AP*

BIOLOGY

4TH EDITION

DEBORAH T. GOLDBERG, M.S.
Former AP Biology Teacher
Lawrence High School
Cedarhurst, New York

BARRON'S

*AP is a registered trademark of the College Board, which was not involved in the production of, and does not endorse, this product.

About the Author

Deborah Goldberg earned her B.S. and M.S. degrees at Long Island University. For fourteen years, she studied cells using electron microscopy at NYU Medical Center and New York Medical College. For the following twenty-two years, she taught Chemistry, Biology, and Forensic Science at Lawrence High School on Long Island.

Dedication

I dedicate this book to my children, Michael and Sara Boilen, to my husband, Howard Blue, and to all my AP Biology students at Lawrence High School, who are every teacher's dream.

Acknowledgments

I wish to thank—

—My husband, Howard Blue, for his constant love and support

—My sister, Rachel, for her mastery of the English language and her eagerness to share it

—Pat Hunter, my editor, for her expert guidance

—My students at Lawrence High School on Long Island, New York, who make teaching the best job in the world

—Michele Sandifer, for her expert copyediting

—Reviewers, who made some great recommendations

All inquiries should be addressed to:
Barron's Educational Series, Inc.
250 Wireless Boulevard
Hauppauge, New York 11788
www.barronseduc.com

ISBN: 978-0-7641-4692-3 (book only)
ISBN: 978-1-4380-7126-8 (book/CD-ROM package)

ISSN 2168-8095 (book only)
ISSN 2168-8087 (book/CD-ROM package)

PRINTED IN THE UNITED STATES OF AMERICA
9 8 7 6 5 4 3 2 1

10%
POST-CONSUMER WASTE
Paper contains a minimum of 10% post-consumer waste (PCW). Paper used in this book was derived from certified, sustainable forestlands.

CONTENTS

MODEL TESTS

As you review the content in this book and work toward earning that **5** on your AP BIOLOGY exam, here are five things that you **MUST** know above everything else.

Barron's Essential

1

The process of evolution drives the diversity and unity of life.
- Darwin's theory of natural selection is evolution's major driving mechanism.
- Evolutionary theory is supported by evidence from many scientific disciplines.
- Phylogenetic trees and cladograms graphically represent evolutionary history.
- The genetic code is universal because all living things descend from a common ancestor.
- Evolution explains how all life is so similar and also why so much diversity exists.

2

Biological systems utilize free energy and molecular building blocks to grow, to reproduce, and to maintain dynamic homeostasis.
- Reciprocal processes of cellular respiration and photosynthesis cycle H_2O, O_2, and CO_2.
- Surface-to-volume ratios affect the capacity of a biological system to obtain resources and eliminate wastes.
- Selectively permeable plasma membranes regulate the movement of molecules across them, and maintain internal environments that differ from external ones.
- Negative and positive feedback mechanisms maintain dynamic homeostasis.

3

Molecules, cells, and organs coordinate activities for the fitness of the organism as a whole.
- Cells communicate by generating, transmitting, and receiving chemical signals.
- Signal transduction pathways link signal reception with cellular response.
- Animals have nervous systems that sense, transmit, and integrate information.
- Different regions of the brain have different functions.
- Cells of the immune system interact in complex ways.

4

Living systems store, retrieve, transmit, and respond to information essential to life processes.
- Genetic information is passed from parent to offspring via DNA with a high degree of accuracy although some mutations do occur.
- DNA directs the production of polypeptides at the ribosome by an elaborate process.
- The cell cycle is complex with highly regulated checkpoints.
- Most traits derive from gene interactions more complex than what Mendel described.
- The expression of genes is controlled by cell signaling, transcription factors, alternate splicing of pre-RNA, and environmental factors.

5

Biological systems interact, and these systems possess complex properties.
- Populations, communities, and ecosystems interact and respond to changes in the environment.
- Mathematic operations can be used to quantify interactions among living things in the environment.
- Interactions between living organisms and their environments result in the recycling of matter and the movement of energy through food chains.

About the New Exam

The College Board revised the AP Biology curriculum for 2012–2013 to reduce the amount of memorization and to focus on understanding broad concepts and to train students to think scientifically. Labs have been changed to encourage students to do real inquiry-based investigations that they themselves devise instead of conducting experiments merely to confirm what they already learned.

WHAT IS INCLUDED IN THE NEW CURRICULUM?

The new curriculum is organized around four major concepts. They are:

1. Evolution
2. Cellular processes: energy and communication
3. Genetics
4. Interactions between molecules, cells, organs, and components of ecosystems

Even though the curriculum has been reorganized, many textbooks have not been changed and your teacher may not change what is taught. However, the structure, style of question, and emphasis of the AP Biology exam is very different.

WHAT ARE THE NEW EXAM QUESTIONS LIKE?

Although the new test questions differ from past questions, they are consistent and predictable. The format of the question itself is almost always the same:

"Which of the following statements best explains. . .?"

Here is a comparison of a sample old- and new-style question.

Old-style question:

An animal that contains both male and female organs is

(A) Known as sexually dimorphic
(B) An example of parthenogenesis
(C) Found only in primitive animals
(D) Hermaphroditic
(E) An example of asexual reproduction

The answer is D. This is a question that focuses on vocabulary and not on underlying concepts.

New-style question:

Hermaphrodites are animals that serve as both male and female by producing both sperm and eggs. Hermaphrodites are often animals that are fixed to a surface (sessile) and are less often motile (free-moving) animals. Which of the following statements is the best explanation for this phenomenon?

(A) As in all examples of asexual reproduction, large numbers of offspring can be produced.

(B) This is a novel adaptation that evolved to meet the challenge of finding a mate without having the ability to move.

(C) Both the male and the female can produce offspring simultaneously.

(D) Hermaphroditism is a degenerate form of sexual reproduction and is found in primitive animals such as parasites.

The answer is B. The new-style question:

- Focuses on reasoning and analysis; not only on vocabulary.
- Contains information in the body of the question so that students do not have to recall details, but can demonstrate an understanding of a concept. In this case, the concept being tested is evolutionary adaptation, not reproduction.
- Contains only four answers from which to choose.
- Takes longer to read, think about, and answer. That's why there are only 63 multiple-choice questions on the exam.

WHAT'S IN THIS REVIEW BOOK?

1. **This book includes every topic in the new AP Biology curriculum** plus some additional ones that probably will not be directly tested on by the College Board. These supplementary topics are identified by an **OWL icon** in the margins. They are included in this book for three reasons:

   - Some teachers will continue to teach these topics because they believe they are important.
   - A student who has familiarity with these topics will be able to demonstrate a depth of knowledge and use them as illustrative examples when answering an essay question.
   - Although the College Board considers these topics background material, some students do not take a pre-AP biology course in high schoool.

2. **Many details have been removed from some topics** in accordance with College Board guidelines. For example, names of specific electron carriers in the electron transport chain or the names or structure of compounds in glycolysis and the citric acid cycle have been removed from review chapters.

3. **Some topics, like Molecular Biology and the Immune System, have been expanded** and include the most up-to-date scientific information.

4. **Thirteen new inquiry-based lab investigations from the College Board are featured.**

5. **Questions at the end of every review chapter test factual recall** and help you prepare for the AP exam. Although these review questions differ from the types of questions on the new exam, they nonetheless provide helpful practice. These questions are designed to reinforce the basics of each topic as you review it.

6. **Two special chapters, "Five Themes to Help You Write a Great Essay" and "Learn How to Grade an Essay"** will help you write better essays on the exam.

7. **Two practice exams mimic the questions that you will encounter on the real AP exam.**

EXAM FORMAT

The AP Biology exam is approximately 3 hours in length and is made up of two sections.

Section I is 90 minutes and worth 50% of the total score. It consists of 63 multiple-choice questions plus 6 grid-in questions, which are short and require math. You may use a simple calculator with a square root function, but you cannot use a graphing calculator.

Section II is 90 minutes and accounts for 50% of the total score. It consists of 2 long free-response questions and 6 short free-response questions. It begins with a 10-minute reading period. The remaining 1 hour and 20 minutes is for writing. Allow 20 minutes for each of the two long free-response questions, and about 6 minutes each for the short free-response questions.

Math

Math is included on the new exam. A table of equations and formulas, including standard deviation or chi-square, is part of the exam. You simply plug numbers into formulas as needed. In previous exams, a Hardy-Weinberg question might have required that you take the square root of 16 to find the value of q. Of course, you did that in your head. Now, you might actually have to use a calculator to find the value of q when q^2 is 23. No big deal!

GRADES ON THE EXAM

Advanced placement and/or college credit is awarded by the college or university you will attend. Different institutions observe different guidelines about awarding AP credit. Success on the AP exam may allow you to take a more advanced course and bypass an introductory course, or it might qualify you for 8 credits of advanced standing and tuition credit. The best source of specific up-to-date information about an institution's policy is its catalog or web site.

Exams are graded on a scale from 1 to 5, with 5 being the best. The total raw score on the exam is translated to the AP's 5-point scale.

AP Grade	Qualification
5	Extremely Well Qualified
4	Well Qualified
3	Qualified
2	Possibly Qualified
1	No Recommendation

Here are the grade distributions for all the students who recently sat for the exam. These numbers tend to be consistent from year to year.

Exam Grade	Student Scoring that Grade
5	19.5%
4	15.5%
3	15.8%
2	15.1%
1	34%

TIPS FOR TAKING SECTION I

Section I consists of 63 multiple-choice questions and 6 grid-in questions. It takes 90 minutes.

Be Neat

Improperly erased pencil marks can cause the machine to misgrade your paper. On the other hand, you may write or draw anywhere in the question booklet.

Pace Yourself

Bring a watch and budget your time.

Answer Every Question

There is no penalty for incorrect answers or for leaving an answer blank. So always guess—even if you don't know the answer.

Read Carefully

Questions with EXCEPT or NOT often trip up students.

TIPS FOR TAKING SECTION II

Section II consists of 8 questions. Questions 1 and 2 are long free-response questions that should take about 20 minutes each to answer. Questions 3–8 are short free-response questions that should take about 6 minutes to answer.

Just as an Olympic athlete must anticipate what the judges want to see, you must be prepared to give the exam readers what they want to read. If you can do that on the AP Exam, you will get a high score.

Here are things the readers **do not** particularly care about:

Spelling
Penmanship (unless they cannot read the paper)
Grammar
Wrong information—**You do not get points off for incorrect statements**.

Here are the things the graders **do** care about:

> The answer must be in essay form, not an outline.
>
> The readers want to see lots of correct information—so write, write, and write!

You Do Not Lose Points for Giving Incorrect Information

You begin with zero points and you gain points as you state correct things that answer the question. The reader does not take off points for any reason, not even incorrect information. He or she only adds points. The reader is like the person who stands at the entrance to a concert and uses a clicker to count the number of people entering. He or she reads your essay. Every time you state a correct piece of information that answers the question, you get a click; that is, you get credit. It is straightforward.

Watch the Point Count!

If one essay is divided into two parts, each part is worth 5 points. If you write a ten-page masterpiece on the first part of the essay but you leave out the second part of the essay, you still get only 5 points. If there are three parts to the essay, each part will be worth 3 points with an extra point given to any section where you have demonstrated extra depth of understanding, for a total of 10 points. So be sure you address every part in order to maximize your points.

Bring a Watch and Budget Your Time

You have 90 minutes total for Section II. The exam proctors will **not** announce when it is time to move from one essay to another. You must monitor the time.

Pay Attention to Words

Read the question and determine what you must do: "Describe," "Explain," "Compare," or "Contrast." Be especially mindful of the word *or*. Some years ago, the students were asked to discuss "*either* the nitrogen cycle *or* the carbon cycle." Don't try to answer both. Focus on one topic and do your best to provide a full response.

Organize Your Thoughts

Before you are allowed to begin to write your essays, you have a 10-minute reading period to think, analyze, and generally prepare to write the essay. Brainstorm and write down all the **key words** you can think of that relate to each topic. Then look over the key words, eliminate the ones that are not related, and prioritize the ones you will be writing about. Present your ideas, in order, *from the general to the particular*. After the reading period is over, you may begin to write your essay.

Do Not Leave Out Basic Material

Many students think that a college-level essay should contain only the most complex ideas. This is incorrect. Include everything you can think of that is related to the topic and that answers the question. Remember, you are trying to accumulate points by presenting relevant, correct statements.

Do Not Contradict Yourself

No points will be given if you give contradictory information. For example, you will receive no credit if you state, "The Calvin cycle occurs in the stroma of chloroplasts," and you also write, "the Calvin cycle occurs in the grana of chloroplasts."

Label Your Answers

Number each essay and label all parts, such as 1a, 1b, 1c, 1d. For readability, leave at least one line between essays. If the reader cannot find your answer, you cannot get any credit.

Include Drawings If You Want

Drawings must be *titled* and *labeled* and must be near the text they relate to. However, you may *not* use drawings (or an outline) instead of writing an essay.

Do Not Write Formal Literary-Style Essays

You do not need to include an introduction, a body, and a conclusion. Doing so is not expected and may take up too much time. Jump right in and answer the question.

Answer the Question, Then Move On

You do not have to include every piece of information about the topic to get full credit. Usually, each essay question is very broad, and there are plenty of ways to get full credit. Remember the reader with the clicker, so just write, write, write!

SUBJECT
AREA
REVIEW

Biochemistry

2

INTRODUCTION

You may think, "This is a biology course, not a chemistry course. Why am I studying chemistry!" Well, the reason you have to know some chemistry is because cells are sacs of chemicals. The fact that sweating cools the skin is a function of the strong forces of attraction between water molecules. Your body maintains your blood at one critical pH because of the bicarbonate buffering system. To prevent heart attacks, you must begin with an understanding about the structure of fatty acids. Mad cow disease is caused by a misfolded protein. Biochemistry affects every aspect of our lives. Here is a review of basic biochemistry.

ATOMIC STRUCTURE

Atoms are the building blocks of all matter. Atoms consist of subatomic particles: **protons**, **neutrons**, and **electrons**.

An atom in the elemental state always has a neutral charge because the number of protons (+) equals the number of electrons (—). Electron configuration is important because it determines how a particular atom will react with atoms of other elements. If all the electrons in an atom are in the lowest available energy levels, the atom is said to be in the **ground state**. When an atom absorbs energy, its electrons move to a higher energy level, and the atom is said to be in the **excited state**. For example, when a chlorophyll molecule in a photosynthetic plant cell absorbs light energy, the molecule becomes excited and electrons get boosted to a higher energy level.

Isotopes are atoms of one element that vary only in the number of neutrons in the nucleus. *Chemically, all isotopes of the same element are identical because they have the same number of electrons in the same configuration.* For example: carbon-12 and carbon-14 are isotopes of each other and are chemically identical. Some isotopes, like carbon-14, are radioactive and decay at a known rate called the **half-life**. Knowing the half-life enables us to measure the age of fossils or to estimate the age of the earth. **Radioisotopes** (radioactive isotopes) are useful in many other ways. For example, **radioactive iodine** (I-131) can be used both to diagnose and to treat certain diseases of the thyroid gland. Additionally, **a tracer** such as radioactive carbon can be incorporated into a molecule and used to trace the path of carbon dioxide in a metabolic pathway.

BONDING

A bond is formed when two atomic nuclei attract the same electron/s. *Energy is released when a bond is formed because atoms acquire a more stable configuration by bonding. They form a completed outer shell. Energy must be supplied to break a bond.* There are two main types of bonds: **ionic** and **covalent**.

Ionic bonds result from the **transfer** of electrons. An atom that gains electrons becomes an **anion** (**a** negative **ion**). An atom that loses an electron becomes a **cation**, a positive ion. The Na^+, Ca^{++}, and Cl^- ions are all necessary for normal nerve function.

Covalent bonds form when atoms **share** electrons. The resulting structure is called a **molecule.** A single covalent bond (–) results when two atoms share a pair of electrons. A double covalent bond (=) results when two atoms share two pairs of electrons. A triple covalent bond (≡) results when two atoms share three pairs of electrons. If electrons are shared *equally* between two identical atoms, the bond is a **nonpolar bond**. This is the type of bond found in **diatomic molecules**, such as H_2 (H–H) and O_2 (O=O). If electrons are shared *unequally*, the bond is referred to as a **polar covalent bond**. This is the case between any two different atoms, such as between atoms of carbon and oxygen in CO_2.

This bond is nonpolar or balanced:	H–H ↑	This bond is polar or unbalanced:	C–H ↑

POLAR AND NONPOLAR MOLECULES

When two or more atoms form a bond, the entire resulting molecule is either **nonpolar** (balanced or symmetrical) or **polar** (unbalanced). CO_2 forms a linear molecule that is symmetrical or balanced and therefore nonpolar. It looks like this: O = C = O. H_2O is asymmetrical and a highly polar molecule. See Figure 2.1.

HYDROPHOBIC AND HYDROPHILIC SUBSTANCES

Hydrophilic means "water loving." Substances that are polar such as hydrochloric acid (HCl), that carry a charge like the hydronium ion (H_3O^+), or that are ionic like table salt (NaCl) will dissolve in water. Since so many substances dissolve in water, water is known as the "universal solvent."

Hydrophobic means "water hating" and applies to nonpolar substances, which are miscible with or will dissolve in lipids. Salad dressing settles upon standing because oil (hydrophobic) and vinegar (hydrophilic, acetic acid solution in water) are not miscible.

Since the plasma membrane is a phospholipid bilayer, only nonpolar substances can readily dissolve through the plasma membrane. Large polar molecules must travel across a membrane in special hydrophilic (protein) channels.

PROPERTIES OF WATER

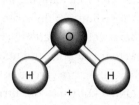

Figure 2.1 H₂O: A Polar Molecule

STUDY TIP

Water is a very polar molecule.

Because the oxygen atom in a molecule of **water** exerts a greater pull on the shared electrons than do the hydrogen atoms, one side of the molecule has a negative charge and the other side has a positive charge. The molecule is therefore asymmetrical and **highly polar**. See Figure 2.1. In addition, the positive hydrogen of one molecule is attracted to the negative oxygen of an adjacent molecule. The two molecules are held together by this **hydrogen bonding**. The strong hydrogen attraction that water molecules have for each other is responsible for the special characteristics of water that are important for life on earth. See Figure 2.2, which shows the hydrogen atom of one molecule of water attracted to the oxygen atom of an adjacent molecule.

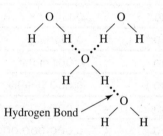

Figure 2.2 Water Molecules with Hydrogen Bonds

- **WATER HAS A HIGH SPECIFIC HEAT.** *Specific heat is the amount of heat a substance must absorb to increase 1 gram of the substance by 1°C.* Because water has a high specific heat, large bodies of water resist changes in temperature and provide a stable environmental temperature for the organisms that live in them. Furthermore, they moderate the climate of the nearby land. High specific heat is responsible for the fact that the marine biome has the most stable temperatures of any biome.

- **WATER HAS A HIGH HEAT OF VAPORIZATION.** Evaporating water requires the absorption of a relatively great amount of heat, so evaporation of sweat significantly cools the body surface.

- **WATER IS THE UNIVERSAL SOLVENT.** Because water is a highly polar molecule, it dissolves all polar and ionic substances.

- **WATER EXHIBITS STRONG COHESION TENSION.** This means that molecules of water tend to attract one another, which results in several biological phenomena.

  1. Water moves up a tall tree from the roots to the leaves without the expenditure of energy by what is referred to as **transpirational-pull cohesion tension**. As one molecule of water is lost from the leaf by transpiration, another molecule is drawn in at the roots.
  2. **Capillary action** results from the combined forces of cohesion and adhesion.
  3. **Surface tension** allows insects to walk on water without breaking the surface.

■ **ICE FLOATS BECAUSE IT IS LESS DENSE THAN WATER.** In a deep body of water, floating ice insulates the liquid water below it, allowing fish and other organisms to survive beneath the frozen surface during winter. In the spring, the ice melts, becomes denser, and sinks to the bottom of the lake, causing water to circulate throughout the lake. Oxygen from the surface is returned to the depths, and nutrients released by the activities of bottom-dwelling bacteria during winter are carried to the upper layers of the lake. This cycling of the nutrients in the lake is known as the **spring overturn** and is an important part of the life cycle of a lake.

pH

pH is a measure of the acidity and alkalinity of a solution. Anything with a pH of less than 7 is an acid, and anything with a pH value greater than 7 is alkaline or basic. A pH of 7 is neutral. *The value of the pH is the negative log of the hydrogen ion concentration in moles per liter.* See Table 2.1, which shows pH values compared with molarity.

Table 2.1

pH Compared with Molarity		
pH	Concentration of H+ ions in Moles per Liter	
1	1×10^{-1} =	0.1 molar
2	1×10^{-2} =	0.01 molar
3	1×10^{-3} =	0.001 molar
4	1×10^{-4} =	0.000 1 molar
7	1×10^{-7} =	0.000 000 1 molar
13	1×10^{-13} =	0.000 000 000 000 1 molar

A substance with a pH of 3 has 1.0×10^{-3} or 0.001 mole per liter of hydrogen ions in solution, while a substance of pH 4 has a H^+ concentration of 1.0×10^{-4} or 0.0001 mole per liter of hydrogen ions in solution. Therefore, a solution of pH 3 is 10 times more acidic than a solution with a pH of 4. A solution with a pH of 6 is 1,000 times more acidic than a solution with a pH of 9; see Figure 2.3.

IT'S TRICKY

As the H+
concentration
increases, the
pH decreases.

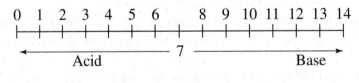

Figure 2.3

The pH of some common substances is as follows:

Stomach acid   2
Human blood   7.4
Acid rain   1.5–5.4

The internal pH of most living cells is close to 7. Even a slight change can be harmful. Biological systems regulate their pH through the presence of **buffers**, substances that resist changes in pH. A buffer works by either absorbing excess hydrogen ions or donating hydrogen ions when there are too few. The most important buffer in human blood is the **bicarbonate ion**.

H+ donor (acid)	Response to rise in pH	H+ acceptor (base)		Hydrogen ion
H_2CO_3	⇌	HCO_3^-	+	H^+
Carbonic acid	Response to drop in pH	Bicarbonate ion		

ISOMERS

Isomers are organic compounds that have the same molecular formula but different structures. Because their structures are different, isomers have different properties. There are three types of isomers: **structural isomers**, *cis-trans* isomers, and **enantiomers**.

Structural isomers differ in the arrangement of their atoms. ***Cis-trans*** **isomers** differ only in spatial arrangement around double bonds, which are not flexible like single bonds are. **Enantiomers** are molecules that are mirror images of each other. The mirror images are called the L- (left-handed) and D- (right-handed) versions. Knowledge of enantiomers is important in the pharmaceutical industry because the two mirror images may not be equally effective. For example, L-dopa is a drug used in the effective treatment of Parkinson's disease. However, D-dopa, its enantiomer, is biologically inactive and useless in the treatment of the disease. For some reason that we do not understand, all the amino acids in cells are left-handed. See Figure 2.4.

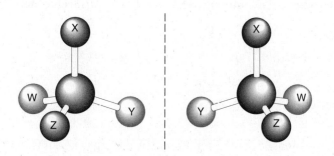

Figure 2.4 Mirror image

ORGANIC COMPOUNDS

All living organisms are made up of organic compounds, compounds that contain carbon. The number of different carbon compounds is vast. There are four classes of organic compounds: **carbohydrates**, **lipids**, **proteins**, and **nucleic acids**.

Carbohydrates

The body uses carbohydrates for fuel and as building materials. Carbohydrates consist of three elements: carbon, hydrogen, and oxygen. The ratio of the number of hydrogen atoms to the number of oxygen atoms in all carbohydrates is always 2 to 1. The empirical formula for all carbohydrates is C_nH_2O. The body uses carbohydrates for quick energy; 1 gram of any carbohydrate will release 4 calories when burned in a calorimeter. Dietary sources include rice, pasta, bread, cookies, and candy. There are three classes of carbohydrates you should know: monosaccharides, disaccharides, and polysaccharides.

Monosaccharides have a chemical formula of $C_6H_{12}O_6$. Three examples are **glucose**, **galactose**, and **fructose**, which are isomers of each other. The structural formula of glucose is shown in Figure 2.5. Notice the conventional numbering of the carbons in the rings. The numbering begins to the right of the oxygen.

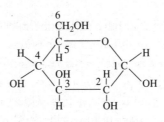

Figure 2.5 Glucose

Disaccharides have the chemical formula $C_{12}H_{22}O_{11}$. They consist of two monosaccharides joined together, with the release of one molecule of water, by the process known as **dehydration synthesis** or **condensation**. Here are the three condensation reactions of monosaccharides that produce the three disaccharides.

monosaccharide	+	**monosaccharide**	→	**disaccharide**	+	**water**
$C_6H_{12}O_6$	+	$C_6H_{12}O_6$	→	$C_{12}H_{22}O_{11}$	+	H_2O
glucose	+	glucose	→	**maltose**	+	water
glucose	+	galactose	→	**lactose**	+	water
glucose	+	fructose	→	**sucrose**	+	water

Hydrolysis is the breakdown of a compound by adding water. It is the reverse of condensation synthesis.

$$\text{sucrose} + \text{water} \rightarrow \text{glucose} + \text{fructose}$$

Polysaccharides are macromolecules, are polymers of carbohydrates, and are formed as many monosaccharides join together by dehydration reactions. There are four important polysaccharides, as shown in Table 2.2.

Table 2.2

	Structural	Storage
Structural and Storage Polysaccharides		
	Structural	**Storage**
Found in plants:	Cellulose	Starch
	Makes up plant cell walls	Two forms are amylose and amylopectin
Found in animals:	Chitin	Glycogen
	Makes up the exoskeleton in arthropods (and cell walls in mushrooms)	"Animal starch." In humans, this is stored in liver and skeletal muscle

Lipids

Lipids are a diverse class of organic compounds that include **fats**, **oils**, **waxes**, and **steroids**. They are grouped together because they are all hydrophobic, meaning that they are not soluble in water. Structurally, most lipids consist of 1 glycerol and 3 fatty acids; see Figure 2.6.

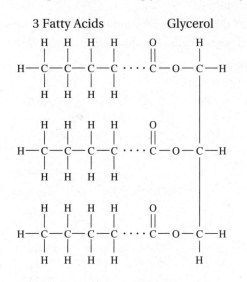

Figure 2.6 Lipid

Glycerol is an alcohol and exists only as shown in Figure 2.7.

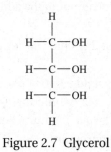

Figure 2.7 Glycerol

A **fatty acid** is a hydrocarbon chain with a carboxyl group at one end. Fatty acids exist in two varieties, **saturated** and **unsaturated**, as shown in Figure 2.8.

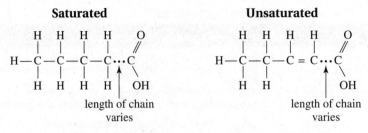

Saturated   **Unsaturated**

length of chain
varies

length of chain
varies

Figure 2.8 Fatty Acids

In general, saturated fats come from animals, are solid at room temperature, and when ingested in large quantities, are linked to heart disease. An example of a saturated fat is butter. Saturated fatty acids contain only *single bonds* between carbon atoms. Unsaturated fatty acids, in general, are extracted from plants, are liquid at room temperature, and are considered to be good dietary fats. Unsaturated fatty acids have at least one *double bond* formed by the removal of hydrogen atoms in the carbon skeleton. As a result, they hold fewer hydrogen atoms than saturated fatty acids. One exception is the group of tropical oils such as coconut and palm oil that are saturated, somewhat solid at room temperature, and are as unhealthy as are fats extracted from animals.

Steroids are lipids that do not have the same general structure as other lipids. Instead, they consist of four fused rings. Figure 2.9 shows the steroid cholesterol. Other steroids are testosterone and estradiol.

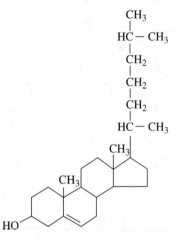

Figure 2.9   The Steroid Cholesterol

Lipids serve many functions.

- **ENERGY STORAGE:** One gram of any lipid will release 9 calories per gram when burned in a calorimeter.
- **STRUCTURAL:** Phospholipids (a lipid where a phosphate group replaces one fatty acid) are a major component of the cell membrane. One steroid, cholesterol, serves as an important component of the plasma membrane of animal cells.
- **ENDOCRINE:** Some steroids are hormones.

Phospholipids

Phospholipids are modified lipids. They consist of only two fatty acids attached to the glycerol backbone, forming two hydrophobic "tails." The third hydroxyl group of the glycerol attaches to a phosphate group, which is charged and therefore hydrophilic. This hydrophilic "head" attracts water. When phospholipids are added to water, they self-assemble into a double-layered structure called a "bilayer." The phosphate is on the outside and the hydrophobic tails are on the inside, shielded from water. This phospholipid bilayer is the structural basis of all plasma membranes. It forms the boundary between the inside of the cell and the external environment.

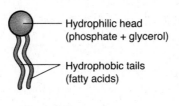

Figure 2.10

Proteins

Proteins are complex, unbranched macromolecules that carry out many functions in the body including:

- Growth and repair
- Signaling from one cell to another
- Regulation: Hormones such as insulin lower blood sugar
- Enzymatic activity: Catalyzing chemical reactions
- Movement: Actin and myosin are protein fibers responsible for muscle contractions.

Dietary sources of proteins include fish, poultry, meat, and certain plants like beans and peanuts. One gram of protein burned in a calorimeter releases 4 calories. Proteins consist of the elements S, P, C, O, H, and N. They are **polymers** or **polypeptides** consisting of units called **amino acids**, which are joined by **peptide bonds**.

Amino acids consist of a **carboxyl group**, an **amine group**, and a **variable (R)** all attached to a central asymmetric carbon atom. The R group, also called the side chain or variable, differs with each amino acid. With only 20 different amino acids, cells can build thousands of different proteins; see Figure 2.11.

See Figure 2.12, showing two amino acids combining to form a dipeptide. A **dipeptide** is a molecule consisting of two amino acids connected by one peptide bond.

Each protein has a unique shape or **conformation** that determines what job it performs and how it functions. There are four levels of protein structure that are responsible for a protein's unique conformation. They are **primary**, **secondary**, **tertiary**, and **quaternary structures**; see Figure 2.13.

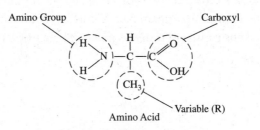

Figure 2.11 Amino Acid

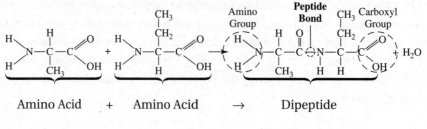

Amino Acid $+$ Amino Acid $\rightarrow$ Dipeptide

Figure 2.12

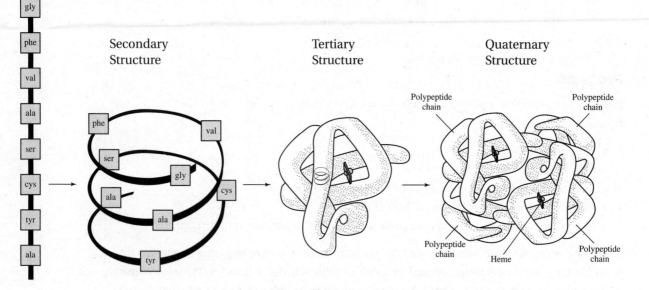

Figure 2.13 Four Levels of Protein Structure

The **primary structure** of a protein refers to the unique linear sequence of amino acids. The slightest change in the amino acid sequence of a protein can have major consequences. Such is the case with sickle cell anemia, a life-threatening condition that results from a substitution of one amino acid (valine) for another (glutamic acid) in a molecule of hemoglobin.

While working in the 1940s and 1950s, Fred Sanger was the first to sequence a protein. That protein was **insulin**, and he received the Noble Prize for his work.

The **secondary structure** of a protein results from **hydrogen bonding** within the polypeptide molecule. It refers to how the polypeptide coils or folds into two distinct shapes: an **alpha helix** or a **beta pleated sheet**.

Proteins that exhibit either alpha helix or beta pleated sheet or both are called **fibrous proteins**. Examples of fibrous proteins are wool, claws, beaks, reptile scales, collagen, and ligaments. The protein that makes up human hair, **keratin**, is composed mostly of alpha helixes, while silk and spider webs consist of proteins made of beta pleated sheets.

Tertiary structure is the intricate three-dimensional shape or conformation of a protein that is superimposed on its secondary structure. Tertiary structure determines the protein's **specificity**. The following intramolecular factors contribute to the tertiary structure:

- Hydrogen bonding between R groups of amino acids
- Ionic bonding between R groups
- Hydrophobic interactions
- Van der Waals interactions
- Disulfide bonds between cysteine amino acids

Quaternary structure refers to proteins that consist of more than one polypeptide chain. Hemoglobin exhibits quaternary structure because it consists of four polypeptide chains, each one forming a heme group.

The Protein-Folding Problem

Under normal cellular conditions, the primary structure of a protein determines how it folds into its particular three-dimensional shape. Protein structure also depends on physical and chemical conditions in the environment, such as pH, salt concentration, and temperature. Adverse conditions alter the weak intramolecular forces, causing the protein to lose its characteristic shape as well as its function, a phenomenon known as **denaturation** or **denaturing**.

The concept that the shape or **conformation** of a protein determines how it functions is a basic concept of modern biology. Scientists have yet to discover all the rules about how proteins spontaneously fold into their proper conformation. However, one important recent discovery is that molecules called **chaperone proteins** or **chaperonins** assist in folding other proteins. A misfolded protein, one with an incorrect shape, can present a serious problem for a cell. Learning more about protein folding is important. Many serious diseases, such as Alzheimer's, Parkinson's, and mad cow disease, result from the accumulation of misfolded proteins, called **prions,** in brain cells.

Three complementary techniques are used to reveal the three-dimensional shape or conformation of proteins. These are **X-ray crystallography**, nuclear magnetic resonance (NMR) spectroscopy and a new field, **bioinformatics**. Bioinformatics uses computers and mathematical modeling to integrate the huge volume of data generated from the analysis of an amino acid sequence of a protein to predict the three-dimensional structure of the resulting protein molecule. Currently, the three-dimensional shape of more than 20,000 proteins has been determined.

Nucleic Acids

The two nucleic acids are **ribonucleic acid (RNA)** and **deoxyribonucleic acid (DNA)**. They encode all hereditary information. Through RNA intermediates, the information encoded in DNA is used to specify the amino acid sequences of all proteins. Nucelic acids are polymers, polynucleotides that consist of repeating units called **nucleotides**. A nucleotide consists of a **phosphate**, a **5-carbon sugar—deoxyribose** or **ribose**, and a **nitrogen base—adenine, cytosine**, **guanine**, or either **thymine** (in DNA) or **uracil** (in RNA). The carbon atoms of deoxyribose are numbered from 1 to 5. Nucleic acids are also discussed in the genetics section later on. In Figure 2.14, P stands for phosphate.

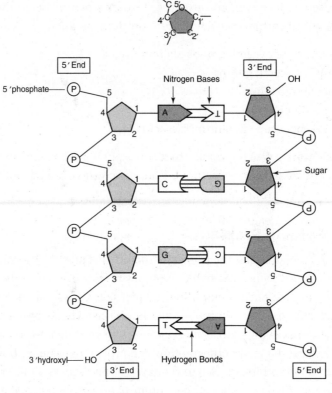

Figure 2.14 DNA

Functional Groups

The components of organic molecules that are most often involved in chemical reactions are known as **functional groups**. These groups are attached to the carbon skeleton, replacing one or more hydrogen atoms that would be present in a hydrocarbon. For example, the difference between testosterone and estradiole is the functional group attached to the carbon skeleton. Each functional group behaves in a consistent fashion from one organic molecule to another. Table 2.3 shows some common functional groups.

Table 2.3

Functional Groups in Organic Compounds		
Group	**Formula**	**Name of Compound**
Amino	$R-N{<}^H_H$	Amine
Carboxyl	$R-C{<}^O_{OH}$	Carboxyl (acid)
Hydroxyl	$R-OH$	Alcohol
Phosphate	$R-O-P(=O)(O^-)-O^-$	Phosphate

ENERGY, ENZYMES, AND METABOLISM

Living organisms require a constant input of energy. Biological systems constantly transform this energy from one form to another in order to carry out all life functions. The laws of thermodynamics govern all energy transformations. The **first law of thermodynamics** states that energy cannot be created or destroyed, only transformed from one form to another. This is also known as the law of conservation of energy. The **second law of thermodynamics** states that during energy conversions, the universe becomes more disordered. In other words, entropy increases. We can determine how much free energy is available to do work within a cell by calculating the **Gibb's free energy**, or just free energy, represented by the letter *G*. During the course of a reaction, if energy is released, the reaction is said to be **exergonic** or exothermic. If in a chemical reaction energy is absorbed, the reaction is **endergonic** or endothermic. In complex cellular reactions, exergonic and endergonic chemical reactions are coupled. *Exergonic reactions power the endergonic ones.*

Metabolism is the sum of all the chemical reactions that take place in cells. Some reactions break down molecules (**catabolism**); other reactions build up molecules (**anabolism**). Metabolic reactions take place in a series, called **pathways**, each of which serves a specific function. These multistep pathways are controlled by enzymes and enable cells to carry out their chemical activities with remarkable efficiency.

Enzymes do not provide energy for a reaction, and they do not enable a reaction to occur that would not occur on its own. Enzymes serve as **catalytic proteins** that speed up reactions by lowering the **energy of activation,** E_A, the amount of energy needed to begin a reaction. In the potential energy diagram in Figure 2.15, the potential energy of the products is less than the potential energy of the reactants, so energy is released and the reaction is exergonic. The dotted line shows the same reaction when an enzyme is introduced. The enzyme serves to lower the energy of activation, and the reaction can proceed more quickly.

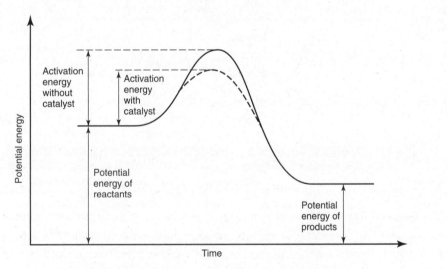

Figure 2.15 Progress of an Exergonic Reaction

In the potential energy diagram in Figure 2.16, the potential energy of the products is greater than the potential energy of the reactants, so energy is absorbed and the reaction is endergonic. The dotted line shows the same reaction when an enzyme is introduced. The enzyme serves to lower the energy of activation, and the reaction can proceed more quickly.

Notice that in both reactions, the only factor altered by the presence of an enzyme is the energy of activation and the activated complex. The **transition state** is the reactive (unstable) condition of the substrate after sufficient energy has been absorbed to initiate the reaction.

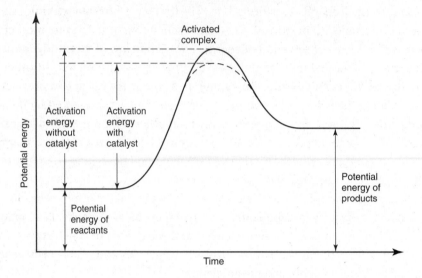

Figure 2.16 Endergonic Reaction

Characteristics of Enzymes

- Enzymes are **globular proteins** that exhibit **tertiary structure**.
- Enzymes are substrate specific. In Figure 2.17, only substrate A will bind to the enzyme.

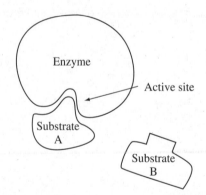

Figure 2.17 Enzymes are Specific

- The **induced-fit model** describes how enzymes work. As the substrate enters the active site, it induces the enzyme to alter its shape slightly so the substrate fits better. The old lock and key model was abandoned because it falsely implied that the lock and the key were unchanging.
- The enyzme binds to its substrate(s), forming an enzyme-substrate complex.
- Enzymes are not destroyed during a reaction. They are reused.
- Enzymes are named after their substrate, and the name ends in the suffix "**ase**." For example, sucrase is the name of the enzyme that hydrolyzes sucrose, and lactase is the name of the enzyme that hydrolyzes lactose.

- Enzymes catalyze reactions in both directions:

Lactase

lactose ⟷ glucose + galactose

- Enzymes often require assistance from **cofactors** (inorganic) or **coenzymes** (vitamins).
- The efficiency of the enzyme is affected by temperature and pH. Average human body temperature is 37°C, near optimal for human enzymes. When body temperature is too high, enzymes will begin to denature and lose both their unique conformation and their ability to function. Gastric enzymes become active at low pH, when mixed with stomach acid, while intestinal amylase works best in an alkaline environment; see Figures 2.18 and 2.19.

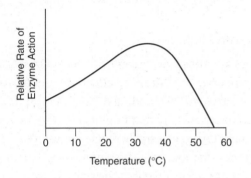

Figure 2.18

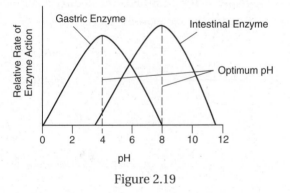

Figure 2.19

Control of Enzyme Activity

Cell metabolism is tightly regulated by controlling when and where different enzymes are active. This is done either by switching on and off the genes that code for enzymes or by regulating the enzymes once they are made. Enzymes that have already been produced are regulated by competitive or noncompetitive inhibition.

COMPETITIVE INHIBITION

In **competitive inhibition**, some compounds resemble the substrate molecules and compete for the same active site on the enzyme. These mimics or **competitive inhibitors** reduce the productivity of enzymes by preventing or limiting the substrate from binding to the enzyme. This kind of inhibition can be overcome by increasing the concentration of substrate. See Figure 2.20.

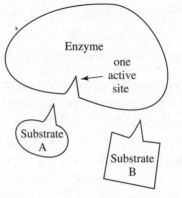

Figure 2.20 Competitive Inhibition

NONCOMPETITIVE INHIBITION

In contrast, some enzymes are **allosteric**; meaning a change in shape alters their efficiency. Molecules called **noncompetitive inhibitors** or **allosteric regulators** bind to a site distinct from the active site of the enzyme. This binding of the inhibitor to the alternate site causes the enzyme to change its shape in a way that inhibits the enzyme from catalyzing substrate into product. Other allosteric enzymes usually toggle between two different conformations (shapes)—one active and the other inactive. The binding of either an activator or an inhibitor locks or stabilizes the allosteric enzyme in either the active or inactive form, respectively. One economical mechanism for regulating a lengthy metabolic pathway is known as **feedback inhibition**, where the *end product* of the pathway is the *allosteric inhibitor* for an enzyme that catalyzes an early step in the pathway; see Figure 2.21.

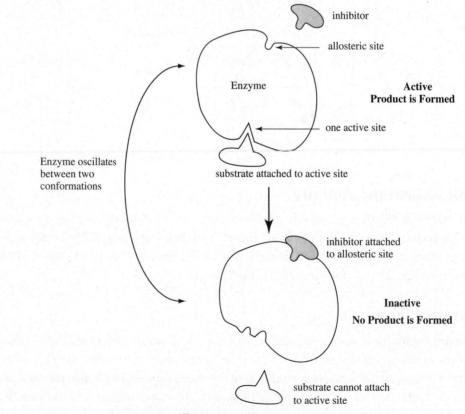

Figure 2.21 Allosteric or Noncompetitive Inhibition

COOPERATIVITY

Cooperativity is another type of *allosteric activation*. The binding of one substrate molecule to one active site of one subunit of the enzyme causes a change in the entire molecule and locks all subunits in an active position. This mechanism amplifies the response of an enzyme to its substrates.

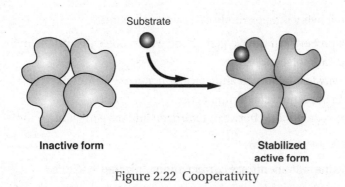

Figure 2.22 Cooperativity

MULTIPLE-CHOICE QUESTIONS

1. Which of the following is correct about isotopes of carbon?

    (A) They are all radioactive.
    (B) They contain the same number of neutrons but a different number of protons.
    (C) They contain the same number of electrons but are chemically different because the number of neutrons is different.
    (D) They are chemically identical because they have the same number of electrons.

2. All of the following are characteristics of water EXCEPT

    (A) water has a relatively high boiling point
    (B) water molecules have little attraction for each other
    (C) water is a universal solvent
    (D) ice is less dense than water

3. The pH of blood

    (A) is normally close to 7.4
    (B) is strongly acidic
    (C) varies with the needs of the cells
    (D) is about 6.9

4. Chaperonins are involved with

    (A) holding the two strands of DNA together
    (B) maintaining the normal pH level of blood
    (C) proper folding of proteins
    (D) enantiomers

5. Which of the following is stored in the human liver for energy?

   (A) glucose
   (B) glycogen
   (C) glycerol
   (D) glucagon

6. Which of the following is an example of a hydrogen bond?

   (A) the attraction between the oxygen of one water molecule and the hydrogen of an adjacent molecule
   (B) the bond between hydrogen and carbon in glucose or any sugar
   (C) the peptide bond between amino acids
   (D) the intramolecular bond between hydrogen and oxygen within a molecule of water

7. All of the following statements are correct about enzymes EXCEPT

   (A) they raise the energy of activation of all reactions
   (B) they enable reactions to occur at a relatively low temperature
   (C) they remain unchanged during a reaction
   (D) they are often located within the plasma membrane of the cell

8. Which statement is correct about pH?

   (A) There are no hydrogen ions in a strong basic solution.
   (B) Pure water has a neutral pH of 7 because the concentration of hydrogen ions equals the concentration of hydroxyl ions.
   (C) The concentration of a solution with a pH of 5 is 5 times more acidic than a solution with pH 1.
   (D) The concentration of hydrogen ions in a solution with a pH of 2 is 2,000 times more acidic than a solution with pH of 4.

9. Which of the following can be used to determine the rate of an enzyme-catalyzed reaction?

   (A) the rate of substrate formed
   (B) the decrease in temperature in the system
   (C) the rate of enzyme used up
   (D) the rate of substrate used up

10. Which of the following best describes the reaction shown below?

    A + B → AB + energy

    (A) hydrolysis
    (B) an exergonic reaction
    (C) an endergonic reaction
    (D) catabolism

Questions 11–12

The graph below demonstrates two chemical reactions. One is catalyzed by an enzyme, one is not.

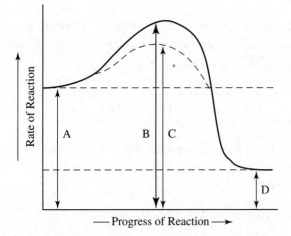

11. Which letter shows the energy of activation for the enzyme-catalyzed reaction?

    (A) A
    (B) B
    (C) C
    (D) D

12. Which letter shows the potential energy of the product?

    (A) A
    (B) B
    (C) C
    (D) D

13. Which level of protein structure is most related to specificity?

    (A) tertiary
    (B) primary
    (C) secondary
    (D) quaternary

Questions 14–20

In a lab experiment, one enzyme is combined with its substrate at time 0. The product is measured in micrograms at 20-second intervals and recorded on the data table below.

Time (s)	0	20	40	60	80	100	120
Product (µg)	0.0	0.25	0.50	0.70	0.80	0.85	0.85

14. What is the initial rate of the enzyme reaction?

15. What is the rate after 100 seconds?

16. Why is there no increase in product after 100 seconds?

17. What would happen if you added only more enzyme after 100 seconds?

18. What would happen if you added only more substrate after 100 seconds?

19. What would happen if you boiled the enzyme for 10 minutes before you did the experiment?

20. What would happen if you added strong acid to the enzyme 1 hour before you did the experiment?

Answers to Multiple-Choice Questions

1. **(D)** Since they have the same number of electrons, isotopes of the same element are chemically identical. Only some of the isotopes of carbon (such as C-14) are radioactive. Isotopes vary only in the number of neutrons.

2. **(B)** Water molecules have strong attraction for each other because they are polar and because of strong hydrogen bonding between molecules.

3. **(A)** The pH of blood remains very close to 7.4 at all times and is maintained by the bicarbonate buffering system. This is an example of how an organism maintains homeostasis or internal stability.

4. **(C)** Molecules called chaperone proteins or chaperonins assist in the proper folding of other proteins.

5. **(B)** Glycogen is a polysaccharide that stores sugars in the liver and skeletal muscle. Glucose is not stored; it is produced and used up constantly. Glycerol and fatty acids make up lipids. Glucagon is a hormone that is responsible for breaking down glycogen into glucose. Glycine is the simplest amino acid.

6. **(A)** Hydrogen bonding is actually an intermolecular attraction between two molecules, not a covalent bond *within* one molecule. Choice B describes a covalent bond within a molecule. The bond between Na^+ and Cl^- is ionic. A peptide bond exists between the amino group of one amino acid and the carboxyl group of an adjacent amino acid.

7. **(A)** Enzymes lower the energy of activation, thus speeding up the reaction.

8. **(B)** The pH of a solution is neutral when the hydrogen ion (H^+) concentration equals the hydroxyl ion (OH^-) concentration. Pure water has a pH of 7 because it consists of equal amounts of H^+ and OH^-. Strong basic solutions contain mostly hydroxyl ions, but there are some hydrogen ions in it. The concentration of a solution of pH 5 is 10,000 times more acidic than a solution with pH 1. The concentration of hydrogen ions in a solution with a pH of 2 is 100 times more acidic than a solution with pH of 4. A solution with a pH of 5 means there are 1×10^{-5} moles of hydrogen ions in solution.

9. **(D)** Substrate is used up as the product is formed. Since enzymes are never used up, they are reused. So enzyme levels cannot be used to monitor the progress of a reaction. Enzymes lower the energy needed to begin the reaction (the E_A), but they do not affect the temperature of the system.

10. **(B)** The reaction is exergonic because it releases energy. Endergonic and endothermic are synonyms for a reaction that requires energy. Catabolism is the breaking down of a substance; this reaction is anabolic, a building-up process.

11. **(C)** Enzymes lower the energy of activation.

12. **(D)** The potential energy (PE) of the product is the same for both reactions.

13. **(A)** Tertiary structure dictates the three-dimensional shape and function of a protein.

Answers to Lab Questions

14. $0.25\,\mu g$ per 20 seconds or $0.0125\,\mu g/s$. Rate is derived by taking the change in amount divided by the time. $(0.25\,\mu g - 0.00\,\mu g)/20\,s = 0.25\,\mu g/20\,s$.

15. Zero. This is true because no new product is formed. The calculation is $0.85 - 0.85 = 0$.

16. There is no increase in product after 100 seconds because all the enzymes are saturated and are already catalyzing reactions as fast as they can.

17. Assuming there is excess substrate that enzyme is not colliding with, an addition of enzyme after 100 seconds would increase the rate of reaction until the enzyme, once again, became saturated.

18. If you add more substrate after 100 seconds, there would be no change because the enzyme is saturated and reacting with as much substrate as it can. The enzyme cannot handle any more substrate. If you added both substrate and enzyme after 100 seconds, the reaction rate would increase temporarily until the enzyme became saturated once again.

19. If you boiled the enzyme prior to doing the experiment, you would get no product because the enzyme would probably have been denatured by the high heat and would not catalyze the reaction.

20. If you expose the enzyme to strong acid prior to doing the experiment, there would probably be no product because the enzyme would have been denatured by the acid.

Directions: Answer all questions. You must answer the question in essay—**not** outline—form. You may use labeled diagrams to supplement your essay, but diagrams alone are *not* sufficient. Before you start to write, read each question carefully so that you understand what the question is asking.

1. Describe the structure and function of enzymes.

2. The unique properties of water make life on earth possible. Select four characteristics of water and:

   a. Identify one characteristic of water, and explain how the structure of water relates to this property.
   b. Describe one example of how this characteristic affects living organisms.

Typical Free-Response Answers

Note: Key words have been written in bold merely to help you focus on what you should be discussing in an essay on enzyme function. You may use sketches to help explain your ideas, but they must be labeled and drawn near the text they explain.

1. Enzymes act as **catalysts**, lowering the **energy of activation**, thus increasing the rate at which reactions occur. Enzymes are large protein molecules that exhibit **tertiary structure** and are folded into a particular shape or **conformation** that results from many intramolecular interactions of the amino acids that make up the molecule. Enzymes are specific and act only on certain **substrates**. The shape of an enzyme determines which substrate an enzyme will act upon. Enzymes fold in such a way that there is one or more active sites where a substrate can bind. Special proteins called **chaperone proteins** assist enzymes in folding in their unique way. Many enzymes require the assistance of **coenzymes** (**vitamins**) or **cofactors** (**minerals**) to function properly. The shape of an enzyme can be altered or **denatured** and the enzyme will no longer function as it is supposed to. High heat and extremes in pH can denature an enzyme. The way in which an enzyme and a substrate form an enzyme-substrate is known as **induced fit**.

   Enzyme activity can be regulated in a number of different ways. One is by **competitive inhibition**, where two substrates compete for the same active site. Another is called **noncompetitive inhibition**, in which an **allosteric inhibitor** alters the shape of the allosteric site and prevents the substrate from binding to its active site.

2a. Water is a highly **polar** molecule with strong **hydrogen bonding** between adjacent water molecules. Because there are such strong attractions between molecules, water has certain properties that make life on earth possible. These properties are high cohesion tension, **high specific heat**, **high heat** of **vaporization**, and the fact that water is an **excellent solvent**. Another property of water is that when it is frozen, it is less dense than water. The reason for this is that the bonding between the water molecules in ice holds the molecules rigidly and farther apart than in liquid water. Since the water molecules are farther apart in ice than in water, ice is less dense than water.

2b. Because ice is less dense than water, it floats. This has important consequences for living things. In a deep body of water, such as a lake, floating ice insulates the liquid water below it, allowing life to exist beneath the frozen surface during cold seasons. The fact that ice covers the surface of water in the cold months and melts in the spring results in a stratification of the lake in winter and considerable mixing in the spring when the ice melts. In the spring, the ice melts, becomes denser water, and sinks to the bottom of the lake, causing water to circulate throughout the lake. Oxygen from the surface is returned to the depths, and nutrients released by the activities of bottom-dwelling bacteria during the winter are carried to the upper layers of the lake. This cycling of the nutrients in the lake is known as the spring overturn and is a necessary part of the life cycle of a lake.

The Cell

→ **STRUCTURE AND FUNCTION OF THE CELL**

→ **TRANSPORT INTO AND OUT OF THE CELL**

→ **CELL COMMUNICATION**

INTRODUCTION

All organisms on earth are believed to have descended from a common ancestral cell about three and a half billion years ago. According to the **theory of endosymbiosis**, eukaryotic cells emerged when mitochondria and chloroplasts, once free-living prokaryotes, took up permanent residence inside other larger cells, about one and a half billion years ago. Here was the advent of the radically more complex **eukaryotic cell** with internal membranes that compartmentalized the cell and led to the rapid evolution of multicelled organisms.

Modern cell theory states that all organisms are composed of cells and that all cells arise from preexisting cells. Most animal and plant cells have diameters between 10–100 μm (microns or micrometers), although many, like human red blood cells with a diameter of only 8 μm, are smaller.

All cells share certain characteristics. They are all enclosed by a protective and selective barrier called a plasma membrane. They all contain a semifluid substance called cytosol in which subcellular components are suspended. Finally, all cells contain ribosomes and genetic material in the form of DNA.

There are two types of cells: prokaryotes and eukaryotes. Prokaryotes are simple cells containing no nuclei or other internal membranes. All bacteria are prokaryotic cells. Instead of a nucleus, they have a nucleoid, which is a non-membrane-bound region where the chromosome is located. Eukaryotic cells have a nucleus bound by a double membrane. Eukaryotic cells also have organelles and internal membranes that compartmentalize each cell so that complex chemical reactions can be carried out efficiently in separate regions of a cell. All cells of the human body are eukaryotic cells; see Figure 3.1 and Table 3.1.

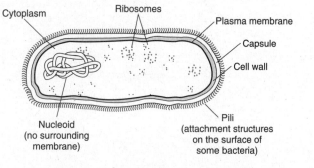

Figure 3.1 Typical Prokaryotic Cell

Table 3.1

Comparison of Cell Types	
Prokaryotes	Eukaryotes
No internal membranes: no nuclear membrane, E.R., mitochondria, vacuoles, or other organelles	Contain distinct organelles
Circular, naked DNA	DNA wrapped with histone proteins into chromosomes
Ribosomes are very small	Ribosomes are larger
Metabolism is anaerobic or aerobic	Metabolism is aerobic
Cytoskeleton absent	Cytoskeleton present
Mainly unicellular	Mainly multicellular with differentiation of cell types
Cells are very small: 1–10 μm	Cells are larger: 10–100 μm

STRUCTURE AND FUNCTION OF THE CELL

A major theme in biology (and therefore a common essay question) is that *function dictates form* and vice versa. As a result, one would expect that all cells do not look alike. And they do not. The nerve cell, whose purpose is to send electrical impulses, is long and spindly. Cells that store fat are rounded, large, and distended. Cells that make up a tough peach pit resemble square building blocks. Figure 3.2. shows a sketch of different cell types, each with a different overall appearance suited for each different function.

KEEP THIS IN MIND

Form and function go together.

Look for examples as you read this book.

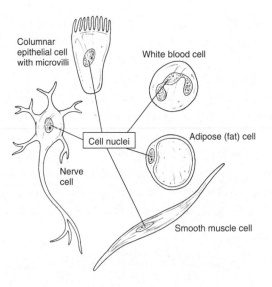

Figure 3.2

Why Cells Are Small

Although various cells look different, they are all tiny. This is true because of surface area to volume ratio. As an object increases in size, both its surface area and volume increase. However, surface area increases more slowly than volume because surface area is a function of the *square* of the radius while volume is a function of the *cube* of the radius.

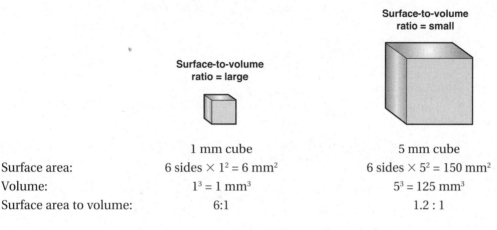

	1 mm cube	5 mm cube
Surface area:	6 sides $\times$ 1^2 = 6 mm^2	6 sides $\times$ 5^2 = 150 mm^2
Volume:	1^3 = 1 mm^3	5^3 = 125 mm^3
Surface area to volume:	6:1	1.2 : 1

Figure 3.3

The volume of a cell determines the amount of metabolic activity the cell carries on. However, the surface area of the plasma membrane limits the amount of material that can enter and leave the cell. You can see from Figure 3.3 that as a cell grows larger and its metabolism increases, the surface area (plasma membrane) does not "keep up." So complex organisms consist of millions of tiny cells carrying out different functions instead of one gigantic all-purpose cell.

Although eukaryotic cells have many organelles in common, they also have organelles that are unique to each cell type, such as cell walls in plant cells. Figure 3.4 shows a typical plant cell.

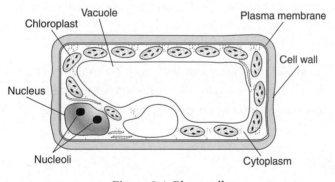

Figure 3.4 Plant cell

The human body consists of approximately two hundred different cell types, each with a different function and, therefore, a different form. Although different cell types have different appearances, they all contain the same organelles. See Figure 3.5, a sketch of a typical animal cell.

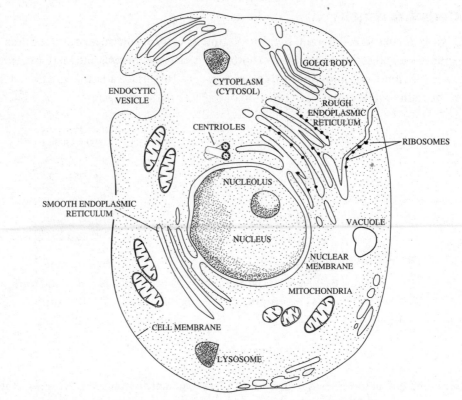

Figure 3.5 Diagram of a Eukaryotic Animal Cell

Nucleolus

The nucleus of a nondividing cell contains one or more prominent nucleoli, where ribosomal RNA (rRNA) is synthesized according to instructions from the DNA. Large and small subunits of ribosomes are also assembled there. A nucleolus combines proteins imported from the cytoplasm with rRNA made in the nucleolus. Nucleoli are not membrane-bound structures but are actually a tangle of chromatin and unfinished ribosomal precursors.

Ribosomes

Ribosomes are protein factories. They can be found free in the cytoplasm or bound to the endoplasmic reticulum. Free ribosomes are associated with protein produced for the cell's own use, while ribosomes attached to the endoplasmic reticulum are meant for export out of the cell.

Peroxisomes

Peroxisomes are found in both plant and animal cells and perform a specialized function. They contain **catalase**, which converts hydrogen peroxide (H_2O_2), a waste product of respiration in the cell, into water with the release of oxygen atoms. They also detoxify alcohol in liver cells.

Endomembrane System

The endomembrane system regulates protein traffic and performs metabolic functions in cells. It includes the nuclear envelope, endoplasmic reticulum, Golgi apparatus, lysosomes, vesicles, vacuoles, and plasma membrane.

Nucleus

The nucleus contains chromosomes which are wrapped with special proteins into a **chromatin network**. The nucleus is surrounded by a selectively permeable **nuclear envelope** that separates the contents of the nucleus from the cytoplasm. The nuclear envelope contains **pores** to allow for the transport of molecules, like messenger RNA (mRNA), which are too large to diffuse directly through the envelope.

Endoplasmic Reticulum

The endoplasmic reticulum (E.R.) is a membranous system of channels and flattened sacs that traverse the cytoplasm and account for more than half the total membranes in a eukaryotic cell. There are two types.

- **Rough E.R.** is studded with ribosomes and produces proteins.
- Smooth E.R. has three diverse functions:

    1. Assists in the synthesis of steroid hormones, like sex hormones, and of other lipids
    2. Stores Ca^{++} ions in muscle cells to facilitate normal muscle contractions
    3. Detoxifies drugs and poisons from the body

Golgi Apparatus

The **Golgi apparatus** lies near the nucleus and consists of flattened membranous sacs stacked next to one another and surrounded by vesicles. They **process** and **package** substances produced in the rough endoplasmic reticulum and **secrete** the substances to other parts of the cell or to the cell surface for export.

Lysosomes

Lysosomes are sacs of **hydrolytic** (digestive) **enzymes** surrounded by a single membrane. They are the principal site of **intracellular digestion**. With the help of the lysosome, the cell continually renews itself by breaking down and recycling cell parts, a process called **autophagy**. Programmed destruction of cells (**apoptosis**) by their own hydrolytic enzymes is a critical part of the development of multicelled organisms. Lysosomes are not generally found in plant cells.

Mitochondria

The mitochondria are the site of cellular respiration. All cells have many mitochondria; a very active cell could have 2,500 of them. Mitochondria have an outer double membrane and an inner series of membranes called **cristae**. Mitochondria also contain their own DNA. Mitochondria constantly divide and fuse with each other in order to exchange DNA and compensate for one another's defects. This is necessary for normal mitochondrial function, including respiration, cell development, and apoptosis.

The fact that mitochondria contain their own DNA is a major support for the **endosymbiotic theory**. This theory postulates that mitochondria were once free-living prokaryotic cells that took up residence inside larger prokaryotic cells billions of years ago.

Vacuoles

Vacuoles are membrane-bound structures used for storage. They are large vesicles derived from the E.R. and Golgi apparatus. Mature plant cells generally have a single large central vacuole. Many freshwater protists have **contractile vacuoles** to pump out excess water. **Food vacuoles** are formed by the **phagocytosis** of foreign material.

Chloroplast

Chloroplasts contain the green pigment chlorophyll that, along with enzymes, absorbs light energy and synthesizes sugar. They are found in plants and algae. In addition to a double outer membrane, they have another inner membrane system called **thylakoids**. (See page 76 for details.)

According to the **theory of endosymbiosis**, chloroplasts were once tiny, free-living prokaryotic cells that were engulfed by a larger prokaryotic cell. Eventually, the engulfed cell became a permanent resident, and the two became one entity. A major piece of evidence for this theory is that chloroplasts have their own DNA that resembles bacterial DNA, not eukaryotic nuclear DNA.

Cytoskeleton

The cytoskeleton of the cell is a complex mesh of protein filaments that extends throughout the cytoplasm. The cytoskeleton has several important roles.

1. It maintains the cell's shape.
2. It controls the position of organelles within the cell by anchoring them to the plasma membrane.
3. It is involved with the flow of the cytoplasm, known as **cytoplasmic streaming**.
4. It anchors the cell in place by interacting with extracellular elements.

The cytoskeleton includes microtubules and microfilaments.

- **Microtubules** are hollow tubes made of the protein tubulin that make up the **cilia, flagella**, and **spindle fibers**. Cilia and flagella, which move cells from one place to another consist of 9 pairs of microtubules organized around 2 singlet microtubules; see Figure 3.6. **Spindle fibers** help separate chromosomes during mitosis and meiosis and consist of microtubules organized into 9 triplets with no microtubules in the center. Flagella, when present in prokaryotes, are not made of microtubules.

- **Microfilaments** are assembled from **actin filaments** and help support the shape of the cell. They enable

    1. Animal cells to form a **cleavage furrow** during cell division
    2. Ameoba to move by sending out **pseudopods**
    3. Skeletal muscle to contract as they slide along myosin filaments

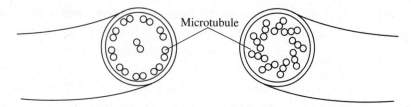

Cilia and Flagella   Spindle Fibers and Centrioles

Figure 3.6

Centrioles, Centrosomes, and the Microtubule Organizing Centers

Centrioles, **centrosomes**, or **microtubule organizing centers** (**MTOCs**) are nonmembranous structures that lie outside the nuclear membranes. They organize spindle fibers, and give rise to the spindle apparatus required for cell division (see Figure 3.7). Two centrioles oriented at right angles to each other make up one centrosome and consist of 9 triplets of microtubules (just like spindle fibers) arranged in a circle. Plant cells lack centrosomes, but have MTOCs. In animal cells, the MTOC is synonymous with centrosome.

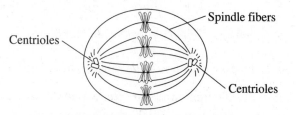

Figure 3.7 Spindle Fibers

Cell Wall

The **cell wall** is one cell structure not found in animal cells. Plants and algae have cell walls made of **cellulose**. The cell walls of fungi are usually made of **chitin**. Those of prokaryotes consist of other polysaccharides and complex polymers. The **primary cell wall** is immediately outside the plasma membrane. Some cells produce a second cell wall underneath the primary cell wall, called the **secondary cell wall**. When a plant cell divides, a thin gluey layer is formed between the two new cells, which becomes the **middle lamella**.

Plasma Membrane

The **cell** or **plasma membrane** is a **selectively permeable** membrane that regulates the steady traffic that enters and leaves the cell. S. J. Singer is famous for his description of the cell membrane in 1972, which he called the **fluid mosaic model**. The eukaryotic plasma membrane consists of a **phospholipid bilayer** with proteins dispersed throughout the layers. A phospholipid is **amphipathic**, meaning it has both a hydrophobic and hydrophilic region. **Integral proteins** have nonpolar regions that completely span the hydrophobic interior of the membrane. **Peripheral proteins** are loosely bound to the surface of the membrane. **Cholesterol molecules** are embedded in the interior of the bilayer to stabilize the membrane. The average membrane has the consistency of olive oil and is about 40 percent lipid and 60 percent protein. Phospholipids move along the plane of the membrane rapidly. Some proteins are kept in place by attachment to the cytoskeleton, while others drift slowly. Extending from the external surface of the plasma membrane are glycolipids (carbohydrates covalently bonded to lipids), and glycoproteins (carbohydrates covalently bonded to proteins). Both structures may serve as signaling molecules that distinguish one cell type from another. Glycoproteins on the surface of red blood cells are responsible for ABO and Rh blood types; see Figure 3.8.

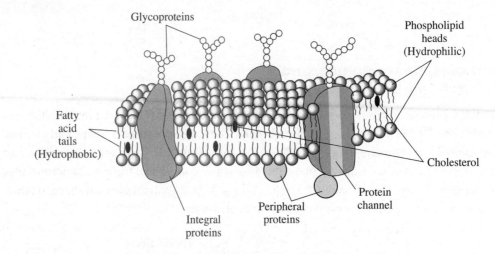

Figure 3.8 Detail of the Plasma Membrane

Proteins in the plasma membrane provide a wide range of functions.

- **TRANSPORT:** Molecules, electrons, and ions are carried through channels, pumps, carriers, and electron transport chains, which manufacture ATP.
- **ENZYMATIC ACTIVITY:** One membrane-bound enzyme is adenylate cyclase, which synthesizes cyclic AMP (cAMP) from ATP.
- **SIGNAL TRANSDUCTION:** Binding sites on protein receptors fit chemical messengers (signaling molecule) like hormones. The protein changes shape and relays the message to the inside of the cell.
- **CELL-TO-CELL RECOGNITION:** Some glycoproteins serve as identification flags that are recognized by other cells.
- **CELL-TO-CELL ATTACHMENTS:** Desmosomes, gap junctions, and tight junctions are examples.
- **ATTACHMENT TO THE CYTOSKELETON AND EXTRACELLULAR MATRIX:** This helps maintain cell shape and stabilizes the location of certain membrane proteins.

TRANSPORT INTO AND OUT OF THE CELL

Transport is the movement of substances into and out of a cell. Transport can be either active or passive. Active transport requires energy (ATP). Passive transport requires no energy. Figure 3.9 shows an overview of passive and active transport.

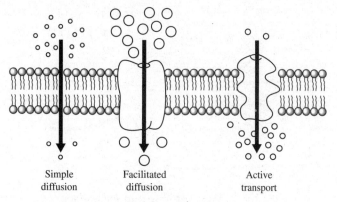

Figure 3.9 Overview: Active and Passive Transport

Simple
diffusion

Facilitated
diffusion

Active
transport

Passive Transport

Passive transport is the movement of molecules *down a concentration gradient from a region of high concentration to a region of low concentration* until equilibrium is reached. Examples of passive transport are **diffusion** and **osmosis**.

There are two types of **diffusion: simple** and **facilitated**. Simple diffusion does not involve protein channels, but facilitated diffusion does. An example of simple diffusion is found in the glomerulus of the human kidney, where solutes dissolved in the blood diffuse into Bowman's capsule of the nephron. Facilitated diffusion requires a **hydrophilic protein channel** that will passively transport specific substances across the membrane; see Figure 3.10. One type of channel transports single ions such as Na^+, K^+, Ca^{2+}, and Cl^-. (Neither simple nor facilitated diffusion requires energy.)

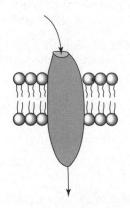

Figure 3.10 Protein Channel

A special case of simple diffusion is called **countercurrent exchange**—the flow of adjacent fluids in opposite directions that maximizes the rate of simple diffusion. One example of countercurrent exchange can be seen in fish gills. Blood flows toward the head in the gills, while water flows over the gills in the opposite direciton. This process maximizes the diffusion of respiratory gases and wastes between the water and the fish; see Figure 3.11.

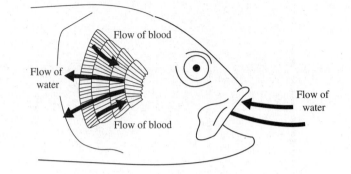

Figure 3.11 Countercurrent Exchange

Here is some basic vocabulary to aid in your understanding of passive transport.

- **DIFFUSION:** random movement of molecules or other particles from a higher concentration to a lower concentration.
- **OSMOSIS:** is the term used for a specific type of diffusion of *water across a membrane*.
- **SOLVENT:** the substance that does the dissolving
- **SOLUTE:** the substance that dissolves
- **HYPERTONIC:** having greater concentration of *solute* than another solution
- **HYPOTONIC:** having lesser concentration of *solute* than another solution
- **ISOTONIC:** two solutions containing equal concentration of *solutes*
- **OSMOTIC POTENTIAL:** the tendency of water to move across a permeable membrane into a solution
- **WATER POTENTIAL:** Scientists look at movement of water in terms of water potential. **Water potential**, symbolized by the Greek letter psi, (ψ), results from two factors: solute concentration and pressure. The water potential for *pure water is zero*; the addition of solutes lowers water potential to a value less than zero. Therefore, the water potential inside a cell is a negative value. Water will move across a membrane from the solution with the higher water potential to the solution with the lower water potential.

Figure 3.12 shows a sketch of two containers. Solution *A* is hypertonic to solution *B*.

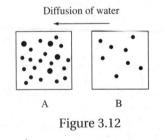

Diffusion of water

A   B

Figure 3.12

Example 1: In Figure 3.13 the cell is in an isotonic solution. Water diffuses in and out, but there is no net change in the size of the cell.

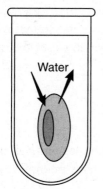

Figure 3.13 Cell in Isotonic Solution

Example 2: The cell in Figure 3.14 is in a **hypotonic** solution. The concentration of solute in the beaker is less than the concentration of solute in the cell. Therefore, *water* will flow into the cell, causing the cell to swell or burst. If the cell is a plant cell, the cell wall will prevent the cell from bursting. The cell will merely swell or become **turgid**. This turgid pressure is what keeps plants like celery crisp. If a plant loses too much water (dehydrates), it loses its **turgor pressure** and wilts.

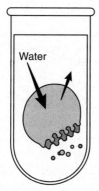

Figure 3.14 Cell in Hypotonic Solution

Example 3: The cell in Figure 3.15 is in a **hypertonic** solution. The concentration of **solute** in the beaker is greater than the concentration of solute in the cell. Therefore, water will flow out of the cell because water flows from high concentration of water to low concentration of water. As a result, the cell shrinks, exhibiting **plasmolysis**.

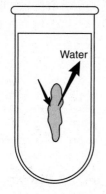

Figure 3.15 Cell in Hypertonic Solution

Aquaporins are special water channel proteins found in certain cells that facilitate the diffusion of massive amounts of water across a cell membrane. These channels do not affect the water potential gradient or the direction of water flow but, rather, the rate at which water diffuses down its gradient. It is possible that aquaporins can also function as **gated channels** that open and close in response to variables such as turgor pressure of a cell. The sudden change in a cell in response to changes in tonicity as seen in Figure 3.15 may be the result of the action of aquaporins.

Active Transport

Active transport is the movement of molecules *against a gradient*, which requires energy, usually in the form of ATP. There are many examples of active transport.

- **Pumps** or **carriers** carry particles across the membrane by active transport.

    1. The **sodium-potassium pump** that pumps Na⁺ and K⁺ ions across a nerve cell membrane to return the nerve to its resting state is an example; see Figure 3.16. The sodium-potassium pump moves Na⁺ and K⁺ ions against a gradient, pumping two K⁺ ions for every three Na⁺ ions. ATP provides the energy for this mechanism.

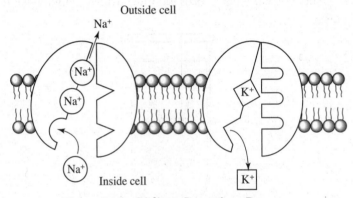

Figure 3.16   Sodium-Potassium Pump

    2. The **electron transport chain** in mitochondria consists of proteins that pump protons across the cristae membrane.

- The **contractile vacuole** in freshwater Protista pumps out excess water that has diffused inward because the cell lives in a hypotonic environment.
- **Exocytosis** in nerve cells occurs as vesicles release neurotransmitters into a synapse.
- **Pinocytosis**, *cell drinking*, is the uptake of large, dissolved particles. The plasma membrane invaginates around the particles and encloses them in a vesicle; see Figure 3.17.

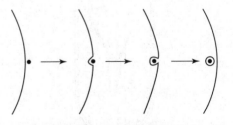

Figure 3.17 Pinocytosis

■ **Phagocytosis** is the engulfing of large particles or small cells by pseudopods. The cell membrane wraps around the particle and encloses it into a vacuole. This is the way human white blood cells engulf bacteria and also the way in which ameoba gain nutrition; see Figure 3.18.

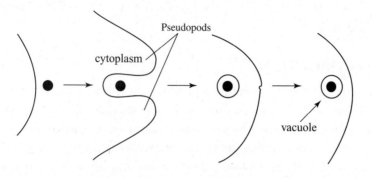

Figure 3.18 Phagocytosis

■ **Receptor-mediated endocytosis** enables a cell to take up large quantities of very **specific substances**. It is a process by which extracellular substances bind to receptors on the cell membrane. Once the **ligand** (the general name for any molecule that binds specifically to a receptor site of another molecule) binds to the receptors, endocytosis begins. The receptors, carrying the ligand, migrate and cluster along the membrane, turn inward, and become a **coated vesicle** that enters the cell. This is the way cells take in **cholesterol** from the blood; see Figure 3.19.

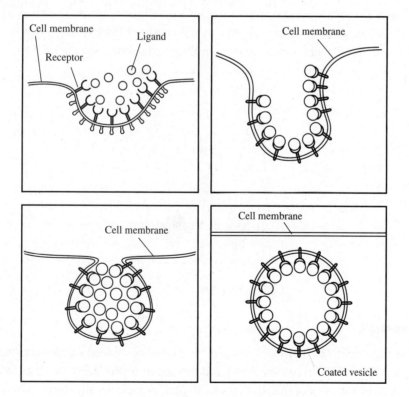

Figure 3.19 Receptor-mediated endocytosis

Bulk Flow

Bulk flow is a general term for the overall movement of a fluid in one direction in an organism. In humans, **blood** moves around the body by bulk flow as a result of blood pressure created by the pumping heart. **Sap** in trees moves by bulk flow from the leaves to the roots due to active transport in the phloem. Bulk flow movement is always from **source** (where it originates) **to sink** (where it is used).

CELL COMMUNICATION

Cells must communicate with each other and with their environment. They do it in a wide variety of ways. Simple bacterial cells secrete molecules that enable them to respond to changes in their population density by a phenomenon called **quorum sensing**. One example is the bacteria *Vibrio fischeri*, which produces the bioluminsescent substance luciferin that makes the bacteria glow. If luciferin were produced by only one single cell, it would not be detected by other animals. By using quorum sensing though, *V. fischeri* produces luciferin when its population is large enough that its bioluminescence can really be noticed.

Complex multicellular organisms contain trillions of cells that must communicate with each other and with the environment in order to maintain *homeostasis*. In plant cells, plasmodesmata (plant cell junctions) singal adjacent cells. Figures 3.20, 3.21, and 3.22 show the structure of three different types of junctions in animal cells: tight junctions, desmosomes, and gap junctions.

Tight Junctions

Tight junctions are belts around the epithelial cells that line organs and serve as a barrier to prevent leakage into or out of those organs. In the **urinary bladder**, they prevent the urine from leaking out of the bladder into the surrounding body cavity; see Figure 3.20.

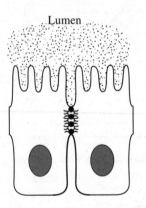

Figure 3.20 Tight junctions

Desmosomes

Desmosomes are found in many tissues and have been compared to **spot welds** that rivet cells together. They occur in tissues that are subjected to severe mechanical stress, such as skin epithelium or the neck of the uterus, which must expand greatly during childbirth; see Figure 3.21.

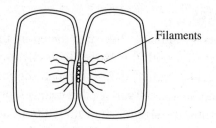

Figure 3.21 Desmosomes

Gap Junctions

Gap junctions permit the passage of materials directly from the cytoplasm of one cell to the cytoplasm of an adjacent cell. In the muscle tissue of the **heart**, the flow of ions through the gap junctions coordinates the contractions of the cardiac cells; see Figure 3.22.

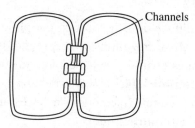

Figure 3.22 Gap junctions

Plasmodesmata

Plasmodesmata connect one plant cell to the next. They are analogous to gap junctions in animal cells.

Signal Transduction Pathways

Cell signals can diffuse locally or can travel long distances. **Autocrine signals** diffuse from one part of a cell to another part of the same cell. **Synaptic signaling**, in which the neurotransmitter is the signaling molecule, occurs in the animal nervous system. **Paracrine signals** send a message to nearby cells by diffusion, while **endocrine signals** (hormones) travel anywhere in the blood to reach their target cells.

How do cells interpret these signals once the cells receive them? To answer that question we focus on the **signal transduction pathway**, which coordinates the activities within, between, and among individual cells and supports the entire organism. Signal transduction pathways begin when very low concentrations of specific signaling chemicals, **ligands,** reach target cells either by local diffusion or by the circulatory system. The ligand binds to a specific **receptor**, either on the cell membrane or inside the cytoplasm of the target cell, in the same way a key fits into a lock. The binding of the ligand to a receptor activates the receptor and changes or *transduces* the signal into a form the cell can recognize. Ultimately, the signal transduction pathway leads to a cellular response such as a change in gene expression, a change in protein activity, or even cell death (apoptosis).

CELL SURFACE RECEPTORS

All **cell membrane receptors** are similar. They span the entire thickness of the membrane and are therefore in contact with both the extracellular environment and the cytoplasm. **Hydrophilic signaling molecules** cannot diffuse through the membrane. Hoewever, they bind to the part of the receptor on the cell surface that changes the shape on the cytoplasmic side of the same receptor. Thus, the signal is transmitted from outside the cell to the cytoplasm. Once inside the cell, the signal is carried by a **second messenger**. The most common second messenger is **cyclic AMP (cAMP)**. Notice that the **ligand**, the **first messenger**, never enters the cell. Cell surface receptors are so prevalent that they make up about 30% of all human proteins.

Three examples of cell surface receptors are **ion channel receptors**, **G-protein-coupled receptors**, and **protein kinase receptors**.

CYTOPLASMIC RECEPTORS

Small, nonpolar ligands can diffuse directly through the plasma membrane and set up a signal transduction pathway when they bind to receptors in the cytoplasm. This **ligand-receptor complex** either activates a pathway in the cytoplasm immediately or migrates to the nucleus where it acts as a **transcription factor** and switches genes on or off. Such *hydrophobic* chemical messengers include **steroid hormones** like **testosterone** and **estrogen** as well as **thyroid hormones** and the gas **NO (nitric oxide)**.

THE CASCADE EFFECT

Signal transduction pathways utilize a small number of extracellular signal molecules to produce a major cellular response. The advantage of this multistep cascade is that *it amplifies the signal and provides multiple opportunities for coordination and regulation.* Many signal transduction pathways have been identified and studied extensively across several kingdoms: in bacteria, yeast, animal cells, and plants. The amazing similarity in all of these pathways and in all these diverse organisms suggests that *signal transduction pathways evolved hundreds of millions of years ago in a common ancestor.*

The numbers on the side of Figure 3.23 show a cascade of reactions. The signal begins with one molecule of epinephrine binding to a receptor. It ends with the response, with thousands of molecules of glucose being released into the bloodstream.

RESPONSE

Ultimately, signal transduction pathways lead to a multitude of responses, either in the cytoplasm or in the nucleus. Regardless of the outcome, you need to remember four things about **signal transduction pathways**.

1. They are characterized by a signal, a transduction, and a response.
2. They are highly specific and regulated
3. One signal molecule can cause a cascade effect, releasing thousands of molecules inside a cell.
4. These pathways evolved millions of years ago in a common ancestor.

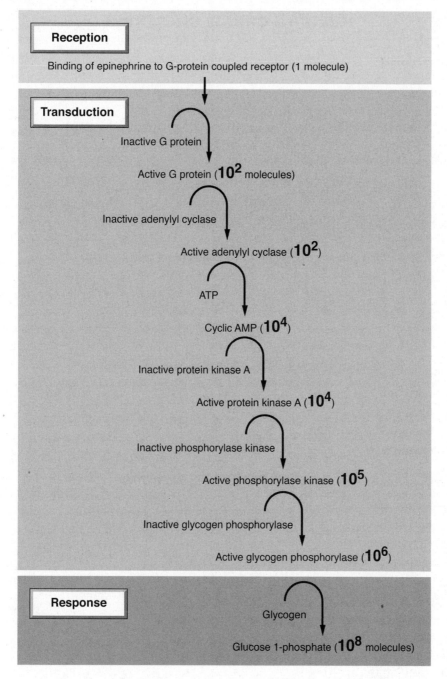

Reception

Binding of epinephrine to G-protein coupled receptor (1 molecule)

Transduction

Inactive G protein

Active G protein (10^2 molecules)

Inactive adenylyl cyclase

Active adenylyl cyclase (10^2)

ATP

Cyclic AMP (10^4)

Inactive protein kinase A

Active protein kinase A (10^4)

Inactive phosphorylase kinase

Active phosphorylase kinase (10^5)

Inactive glycogen phosphorylase

Active glycogen phosphorylase (10^6)

Response

Glycogen

Glucose 1-phosphate (10^8 molecules)

Figure 3.23 Cascade Effect of a Signal Transduction Pathway

Matching Column*

1. Produces ATP

2. Produces proteins

3. Packages and secretes substances

4. Contains hydrolytic enzymes

5. Directly assists with cell division

(A) Golgi apparatus

(B) Microtubules

(C) Rough endoplasmic reticulum

(D) Mitochondria

(E) Lysosomes

6. Which of the following is *not* normally found in a plant cell?

   (A) mitochondria
   (B) endoplasmic reticulum
   (C) plastids
   (D) centrioles

7. Which of the following is *present* in a prokaryote cell?

   (A) mitochondria
   (B) ribosomes
   (C) endoplasmic reticulum
   (D) chloroplasts

8. Membranes are components of all of the following EXCEPT a

   (A) microtubule
   (B) nucleus
   (C) Golgi apparatus
   (D) mitochondrion

9. Which are the following is NOT related to cell-to-cell communication?

   (A) Contact inhibition
   (B) Countercurrent exchange
   (C) Quorum sensing
   (D) Cyclic-AMP

*For study purposes, matching columns contain 5 choices.

10. A scientist has made a homogenate of human liver cells in a blender and then spun that mixture in an ultracentrifuge, as shown below. Which layer would include the most mitochondria?

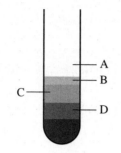

(A) A
(B) B
(C) C
(D) D

11. What is the approximate size of a human red blood cell?

(A) 0.01 micrometer
(B) 8 micrometers
(C) 80 micrometers
(D) 8 nanometers

12. Smooth E.R. carries out all of the following activities EXCEPT

(A) lipid production
(B) detoxification
(C) connects rough E.R. to the Golgi
(D) produces RNA

13. An animal cell in a hypertonic solution would

(A) swell
(B) swell and exhibit turgor
(C) exhibit plasmolysis
(D) shrink and then swell

14. Which one of the following would *not* normally diffuse through the lipid bilayer of a plasma membrane?

(A) CO_2
(B) amino acid
(C) starch
(D) water

15. Which of the following requires ATP?

(A) the uptake of cholesterol by a cell
(B) the facilitated diffusion of glucose into a cell
(C) countercurrent exchange
(D) the diffusion of oxygen into a fish's gills.

16. All of the following cellular activities require ATP EXCEPT

   (A) sodium-potassium pump
   (B) cells absorbing oxygen
   (C) receptor-mediated endocytosis
   (D) amoeboid movement

17. Which of the following best characterizes the structure of the plasma membrane?

   (A) rigid and unchanging
   (B) rigid but varying from cell to cell
   (C) fluid but unorganized
   (D) very active

18. The cytoplasmic channels between plant cells are called

   (A) desmosomes
   (B) middle lamellae
   (C) plasmodesmata
   (D) tight junctions

Questions 19–21

The following questions refer to the figure below, which shows the plasma membrane.

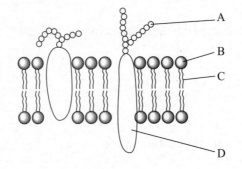

19. Identify the hydrophilic portion of a lipid molecule.

20. Identify the proteins involved in transport.

21. Identify the structure involved in cell-to-cell communication.

22. Which organelle contains DNA?

   (A) ribosomes
   (B) mitochondria
   (C) Golgi body
   (D) lysosomes

23. Which of the following is NOT a receptor on the surface of a cell?

   (A) Ion-channel receptor
   (B) G-protein-coupled receptor
   (C) Cyclic-AMP
   (D) Protein kinase receptor

24. Which of the following is correct about protein kinase receptors?

    (A) They activate G-protein in the cytoplasm of the cell.

    (B) They bind to steroid messengers on the surface of a cell.

    (C) They belong to a class of plasma membrane receptors that exhibit enzyme activity.

    (D) They allow for the passage of ions such as Na^+ ions through the plasma membrane of a cell.

25. Which of the following is correct about signal transduction pathways?

    (A) Signal transduction pathways are found only in the cells of the most complex animals.

    (B) Signal transduction pathways have evolved recently along with the development of enlarged brain size in mammals.

    (C) In the signal transduction pathway, a single molecule can stimulate the release of thousands of molecules of product within a cell.

    (D) Signal transduction pathways are unique in that receptors that stimulate the pathway are located only on the surface of a plasma membrane.

Answers to Multiple-Choice Questions

1. **(D)** Mitochondria release energy from organic molecules and store it in ATP molecules.

2. **(C)** Ribosomes attached to rough endoplasmic reticulum produce proteins.

3. **(A)** The Golgi apparatus receives newly synthesized proteins and lipids from the E.R. and distributes them to the plasma membrane, lysosomes, and secretory vesicles.

4. **(E)** Lysosomes are the principal sites of intracellular digestion.

5. **(B)** Microtubules make up the spindle fibers, which connect to the centromeres of chromosomes and assist in mitosis.

6. **(D)** Plant cells lack centrioles. Instead they have microtubule organizing regions. They also have mitochondria, ribosomes, plastids, and endoplasmic reticulum.

7. **(B)** Prokaryotes have NO internal membranes. Therefore, they lack mitochondria, E.R., chloroplasts, and nuclear membrane. They do have small ribosomes.

8. **(A)** Microtubules are part of the cytoskeletal structure and are made of the protein tubulin. The others all consist of selectively permeable plasma membranes.

9. **(B)** Countercurrent exchange refers to a mechanism that enhances diffusion, such as increased absorption of oxygen across fish gills. It is not an example of how cells communicate. Contact inhibition is the phenomenon whereby cells can sense that they are in a crowded environment at which point they stop dividing. Quorum sensing relates to the fact that cells, such as bacteria, can sense the size of their own population. Cyclic-AMP is a common secondary messenger in the cytoplasm of a cell receiving a signal from outside the cell.

10. **(D)** Centrifugation causes the densest structures to sink to the bottom and the lightest to remain on top. Nuclei are the densest, and mitochondria are the next most dense. The least dense layer would consist of ribosomes.

11. **(B)** Human red blood cells are small, 8 micrometers (μm) or 80 nanometers (nm). An average cell is about 80 micrometers.

12. **(D)** Smooth E.R. connects the rough E.R. to the Golgi, carries out detoxification, and produces lipids like steroids. The nucleolus produces RNA.

13. **(C)** An animal cell in a hypertonic solution would shrink because the concentration of water is greater inside the cell than outside the cell. Since water flows down a gradient, it would flow out of the cell. Plasmolysis means cell shrinking.

14. **(C)** Starch, a polysaccharide, is too large to diffuse through the plasma membrane.

15. **(A)** The uptake of cholesterol occurs by receptor-mediated endocytosis, which requires energy. B is an example of facilitated diffusion, and C and D are examples of countercurrent exchange. B, C, and D are all examples of passive transport and do not require energy.

16. **(B)** Oxygen is absorbed by diffusion. All the other choices are examples of active transport.

17. **(D)** The plasma membrane is organized and made of many small particles that move about readily. Hence the name, fluid mosaic. A membrane is a very active structure. A cell's activity is limited by how fast plasma membranes can take in and get rid of materials.

18. **(C)** Plasmodesmata are functionally like gap junctions in animal cells and are a means of cytoplasmic communication among plant cells. A system of plasmodesmata is called a symplast.

19. **(B)** The phospholipid head is polar and hydrophilic.

20. **(D)** This is a protein channel.

21. **(A)** The glycocalyx is involved in cell-to-cell communication.

22. **(B)** The nucleus, chloroplasts, and mitochondria all contain DNA.

23. **(C)** Cyclic-AMP is a common secondary messenger in the cytoplasm of a cell, not on the plasma membrane of a cell. It is activated when a lipid-soluble molecule (the first messenger) passes through the plasma membrane and enters the cell. All the others are examples of receptors on the cell surface.

24. **(C)** Protein kinase receptors belong to a class of plasma membrane receptors that function as enzymes. One example is RTK, receptor tyrosine kinase. Steroid messengers diffuse directly through the cell membrane and once inside the cell, bind to a second messenger, like c-AMP. G-protein-coupled receptors are another example of a receptor on the surface of the cell membrane. Ion-channel receptors are another example of a receptor on the surface of a cell.

25. **(C)** Signal transduction pathways are highly specific and regulated. Their similarity across different kingdoms speaks to a common ancestry. Receptors that stimulate signal transduction pathways are located both on the surface of the membrane as well as within the cytoplasm.

Directions: Answer all questions. You must answer the question in essay—**not** outline—form. You may use labeled diagrams to supplement your essay, but diagrams alone are *not* sufficient. Before you start to write, read each question carefully so that you understand what the question is asking.

1. Compare and contrast the characteristics of prokaryotic and eukaryotic cells.

2. Living cells are highly organized and regulated.
   a. Describe the structure of the plasma membrane
   b. Explain how the plasma membrane contributes to the regulation of the cell

Typical Free-Response Answers

Note: In the following essay, key words are in bold to make them more visible to you so you can focus on the importance of stating scientific terms when you write an essay.

1. Being the ancestor of all cells, prokaryotic cells have certain things in common with eukaryotic cells. They both have DNA, a cell wall (in the case of plant cells), and cytoplasm. Prokaryotic cells, like eukaryotic cells, carry out a wide range of metabolic processes including glycolysis, photosynthesis, and respiration; but prokaryotes carry them out in a far simpler fashion. Prokaryotes are generally small (1–10 μm in diameter); while eukaryotic cells are 10 times larger (10–100 μm). Prokaryotic cells have circular DNA while eukaryotes have long, linear DNA molecules with many noncoding regions encircled by a nuclear membrane. Prokaryotes are mainly unicellular, while eukaryotes are mostly multicellular. Prokaryotes have no cytoskeleton and no internal membranes. Metabolism in prokaryotes can be anaerobic or aerobic, but in eukaryotes metabolism is aerobic. Eukaryotes have an elaborate cytoskeletal structure and an array of internal membranes such as Golgi, endoplasmic reticulum, mitochondria, chloroplasts, and vacuoles.

2a. In 1972, S. J. Singer elucidated the structure of the cell membrane, which he called the **fluid mosaic model**. The plasma membrane is a selectively permeable and dynamic, fluid structure consisting of a continuous double phospholipid layer about 5nm thick, with proteins dissolved throughout the bilayer. The phospholipids are *amphipathic*, meaning they have polar (**hydrophilic**) heads and nonpolar (**hydrophobic**) tails. The polar heads face outward and the nonpolar tails face toward the interior of the membrane. Proteins are dispersed throughout the lipid layers. **Peripheral proteins** are bound to one face or the other and do not extend into the hydrophobic interior of the lipid bilayer. **Integral membrane proteins** are amphipathic and pass through the lipid layers. In eukaryotes, large quantities of **cholesterol** serve to enhance the flexibility and the mechanical stability of the bilayer.

2b. The lipid bilayer of the plasma membrane is a **selectively permeable membrane**. The lipid layer allows for the passage of water and small, nonpolar molecules. The protein component carries out most of the other functions. **Channel proteins** are large pores that extend across the membrane and allow the passage of specific molecules of appropriate size and charge to flow down a gradient into or out of a cell. **Carrier proteins** behave like membrane-bound enzymes and transport specific molecules across the membrane against a gradient. One example of proteins that work by active transport is the **sodium-potassium pump** in nerve cells. The Na-K pump acts as an *antiport*, carrying two different ions in opposite directions as the protein changes its conformation. Some proteins act as proton pumps, which are responsible for ATP production in mitochondria and the thylakoid membrane of chloroplasts. Other proteins are specific markers that are embedded on the surface of the cell and take part in the cell-cell recognition process. Plasma membranes vary from cell type to cell type depending on their particular function.

Cell Respiration

4

→ **ATP—ADENOSINE TRIPHOSPHATE**
→ **GLYCOLYSIS**
→ **ANAEROBIC RESPIRATION—FERMENTATION**
→ **AEROBIC RESPIRATION: THE CITRIC ACID CYCLE**
→ **STRUCTURE OF THE MITOCHONDRION**
→ **NAD⁺ AND FAD**
→ **AEROBIC RESPIRATION: THE ELECTRON TRANSPORT CHAIN**
→ **OXIDATIVE PHOSPHORYLATION AND CHEMIOSMOSIS**
→ **SUMMARY OF ATP PRODUCTION**

INTRODUCTION

Cell respiration is the means by which cells extract energy stored in food and transfer that energy to molecules of **ATP**. Energy that is temporarily stored in molecules of ATP is instantly available for every cellular activity such as passing an electrical impulse, contracting a muscle, moving cilia, or manufacturing a protein. The equation for the complete aerobic respiration of one molecule of glucose (seen below) is a highly exergonic process (releases energy). See Figure 4.1.

$$C_6H_{12}O_6 + 6O_2 \rightarrow 6CO_2 + 6H_2O + \text{energy}$$
$$\text{free energy} = \Delta G = -686 \text{ kcal/mole}$$

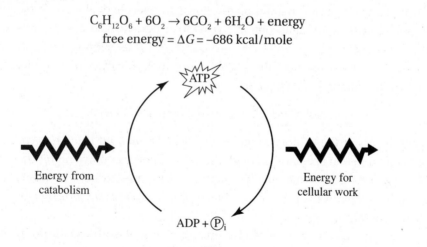

Figure 4.1

There are two types of cell respiration: anaerobic and aerobic. If oxygen is not present (anaerobic respiration), **glycolysis** is followed by either **alcoholic fermentation** or **lactic acid fermentation**. If oxygen is present (aerobic respiration), glycolysis is only the first phase of aerobic respiration. It is followed by the **Citric acid cycle (Krebs cycle)**, the **electron transport chain**, and **oxidative phosphorylation**.

ATP—ADENOSINE TRIPHOSPHATE

A molecule of ATP (adenosine triphosphate) consists of **adenosine** (the nucleotide adenine plus ribose) plus three **phosphates**. ATP is an unstable molecule because the three phosphates in ATP are all negatively charged and repel one another. When one phosphate group is removed from ATP by hydrolysis, a more stable molecule, ADP (adenosine diphosphate), results. *The change, from a less stable molecule to a more stable molecule, always releases energy.* ATP provides energy for all cells activities by transferring phosphates from ATP to another molecule, as seen in Figure 4.1. Figure 4.2 shows the structure of a molecule of ATP.

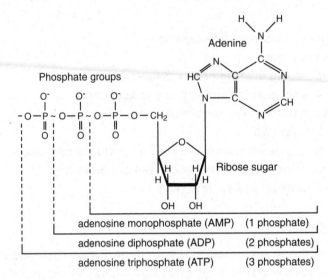

Figure 4.2 The Structure of Adenosine Triphosphate

GLYCOLYSIS

Glycolysis is a ten-step process that breaks down 1 molecule of glucose (a six-carbon molecule) into 2 three-carbon molecules of **pyruvate** or **pyruvic acid** and releases **4** molecules of **ATP**. The energy of activation for glycolysis is 2 ATP. After subtracting 2 ATPs from the 4 ATPs released from the reaction, the glycolysis of 1 molecule of glucose results in a *net gain of 2 ATP*. Here is the simplified equation.

$$2 \text{ ATP} + 1 \text{ Glucose} \rightarrow 2 \text{ Pyruvate} + 4 \text{ ATP}$$

Glycolysis occurs in the cytoplasm and produces ATP *without using oxygen*. Each step is catalyzed by a different enzyme. Although this process releases only one-fourth of the energy stored in glucose (most of the energy remains locked in pyruvate), the reaction is critical. The end product, pyruvate, is the *raw material for the Krebs cycle*, which is the next step in aerobic respiration.

During glycolysis, ATP is produced by **substrate level phosphorylation**—by direct enzymatic transfer of a phosphate to ADP. Only a small amount of ATP is produced this way.

There is one other important thing about glycolysis. The enzyme that catalyzes the third step, phosphofructokinase (PFK), is an **allosteric** enzyme. It inhibits glycolysis when the cell contains enough ATP and does not need to produce any more. If ATP is present in the cell in large quantities, it inhibits PFK by altering the conformation of that enzyme, thus stopping glycolysis. As the cell's activities use up ATP, less ATP is available to inhibit PFK and glycolysis continues, ultimately to produce more ATP. *This is an important example of how a cell regulates ATP production through allosteric inhibition;* see Figure 4.3.

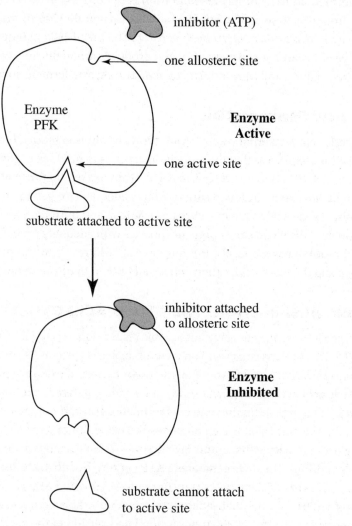

Figure 4.3 Allosteric Inhibition

ANAEROBIC RESPIRATION—FERMENTATION

Anaerobic respiration or **fermentation** is not a synonym for glycolysis. It is an anaerobic, catabolic process that consists of glycolysis *plus* alcohol or lactic acid fermentation. Anaerobic respiration originated millions of years ago when there was no free oxygen in the earth's atmosphere. Even today it is the sole means by which anaerobic bacteria such as *botulinum* (the bacterium that causes a form of food poisoning, botulism) release energy from food. There are two types of anaerobes: facultative and obligate. *Facultative anaerobes* can tolerate the presence of oxygen; they simply do not use it. *Obligate anaerobes* cannot live in an environment containing oxygen.

Fermentation can generate ATP during anaerobic respiration *only as long as there is an adequate supply of NAD+ to accept electrons* during glycolysis. Without some mechanism to convert NADH back to NAD+, glycolysis would shut down. Fermentation consists of glycolysis plus the reactions that regenerate NAD+. Two types of fermentation are **alcohol fermentation** and **lactic acid fermentation**.

Alcohol Fermentation

Alcohol fermentation or simply **fermentation** is the process by which certain cells convert pyruvate from glycolysis into **ethyl alcohol** and **carbon dioxide** in the **absence of oxygen** and, *in the process, oxidize NADH back to NAD+*. The bread-baking industry depends on the ability of yeast to carry out fermentation and produce carbon dioxide, which causes bread to rise. The beer, liquor, and wine industry depends on yeast to ferment sugar into ethyl alcohol.

Lactic Acid Fermentation

During **lactic acid fermentation**, pyruvate from glycolysis is reduced to form **lactic acid** or **lactate**. This is the process that the dairy industry uses to produce yogurt and cheese. Also in the process, *NADH gets oxidized back to NAD+*. **Human skeletal muscles** also carry out lactic acid fermentation when the blood cannot supply adequate oxygen to muscles during strenuous exercise. Lactic acid in the muscle causes fatigue and burning. The lactic acid continues to build up until the blood can supply the muscles with adequate oxygen to repay the oxygen debt. With normal oxygen levels, the muscle cells will revert to the more efficient aerobic respiration and the lactic acid is then converted back to pyruvate in the liver.

AEROBIC RESPIRATION: THE CITRIC ACID CYCLE

When oxygen is present, cells carry out aerobic respiration. It is highly efficient and produces a lot of ATP. This process consists of an anaerobic phase—glycolysis—plus an aerobic phase. The aerobic phase consists of two parts: the Krebs cycle and oxidative phosphorylation.

The **citric acid cycle** is a cyclical series of enzyme-catalyzed reactions also known as the **Krebs cycle**. It takes place in the **matrix of mitochondria** and requires pyruvate, the product of glycolysis. The citric acid cycle completes the oxidation of glucose to CO_2. It turns twice for each glucose molecule that enters glycolysis (once for each pyruvate molecule that enters the mitochondrion). The cycle generates 1 ATP per turn by **substrate-level phosphorylation**, but most of the chemical energy is transferred to NAD+ and FAD. The reduced coenzymes, NADH and $FADH_2$, shuttle high-energy electrons into the electron transport chain in the cristae membrane. Figure 4.4 illustrates a simplified version of the citric acid cycle. Here are some other important points:

- In the first step, acetyl co-A combines with oxaloacetic acid (OAA or oxaloacetate) to produce citric acid; hence the name citric acid cycle.
- Remember that each molecule of glucose is broken down to 2 molecules of pyruvate during glycolysis. Therefore, the respiration of each molecule of glucose causes the Krebs cycle to turn two times.
- Before it enters the Krebs cycle, pyruvate must first combine with coenzyme A (a vitamin) to form **acetyl co-A**, which does enter the Krebs cycle. The conversion of pyruvate to acetyl co-A produces **2 molecules of NADH**, 1 NADH for each pyruvate.
- Each turn of the Krebs cycle releases **3 NADH, 1 ATP, 1 FADH**, and the waste product CO_2, which is exhaled. (Remember, two turns of the Krebs cycle occur per glucose molecule.)
- During the Krebs cycle, ATP is produced by **substrate level phosphorylation**—direct enzymatic transfer of a phosphate to ADP. Very little energy is produced this way compared with the amount produced by oxidative phosphorylation.

TIP

The Citric Acid cycle is also referred to as the Krebs cycle. Don't let the different names confuse you.

STUDY TIP

Your teacher may want you to memorize this, but you won't need to know the specifics for the AP exam.

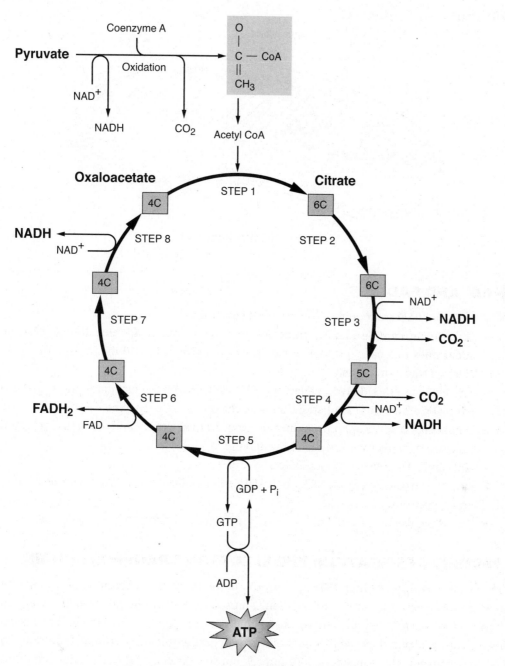

Figure 4.4 The Citric Acid Cycle (Krebs Cycle)

STRUCTURE OF THE MITOCHONDRION

The mitochondrion is enclosed by a double membrane. The outer membrane is smooth, but the inner, or **cristae membrane** is folded. This inner membrane divides the mitochondrion into two internal compartments, the **outer compartment** and the **matrix**. The Krebs cycle takes place in the matrix; the electron transport chain takes place in the cristae membrane. Figure 4.5 shows a diagram of the mitochondrion.

STUDY TIP

Cytoplasm—
Glycolysis

Matrix—Krebs
cycle

Cristae
membrane—
Electron
transport chain

Outer
compartment—
Proton
concentration
builds up

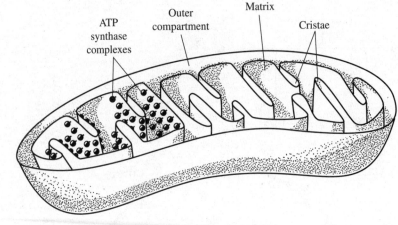

Figure 4.5 Mitochondrion

NAD⁺ AND FAD

- NAD and FAD are required for normal cell respiration.
- NAD⁺ (**nicotinamide adenine dinucleotide**) and FAD (**flavin adenine dinucleotide**) are **coenzymes** that carry protons or electrons from glycolysis and the Krebs cycle to the electron transport chain.
- The enzyme NAD dehydrogenase or FAD dehydrogenase facilitates the transfer of hydrogen atoms from a substrate, such as glucose, to its coenzyme NAD⁺.
- Without NAD⁺ to accept protons and electrons from glycolysis and the Krebs cycle, both processes would cease and the cell would die.
- NAD and FAD are vitamin derivatives.
- NAD⁺ is the oxidized form. NAD_{re} or NADH is the reduced form. NADH carries 1 proton and 2 electrons.
- FAD is the oxidized form. FAD_{re} or $FADH_2$ is the reduced form.

AEROBIC RESPIRATION: THE ELECTRON TRANSPORT CHAIN

The **electron transport chain (ETC)** is a **proton pump** in the mitochondria that couples two reactions, one exergonic and one endergonic. It uses the energy released from the exergonic flow of electrons to pump protons against a gradient from the matrix to the outer compartment. This results in the establishment of a **proton gradient** inside the mitochondrion. The electron transport chain makes no ATP directly but sets the stage for ATP production during **chemiosmosis**; see Figure 4.6.

Important Points

- The ETC is a collection of molecules embedded in the **cristae membrane** of the mitochondrion.
- There are thousands of copies of the ETC in every mitochondrion due to the extensive folding of the cristae membrane.
- The ETC carries electrons delivered by NAD and FAD from glycolysis and the Krebs cycle to oxygen, the **final electron acceptor**, through a series of **redox reactions**. In a redox reaction, one atom gains electrons or protons (**reduction**), and one atom loses electrons (**oxidation**).

- The highly electronegative **oxygen** acts to pull electrons through the electron transport chain.
- **NADH** delivers its electrons to a higher energy level in the chain than does **FADH$_2$**. As a result, NADH provides more energy for ATP synthesis than does FAD. Each NADH produces 3 ATP molecules, while each FADH$_2$ produces 2 ATP molecules.
- The ETC consists mostly of **cytochromes**. These are proteins structurally similar to hemoglobin. Cytochromes are present in all aerobes and are used to trace evolutionary relationships.

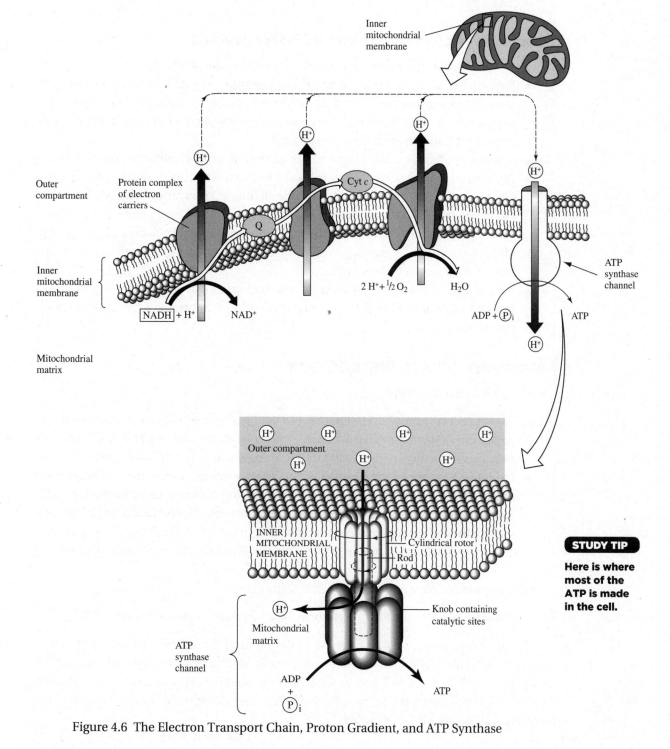

Figure 4.6 The Electron Transport Chain, Proton Gradient, and ATP Synthase

OXIDATIVE PHOSPHORYLATION AND CHEMIOSMOSIS

Most of the energy produced during cell respiration occurs in the mitochondria by a process known by the general name of **oxidative phosphorylation**. This term means the phosphorylation of ADP into ATP by the oxidation of the carrier molecules NADH and $FADH_2$. This energy-coupling mechanism was elucidated in 1961 by Peter Mitchell, who named it the **chemiosmotic theory**. According to the Mitchell hypothesis, chemiosmosis uses potential energy stored in the form of a **proton (H^+) gradient** to phosphorylate ADP and produce ATP (ADP + P → ATP).

Important Points of Oxidative Phosphorylation

- It is powered by the redox reactions of the electron transport chain.
- Protons are pumped from the **matrix** to the **outer compartment** by the electron transport chain. The electron transport chain is an energy converter that couples the exergonic flow of electrons with the endergonic pumping of protons across the cristae membrane and into the outer compartment.
- A proton gradient is created between the outer compartment and the inner matrix.
- Protons cannot diffuse through the cristae membrane; they can flow only down the gradient into the matrix through **ATP synthase channels**. This is **chemiosmosis**, the key to the production of ATP. *As protons flow through the ATP-synthase channels, they generate energy to phosphorylate ADP into ATP.* This process is similar to how a hydroelectric plant converts the enormous potential energy of water flowing through a dam to turn turbines and generate electricity.
- **Oxygen** is the **final hydrogen acceptor**, combining $\frac{1}{2}$ an oxygen molecule with 2 electrons and 2 protons, thus forming **water**. This water is a waste product of cell respiration and is excreted.

SUMMARY OF ATP PRODUCTION

ATP is produced in two ways.

- **Substrate level phosphorylation** occurs when an enzyme, a **kinase**, transfers a phosphate from a substrate directly to ADP. Only a small amount of ATP is produced this way. This is the way energy is produced during glycolysis and the Krebs cycle.
- **Oxidative phosphorylation** occurs during chemiosmosis. This is the way 90 percent of all ATP is produced from cell respiration. During oxidative phosphorylation, NAD and FAD lose protons (become oxidized) to the electron transport chain, which pumps them to the outer compartment of the mitochondrion, creating a steep proton gradient. This electrochemical or proton gradient powers the phosphorylation of ADP into ATP.

During respiration, most energy flows in this sequence:

Glucose → NAD$_{re}$ and FAD$_{re}$ → electron transport chain → chemiosmosis → ATP

Theoretically, about 26–28 ATP can be produced from the aerobic respiration of 1 molecule of ATP. This is hypothetical because some cells are more efficient than others and cells vary in their efficiency at different times. Remember that during glycolysis, 2 pyruvates are formed and each enters the Krebs cycle separately. So following glycolysis, all numbers for $FADH_2$, NADH, and ATP are doubled.

The catabolism (breakdown) of glucose under aerobic conditions occurs in three sequential pathways: glycolysis, pyruvate oxidation, and the citirc acid cycle, which produce the reduced coenzymes NADH and $FADH_2$. These coenzymes are then oxidized by the electron transport chain and ATP is produced by oxidative phosphorylation; see Figure 4.7.

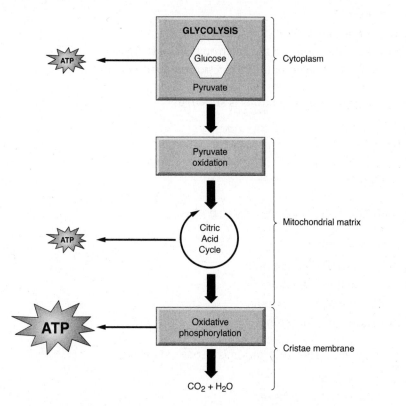

Figure 4.7 The Aerobic Breakdown of Glucose

Figure 4.8 shows an overview of ATP production from cell respiration.

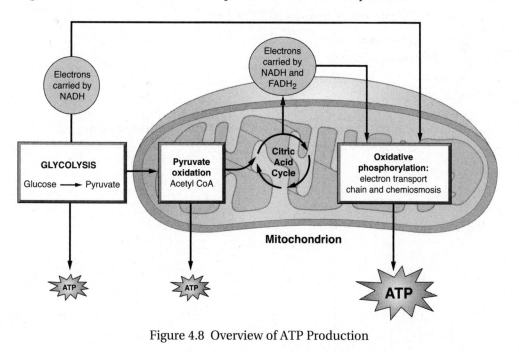

Figure 4.8 Overview of ATP Production

1. The role of oxygen in aerobic respiration is

   (A) to transport CO_2
   (B) most important in the Krebs cycle
   (C) to provide electrons for the electron transport chain
   (D) as the final H_2 acceptor in the electron transport chain

2. The loss of protons and electrons is known as

   (A) dehydration
   (B) hydrogenation
   (C) reduction
   (D) oxidation

3. $C_6H_{12}O_6 + 6O_2 \rightarrow 6H_2O + 6CO_2 + 38$ ATP

   The process shown is

   (A) reduction and is endergonic
   (B) reduction and is exergonic
   (C) oxidation and is endergonic
   (D) oxidation and is exergonic

4. Most energy during cell respiration is harvested during

   (A) the Krebs cycle
   (B) oxidative phosphorylation
   (C) glycolysis
   (D) anaerobic respiration

5. All of the following processes produce ATP EXCEPT

   (A) lactic acid formation
   (B) oxidative phosphorylation
   (C) glycolysis
   (D) the Krebs cycle

6. After strenuous exercise, a muscle cell would contain decreased amounts of _____ and increased amounts of _____.

   (A) glucose; ATP
   (B) ATP; glucose
   (C) ATP; lactic acid
   (D) lactic acid; ATP

Match each process with its correct location

7. Glycolysis

8. Electron transport chain

9. Krebs cycle

(A) The cristae membrane

(B) Cytoplasm

(C) Inner matrix of the mitochondria

The three circles represent three major processes in aerobic respiration.

$$glucose \rightarrow \boxed{process\ A} \rightarrow \boxed{process\ B} \rightarrow \boxed{process\ C} \rightarrow CO_2 + H_2O$$

10. Process C represents

(A) glycolysis
(B) the Krebs cycle
(C) the electron transport chain
(D) substrate level phosphorylation

Questions 11–12

The following questions refer to the sketch of a mitochondrion, shown below.

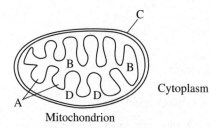

Mitochondrion

11. Identify the site of the Krebs cycle in the sketch of the mitochondrion.

12. Identify the site of the ATP synthase.

_____

13. Each NAD molecule carrying hydrogen to the electron transport chain can produce a maximum of _____ molecules of ATP?

(A) 1
(B) 2
(C) 3
(D) 4

14. Which is true of aerobic respiration but not true of anaerobic respiration?

(A) CO_2 is produced.
(B) ATP is produced.
(C) Water is produced.
(D) Alcohol is produced.

15. Which of the following is the most important thing that happens during aerobic respiration?

   (A) Electrons move down the electron transport chain in a series of redox reactions.
   (B) Acetyl CoA enters the Krebs cycle.
   (C) NAD carries hydrogen to the electron transport chain.
   (D) ATP is produced.

16. The ATP produced during glycolysis is generated by which of the following?

   (A) the electron transport chain
   (B) substrate level phosphorylation
   (C) oxidative phosphorylation
   (D) chemiosmosis

17. In addition to ATP, what is produced during glycolysis?

   (A) pyruvate and NADH
   (B) CO_2 and H_2O
   (C) CO_2 and ethyl alcohol
   (D) CO_2 and NADH

18. Which of the following probably evolved first?

   (A) the Krebs cycle
   (B) oxidative phosphorylation
   (C) glycolysis
   (D) the electron transport chain

19. Which is an example of a feedback mechanism?

   (A) Phosphofructokinase, an allosteric enzyme in glycolysis, is inhibited by ATP.
   (B) Lactic acid gets converted back to pyruvic acid in the human liver.
   (C) ATP is produced in mitochondria as protons flow through the ATP-synthase channel.
   (D) Energy is released from glucose as it decomposes into CO_2 and H_2O.

20. Which process of cell respiration is most closely associated with intracellular membranes?

   (A) oxidative phosphorylation
   (B) the Krebs cycle
   (C) glycolysis
   (D) substrate level phosphorylation

21. During cell respiration, most ATP is formed as a direct result of the net movement of

   (A) electrons flowing against a gradient
   (B) electrons flowing through a channel
   (C) protons flowing through a channel
   (D) protons flowing against a gradient

22. Glycolysis is the first phase of aerobic cellular respiration. It is a complex, enzyme-controlled set of reactions in which glucose molecules are broken down into pyruvate in the absence of oxygen. Although it does not produce much ATP, glycolysis is important because pyruvate is the raw material for the next phase of cellular respiration, which will ultimately produce large amounts of ATP by oxidative phosphorylation. One of the enzymes at the beginning of glycolysis is PFK, phosphofructokinase, an allosteric enzyme. When ATP binds to the allosteric site on PFK, the enzyme changes shape and no longer functions. Which of the following statements best explains the importance of the enzyme PFK in glycolysis?

(A) PFK inhibits glycolysis when oxygen levels are high.

(B) PFK enables glycolysis to continue when no oxygen is present.

(C) PFK inhibits the production of ATP when ATP levels are high.

(D) PFK enhances the production of ATP when ATP levels are high.

Answers to Multiple-Choice Questions

1. **(D)** Oxygen is the final hydrogen and electron acceptor in the electron transport chain. CO_2 is formed and released by cellular respiration. Oxygen does not take part in glycolysis or in the Krebs cycle.

2. **(D)** The loss of hydrogen or loss of electrons is known as oxidation. The gain of hydrogen or of electrons is known as reduction because when electrons are added to an atom, the charge on the atom is reduced.

3. **(D)** This process, cell respiration, is exergonic because energy is being released. It is an oxidation reaction because carbon loses hydrogen atoms and oxidation is defined as the loss of hydrogen atoms.

4. **(B)** A small amount of ATP is produced by substrate level phosphorylation during glycolysis and the Krebs cycle. However, most of the ATP produced during cell respiration occurs by chemiosmosis or oxidative phosphorylation during the electron transport chain.

5. **(A)** Lactic acid is produced from pyruvic acid during fermentation but does not result in the production of energy. All the other choices produce ATP. Oxidative phosphorylation is the process by which ATP is produced during the ETC.

6. **(C)** During sustained strenuous exercise, muscle cells use up ATP and oxygen as they carry out anaerobic respiration. Anaerobic respiration produces lactic acid, which causes weakness and fatigue in the muscle.

7. **(B)** Glycolysis occurs in the cytoplasm.

8. **(A)** The electron transport chain is located within the cristae membrane of the mitochondria.

9. **(C)** The Krebs cycle occurs in the matrix of the mitochondria.

10. **(C)** The order of energy in producing processes in cell respiration is glycolysis, the Krebs cycle, then the electron transport chain.

11. **(B)** The Krebs cycle takes place in the matrix of the mitochondrion.

12. **(A)** The ATP synthase molecules lie within the cristae membrane.

13. **(C)** Each NADH can produce up to 3 ATP, and each FAD produces 2 ATP.

14. **(C)** Water is produced as oxygen combines with hydrogen flowing through the ATP synthase during chemiosmosis of aerobic respiration only. This does not occur in anaerobic respiration. CO_2 is produced in both the aerobic (Krebs cycle) and in anaerobic fermentation. ATP is produced in anaerobic and aerobic respiration. Alcohol is produced only in anaerobic respiration.

15. **(D)** Choices A, B, and C all describe events that occur during cell respiration and lead up to the production of ATP. However, the most important event in respiration is the production of ATP.

16. **(B)** Substrate level phosphorylation is responsible for the production of ATP during glycolysis and the citric acid cycle (also known as the Krebs cycle). It does not produce as much ATP as does chemiosmosis.

17. **(A)** ATP, NADH, and pyruvate are the end products of glycolysis. CO_2 is a by-product of the Krebs cycle. H_2O is a waste product of chemiosmosis and is formed when protons combine with electrons and oxygen. Ethyl alcohol is a by-product of alcoholic fermentation.

18. **(C)** Glycolysis occurs in the first phase of cell respiration and does not require oxygen. The first organisms on earth were probably anaerobes because no free oxygen was available in the atmosphere. Choices A, B, and D all require free oxygen.

19. **(A)** Choices B–D are all correct statements about respiration, but they are not examples of a feedback mechanism. Choice A is an example of negative feedback because when plenty of ATP is available in the cell to meet demand, respiration slows down, conserving valuable molecules and energy for other functions.

20. **(A)** Oxidative phosphorylation occurs as protons flow through the ATP synthase channel within the cristae membrane of the mitochondria. The Krebs cycle occurs within the matrix of the mitochondria. Glycolysis occurs in the cytoplasm of the cell. Substrate level phosphorylation explains how ATP is produced from the Krebs cycle and glycolysis.

21. **(C)** Most of the ATP produced during cell respiration comes from oxidative phosphorylation as protons flow through the ATP synthase channel in the cristae membrane.

22. **(C)** The focus of the question is on PFK and the fact that it is an allosteric enzyme. The stem of the question states that ATP is an inhibitor of PFK. So when ATP levels are high, PFK does not function. That means that when ATP levels are high, PFK stops functioning and glycolysis and the rest of cellular respiration are shut down. This makes sense because the purpose of cellular respiration is to produce ATP. So, if ATP is present, there is no need to produce more. PFK is the key to that feedback—or allosteric—mechanism.

> **Directions:** Answer all questions. You must answer the question in essay—**not** outline—form. You may use labeled diagrams to supplement your essay, but diagrams alone are *not* sufficient. Before you start to write, read each question carefully so that you understand what the question is asking.

Membranes are important structural features of cells.

a. Describe the structure of a membrane.

b. Discuss the role of membranes in ATP synthesis.

Typical Free-Response Answer

> *Note: Key words are in bold to remind you that you must focus on scientific terminology.*

a. The structure of the plasma membrane was elucidated by **S. J. Singer** in 1972, who described the membrane as a **fluid mosaic**, meaning it is made of small pieces that move. The plasma membrane consists of a **phospholipid bilayer** with the **hydrophilic head** of the phospholipid end facing outward and the **hydrophobic tail** facing inward. Protein molecules are dispersed throughout the membrane. Some proteins, **integral proteins**, completely span the membrane. Others, **peripheral proteins**, are loosely bound to the surface of the membrane. **Cholesterol molecules** are embedded in the interior to stabilize the membrane. The average membrane has the consistency of olive oil and is about 40 percent lipid and 60 percent protein. The membrane is **selectively permeable** and therefore controls what enters and leaves the cell. In general only small, uncharged, and hydrophobic molecules can diffuse freely through the membrane. Large, polar molecules cannot diffuse through the membrane; they must pass through special **protein channels**.

b. Cristae membranes in mitochondria play a special role in ATP synthesis. The cristae membrane contains **electron transport chains (ETC)**, collections of molecules embedded in the membrane. Most of these molecules are proteins that carry electrons from higher to lower energy levels. The ETC uses the exergonic flow of electrons to pump protons across the cristae membrane to the outer compartment of the mitochondria to create an **electrochemical** or **proton gradient**. The key here is that the cristae membrane does not allow protons to diffuse through the membrane. Protons can pass through only special protein channels, large enzyme complexes, called **ATP synthase channels**. As protons flow through the ATP synthase channels, like water flowing through a dam, energy is generated to produce ATP by a process known as **chemiosmosis**.

Photosynthesis

5

- → PHOTOSYNTHETIC PIGMENTS
- → THE CHLOROPLAST
- → PHOTOSYSTEMS
- → LIGHT-DEPENDENT REACTIONS—THE LIGHT REACTIONS
- → THE CALVIN CYCLE
- → PHOTORESPIRATION
- → C-4 PHOTOSYNTHESIS
- → CAM PLANTS

INTRODUCTION

Photosynthesis is the process by which light energy is converted to chemical bond energy and carbon is fixed into organic compounds. The general formula is:

$$6CO_2 + 12H_2O \xrightarrow{\text{light}} C_6H_{12}O_6 + 6H_2O + 6O_2$$

There are two main processes of photosynthesis: the **light-dependent** and the **light-independent reactions**. The light reactions use light energy directly to produce ATP that powers the light-independent reactions. The light-independent reactions consist of the Calvin cycle, which produces sugar. To power the production of sugar, the Calvin cycle uses ATP formed during the light reactions. Both reactions occur only when light is present.

PHOTOSYNTHETIC PIGMENTS

Photosynthetic pigments absorb light energy and use it to provide energy to carry out photosynthesis. Plants contain two major groups of pigments, the chlorophylls and carotenoids. **Chlorophyll *a*** and **chlorophyll *b*** are green and absorb all wavelengths of light in the red, blue, and violet range. The **carotenoids** are yellow, orange, and red. They absorb light in the blue, green, and violet range. **Xanthophyll**, another photosynthetic pigment, is a carotenoid with a slight chemical variation. Pigments found in red algae, the **phycobilins**, are reddish and absorb light in the blue and green range. Chlorophyll *b*, the carotenoids, and the phycobilins are known as **antenna pigments** because they capture light in wavelengths other than those captured by chlorophyll *a*. Antenna pigments absorb photons of light and pass the energy along to chlorophyll *a*, which is directly involved in the transformation of light energy to sugars. Figure 5.1 is a graph showing the absorption spectrum for photosynthetic pigments that were extracted from a leaf.

STUDY TIP

Photo—
Light

Synthesis—
Sugar
production:
think of the
Calvin cycle

**Chlorophyll *a*
absorbs light in
the violet, blue,
and red ranges.
It reflects green,
yellow, and
orange light.**

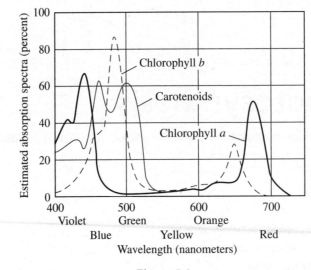

Figure 5.1

Figure 5.2 is an **action spectrum** with wavelengths of light plotted against rate of photo-synthesis. The data are obtained from a living plant—not from isolated photosynthetic pigment as in an absorption spectrum. This figure is different from Figure 5.1 because Figure 5.2 includes data from all photosynthetic pigments in the plants: chlorophyll *a*, chlorophyll *b,* and carotenoids.

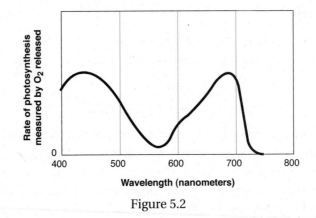

Figure 5.2

Chlorophyll *a* is the pigment that participates directly in the light reactions of photosynthesis. It is a large molecule with a single **magnesium** atom in the head surrounded by alternating double and single bonds. The head, called the porphyrin ring, is attached to a long hydrocarbon tail. The double bonds play a critical role in the light reactions. They are the source of the electrons that flow through the electron transport chains during photosynthesis. Figure 5.3 is a drawing of chlorophyll *a.*

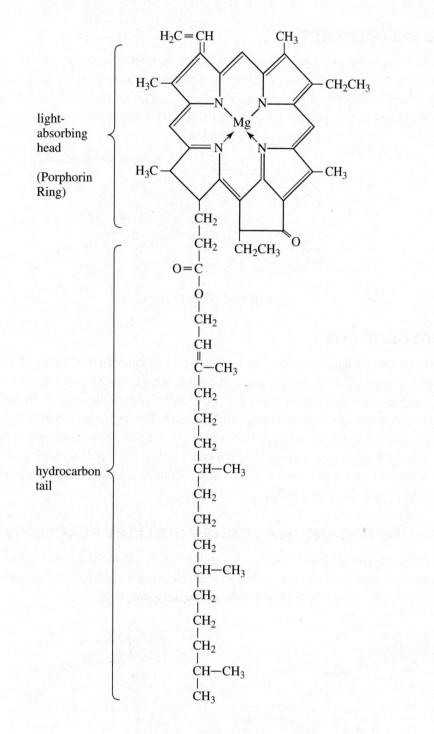

Figure 5.3 Chlorophyll *a*

THE CHLOROPLAST

The chloroplast contains photosynthetic pigments that, along with enzymes, carry out photosynthesis. It also contains **grana**, where the light-dependent reactions occur, and **stroma**, where the light-independent reactions occur. The grana consist of layers of membranes called **thylakoids**, the site of photosystems I and II. The chloroplast is enclosed by a double membrane. Figure 5.4 is a sketch of a chloroplast.

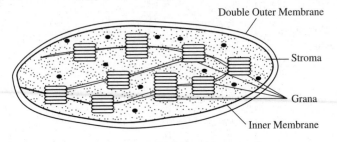

Figure 5.4 Chloroplast

PHOTOSYSTEMS

Photosystems are light-harvesting complexes in the thylakoid membranes of chloroplasts. There are a few hundred photosystems in each thylakoid. Each photosystem consists of a **reaction center** containing chlorophyll *a* and a region containing several hundred antenna pigment molecules that funnel energy into chlorophyll *a*. Two types of photosystems, **PS I** and **PS II**, cooperate in the light reactions of photosynthesis. They are named in the order in which they were discovered. However, PS II operates first, followed by PS I. PS I absorbs light best in the 700 nm range; hence it is also called **P700**. PS II absorbs light best in the 680 nm range; hence it is also called **P680**.

LIGHT-DEPENDENT REACTIONS—THE LIGHT REACTIONS

Light is absorbed by the photosystems (PS II and PS I) in the thylakoid membranes and electrons flow through electron transport chains. As shown in Figure 5.5 there are two possible routes for electron flow: **noncyclic flow** and **cyclic photophosphorylation**.

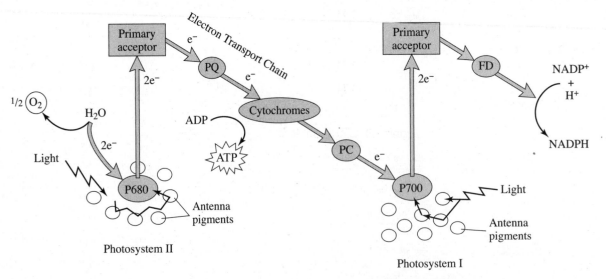

Figure 5.5 Noncyclic Photophosphorylation

Noncyclic Photophosphorylation

REMEMBER

Electrons flow from water to P680 to P700 to NADP, which carries them to the Calvin cycle.

During **noncyclic photophosphorylation**, electrons enter two electron transport chains, and *ATP and NADPH (nicotinamide dinucleotide phosphate) are formed.* The process begins in PS II and proceeds through the following steps.

- **PHOTOSYSTEM II—P680.** Energy is absorbed by P680. Electrons from the double bonds in the head of chlorophyll *a* become energized and move to a higher energy level. They are captured by a **primary electron acceptor**.

- **PHOTOLYSIS.** Water gets split apart, providing electrons to replace those lost from chlorophyll *a* in P680. Photolysis splits water into two electrons, two protons (H^+), and one oxygen atom. Two oxygen atoms combine to form one O_2 molecule, which is released into the air as a waste product of photosynthesis.

- **ELECTRON TRANSPORT CHAIN.** Electrons from P680 pass along an electron transport chain consisting of plastoquinone (PQ), a complex of two cytochromes and several other proteins, and ultimately end up in P700 (PSI). This flow of electrons is exergonic and provides energy to produce ATP by **chemiosmosis**, the same way ATP is produced in mitochondria. Because this ATP synthesis is powered by light, it is called **photophosphorylation**. See Figure 5.6.

- **CHEMIOSMOSIS.** This is the process by which ATP is formed during the light reactions of photosynthesis. Protons that were released from water during photolysis are pumped by the thylakoid membrane from the stroma into the **thylakoid space (lumen)**. ATP is formed as these protons diffuse down the gradient from the thylakoid space, through the **ATP synthase channels**, and into the stroma. The ATP produced here provides the energy that powers the Calvin cycle.

- **NADP** becomes reduced when it picks up the two protons that were released from water in P680. Newly formed **NADPH** carries hydrogen to the Calvin cycle to make sugar in the light-independent reactions.

- **PHOTOSYSTEM I—P700.** Energy is absorbed by P700. Electrons from the head of chlorophyll *a* become energized and are captured by a primary electron receptor. This process is similar to the way it happens in P680. One difference is that the electrons that escape from chlorophyll *a* are replaced with electrons from photosystem II, P680, instead of from water. Another difference is that this electron transport chain contains ferrodoxin and ends with the production of NADPH, not ATP.

Here is an overview of noncyclic photophosphorylation.

light → P680 → ATP produced → P700 → NADPH produced
       oxygen   (NADPH carries
       released H^+ to the Calvin cycle)

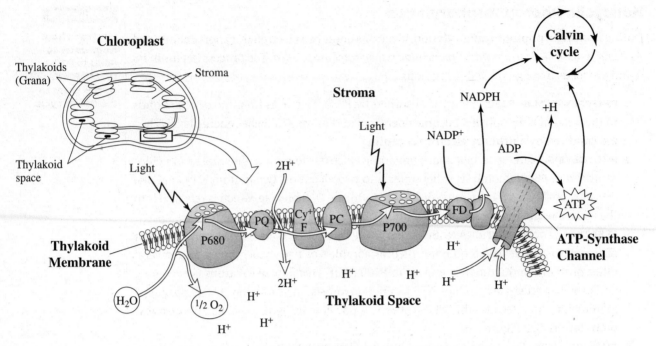

Figure 5.6 The light-dependent reactions

Cyclic Photophosphorylation

The sole purpose of cyclic photophosphorylation is to produce ATP. No NADPH is produced, and no oxygen is released.

The production of sugar that occurs during the Calvin cycle consumes enormous amounts of ATP, so periodically, the chloroplast runs low on ATP. When it does, the chloroplast carries out **cyclic photophosphorylation** to replenish the ATP levels. Cyclic electron flow takes photoexcited electrons on a short-circuit pathway. Electrons travel from the P680 electron transport chain to P700, to a primary electron acceptor, and then back to the cytochrome complex in the P680 electron transport chain. Cyclic photophosphorylation is shown in Figure 5.7.

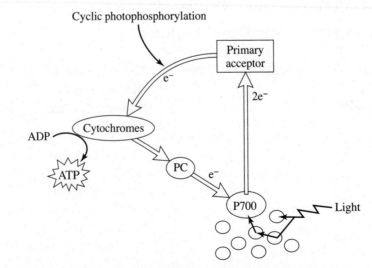

Figure 5.7 Cyclic Photophosphorylation

THE CALVIN CYCLE

The **Calvin cycle** is the main business of the light-independent reactions. It is a cyclical process that produces the 3-carbon sugar **PGAL (phosphoglyceraldehyde)**. Carbon enters the stomates of a leaf in the form of CO_2 and becomes *fixed* or incorporated into PGAL.

Here are the important aspects of the Calvin cycle, as shown in Figure 5.8.

- The process that occurs during the Calvin cycle is **carbon fixation**.
- It is a **reduction reaction** since carbon is gaining hydrogen.
- CO_2 enters the Calvin cycle and becomes attached to a 5-carbon sugar, **ribulose biphosphate (RuBP)**, forming a 6-carbon molecule. The 6-carbon molecule is unstable and immediately breaks down into two 3-carbon molecules of **3-phosphoglycerate (3-PGA)**. The enzyme that catalyzes this first step is ribulose biphosphate carboxylase (**rubisco**).
- The Calvin cycle does not directly depend on light. Instead, it uses the products of the light reactions: ATP and NADPH.
- The Calvin cycle, like the light-dependent reactions, **occurs only in the light**.

STUDY TIP

Your teacher may want you to know this for class, but you will not see all these details on the AP exam. But you do need to understand how it works.

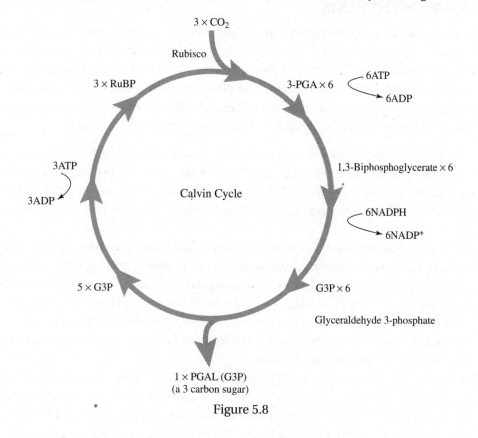

Figure 5.8

PHOTORESPIRATION

In most plants, CO_2 enters the Calvin cycle and is fixed into **3-phosphoglycerate (3-PGA)** by the enzyme rubisco. These plants are called **C-3 plants** because the first step produces the compound 3-PGA, which contains three carbons. This first step is not very efficient because rubisco binds with O_2 as well as with CO_2. When rubisco binds with O_2 instead of CO_2, this process, **photorespiration**, diverts the process of photosynthesis in two ways.

1. Unlike normal respiration, no ATP is produced
2. Unlike normal photosynthesis, no sugar is formed.

Instead, **peroxisomes** break down the products of photorespiration. If photorespiration does not produce any useful products, why do plants carry it out at all? This process is probably a vestige from ancient Earth billions of years ago when the atmosphere had little or no free oxygen to divert rubisco and sugar production.

C-4 PHOTOSYNTHESIS

C-4 photosynthesis is a modification for dry environments. C-4 plants exhibit modified anatomy and biochemical pathways that enable them to minimize excess water loss and maximize sugar production. As a result, C-4 plants thrive in hot and sunny environments where C-3 plants would wilt and die. Examples of C-4 plants are **corn**, **sugar cane**, and **crabgrass**. In C-4 plants, a series of steps precedes the Calvin cycle. These steps pump CO_2 entering the leaf away from the air spaces and deep into the leaf. The details are described:

- CO_2 enters the mesophyll cell of the leaf and combines with a 3-carbon molecule, **PEP (phosphoenolpyruvate)**, to form the 4-carbon molecule **oxaloacetate**. Hence, the plants have the name C-4 plants. The enzyme that catalyzes this reaction, **PEP carboxylase**, does not bind with oxygen and can therefore fix CO_2 more efficiently than rubisco does .
- From oxaloacetate the mesophyll cell produces **malic acid** or **malate**, which it pumps through the **plasmodesmata** into the adjacent **bundle sheath cell**. Once the malate is in the bundle sheath cell, it releases its CO_2, which gets incorporated into PGAL by the Calvin cycle. Because the bundle-sheath cell is deep within the leaf and little oxygen is present, rubisco can fix CO_2 efficiently without being diverted to the dead end of photorespiration. This biochemical pathway is called the **Hatch-Slack pathway**. Its purpose is to remove CO_2 from the air space near the stomate.
- With CO_2 sequestered inside bundle-sheath cells, there is a steep CO_2 gradient between the airspace in the mesophyll of the leaf near the stomates and the atmosphere around the leaf. Thus, C-4 plants can maximize the amount of CO_2 that diffuses into the air space in the leaf and minimize the length of time the stomates must remain open.
- **Kranz anatomy** refers to the structure of C-4 leaves that differs from C-3 leaves. In C-4 leaves, the bundle sheath cells lie under the mesophyll cells, deep within the leaf where CO_2 is sequestered. The light reactions occur in the mesophyll cells and the dark reactions occur in the bundle-sheath cells. In C-3 leaves, all photosynthetic cells have direct access to CO_2. Figure 5.9 is a sketch of both C-3 and C-4 leaves.

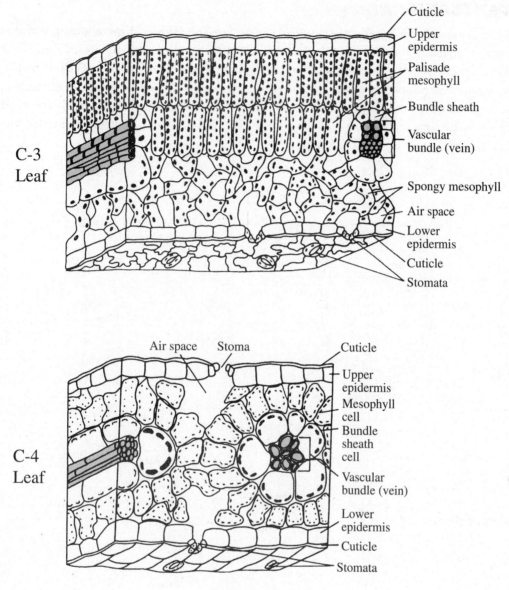

C-3 Leaf

- Cuticle
- Upper epidermis
- Palisade mesophyll
- Bundle sheath
- Vascular bundle (vein)
- Spongy mesophyll
- Air space
- Lower epidermis
- Cuticle
- Stomata

C-4 Leaf

- Air space
- Stoma
- Cuticle
- Upper epidermis
- Mesophyll cell
- Bundle sheath cell
- Vascular bundle (vein)
- Lower epidermis
- Cuticle
- Stomata

Figure 5.9

CAM PLANTS

CAM plants carry out a form of photosynthesis, **crassulacean acid metabolism**, which is another adaptation to dry conditions. These plants keep their stomates closed during the day and open at night, the reverse of how most plants behave. The mesophyll cells store CO_2 in organic compounds they synthesize at night. During the day, when the light reactions can supply energy for the Calvin cycle, CO_2 is released from the organic acids made the night before to become incorporated into sugar. Figure 5.10 shows three sketches comparing carbon fixation in C-3, C-4, and CAM plants.

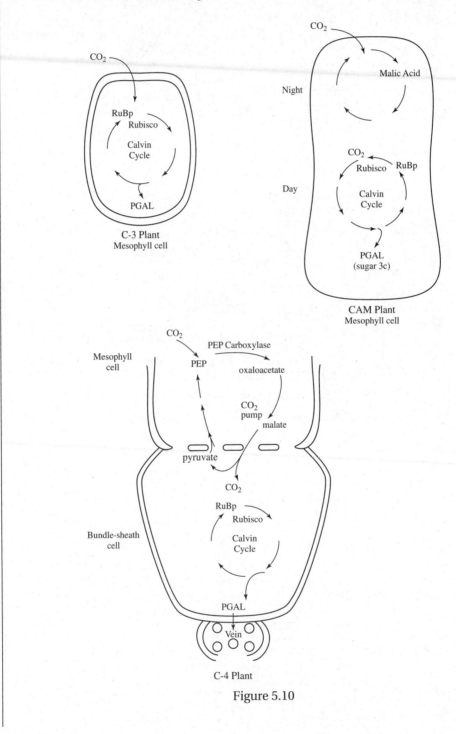

Figure 5.10

1. The graph below shows an absorption spectrum for an unknown pigment molecule.
   What color would this pigment appear?

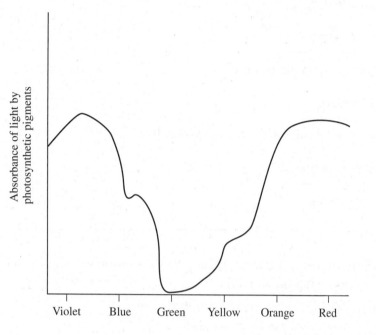

(A) red

(B) yellow

(C) green

(D) blue

2. Cyclic photophosphorylation results in the production of

(A) ATP only

(B) ATP and NADPH

(C) NADPH only

(D) ATP, NADPH, and sugar

3. Plants give off oxygen as a waste product of photosynthesis. This oxygen comes from

(A) the Krebs cycle

(B) the Calvin cycle

(C) photolysis

(D) photorespiration

4. How many turns of the Calvin cycle are required to produce one molecule of glucose?

(A) 1

(B) 2

(C) 3

(D) 6

Questions 5–11

Indicate which of the following events occurs during

(A) light-dependent reactions
(B) light-independent reactions

5. Oxygen is released.

6. Carbon gets reduced.

7. Oxidative photophosphorylation

8. ATP is produced.

9. Electrons flow through an electron transport chain.

10. Oxidation of NADPH

11. Reduction of NADP⁺

_____

12. CAM plants keep their stomates closed during the daytime to reduce excess water loss. They can do this because they

(A) can fix CO_2 into sugars in the mesophyll cells.
(B) can use photosystems I and II at night
(C) modify rubisco so it does not bind with oxygen
(D) can incorporate CO_2 into organic acids at night

13. Which of the following is NOT directly associated with photosystem II?

(A) harvesting light energy by chlorophyll
(B) release of oxygen
(C) splitting of water
(D) production of NADPH

14. Where in the cell is ATP-synthase located?

(A) in the nuclear membrane
(B) in the thylakoid membrane of the chloroplast
(C) in the cristae membrane of mitochondria
(D) both B and C

15. This graph shows the rate of photosynthesis for two plants under experimental conditions.

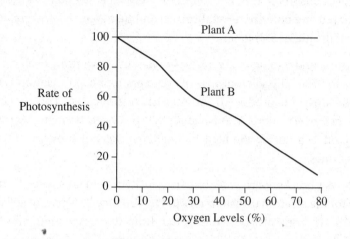

From this graph, what is the best conclusion about the mechanism of photosynthesis in the two plants?

(A) A is a C-3 plant, and B is a C-4 plant.
(B) A is a C-4 plant, and B is a C-3 plant.
(C) A is a CAM plant, and B is a C-4 plant.
(D) A and B are both CAM plants.

16. Which one of the following is NOT required for photosynthesis to occur?

(A) CO_2
(B) O_2
(C) ATP
(D) NADP

17. Which one is NOT correct about where the Calvin cycle occurs?

(A) in the spongy mesophyll in C-3 plants
(B) in the mesophyll in C-4 plants
(C) in the palisade layer in C-3 plants
(D) in the bundle sheath in C-4 plants

Answers to Multiple-Choice Questions

1. **(C)** If light is absorbed, it is not reflected. Only reflected colors are seen. The graph shows that red and blue are most absorbed and that green is most reflected. Therefore the color of the pigment is green.

2. **(A)** Electrons undergoing cyclic phosphorylation move from P680 to P700 and then cycle back to P680. The sole purpose of cyclic photophosphorylation is the production of ATP. No NADPH is produced, and no oxygen is released. This process is necessary when the cell needs more ATP because ATP has been used up by the Calvin cycle. Sugar is produced during the light-independent reactions only, not during the light-dependent ones.

3. **(C)** Photolysis breaks apart water into oxygen, hydrogen ions, and electrons to provide electrons for the electron transport chain. The Krebs cycle is part of cell respiration, not photosynthesis. The Calvin cycle uses CO_2 from the atmosphere and hydrogen from photolysis to make sugar. Cyclic photophosphorylation does not cause photolysis. Photorespiration uses up oxygen; it does not produce oxygen.

4. **(D)** Since glucose is a 6-carbon molecule and each turn of the cycle absorbs 1 molecule of CO_2, 6 turns of the Calvin cycle are needed to produce glucose. Three turns of the Calvin cycle produce 1 PGAL (phosphoglyceraldehyde), a 3-carbon molecule.

5. **(A)** Oxygen is released from the photolysis of water during the light-dependent reactions.

6. **(B)** Carbon is reduced when it enters the Calvin cycle in the dark reactions and combines with hydrogen to yield PGAL.

7. **(A)** Oxidative photophosphorylation is a type of chemiosmosis and explains how ATP is formed during the light-dependent reactions. This is also the way in which energy is produced during aerobic respiration. A steep proton gradient provides the energy for the production of ATP.

8. **(A)** Energy is produced from the process of chemiosmosis, which occurs during the light-dependent reactions of photosynthesis.

9. **(A)** Electrons flow through the electron transport chain during the light reactions, and ATP and NADPH are produced.

10. **(B)** NADP gains a proton (is reduced) during the light-dependent reactions and carries the proton to the Calvin cycle of the light-independent reactions, where it loses the proton (is oxidized).

11. **(A)** $NADP^+$ (also written as NADP) is reduced in the light reactions when it gains hydrogen ions from water.

12. **(D)** During the night, CAM plants fix CO_2 into a variety of organic acids, like malic acid. These acids are stored in vacuoles until the daylight when the CO_2 is released into the Calvin cycle. The Calvin cycle occurs in the bundle-sheath cells, not the mesophyll cells. CAM plants use PEP carboxylase to fix CO_2 into malic acid initially. This is then pumped into the bundle-sheath cells, where CO_2 combines with normal, unmodified rubisco. Lenticels are openings in the stems of woody plants and do not have anything to do with photosynthesis.

13. **(D)** Photosystem II (PS II) is also known as P680. This photosystem absorbs light with an average wavelength of 680nm. When light is absorbed by P680, electrons from chlorophyll *a* become energized and move to a higher energy level and into an electron transport chain within the thylakoid membrane. Water splits apart during photolysis to provide electrons to replace those lost in chlorophyll *a*. Protons that were released from water during photolysis are pumped by the thylakoid membrane from the stroma into the thylakoid space. ATP is formed as protons flow down a steep gradient and through ATP-synthase channels.

14. **(B)** Plant A is not affected by the increase in the concentration of oxygen in the air because in C-4 plants, PEP carboxylase does not react with oxygen. Plant B is a C-3 plant because the increased oxygen levels cause the plant to undergo photorespiration and to carry out less photosynthesis.

15. **(D)** The ATP synthase is the enzyme located within the membranes of mitochondria and chloroplasts. It produces ATP during the light reaction of photosynthesis and during cell respiration.

16. **(B)** All of the choices are required for photosynthesis to occur except oxygen, which is released as a by-product of photosynthesis.

17. **(B)** In C-3 plants, the Calvin cycle occurs in all photosynthetic cells in both the palisade and mesophyll layers. However, in C-4 plants, the light reactions occur in only the mesophyll cells while the Calvin cycle occurs in the bundle-sheath cells.

FREE-RESPONSE QUESTION

Directions: Answer all questions. You must answer the question in essay—not outline—form. You may use labeled diagrams to supplement your essay, but diagrams alone are *not* sufficient. Before you start to write, read each question carefully so that you understand what the question is asking.

The rate of photosynthesis varies with different environmental conditions such as light intensity, wavelength of light, temperature, and so on.

a. Devise an experiment to demonstrate that ONE environmental condition will alter the rate of photosynthesis.
b. State your hypothesis, the procedure, how you would collect the data, the results you expect.
c. The scientific theory behind your expectations.

Typical Free-Response Answer

This experiment will test the hypothesis that wavelengths of light will alter the rate of photosynthesis. The organism chosen for this experiment is elodea, which is inexpensive, readily available, and easy to take care of. Elodea merely requires an aerated, freshwater tank and sunlight.

REMEMBER

1. Make sure you answer the question asked.

2. State why you chose the organism.

3. The experimental and the control must be identical in all ways except the one you are testing.

Set up four 500 mL beakers of freshwater, each containing the same mass of elodea. Place the beakers into four large boxes that do not allow light to enter. This is required to maintain control over the wavelengths of light the elodea is exposed to. Place a light source in each box that is the same distance from each beaker. Include a heat sink between the beaker and the light source to absorb heat and to make sure that the temperature in all beakers remains the same. All light must be of the same intensity so as not to introduce another variable into the experiment. Set up each light source with appropriate light filters so that each beaker is exposed to only one wavelength of light: green, blue, red, or yellow.

The rate of photosynthesis will be measured by counting the number of bubbles of oxygen released in 30-second intervals for a period of 30 minutes. Repeat the experiment five times to verify the results.

I would expect the following results. The elodea exposed to blue light would produce the most bubbling. The elodea exposed to the red and yellow light would produce less bubbling in that order, and the elodea exposed to the green light would produce no bubbles.

The theory behind this experiment is that photosynthetic organisms release oxygen gas from photolysis during the light reactions of photosynthesis. The rate of photosynthesis can therefore be monitored by measuring the amount of oxygen released. The amount of oxygen released is measured by counting the bubbles released. The rate at which photosynthesis occurs is a function of the energy absorbed. The greater the amount of light absorbed, the greater the rate of photosynthesis. Light that is not absorbed is reflected and cannot be used as an energy source.

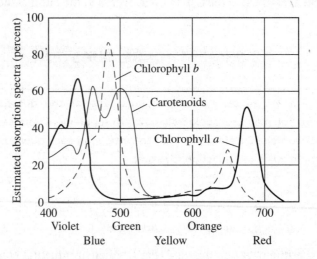

Photosynthetic pigments absorb light of different wavelengths. The figure above shows the absorption spectrum for the photosynthetic pigments chlorophyll *a*, chlorophyll *b*, and the carotenoids. Blue light is absorbed the most, red somewhat less, yellow even less, and green is reflected.

If you wanted the variable to be light intensity, only one thing would vary from the setup above. You must use lightbulbs rated with different lumens, lightbulbs of different intensity. However, the color (wavelength of light) must be the same for all the bulbs. You would not need the filters.

If you wanted temperature to be the variable, everything would be the same as the setup above except for two factors. The lightbulbs would all be identical, but you would control the temperature of each box that contained the light and plants.

Cell Division

6

INTRODUCTION

Cell division functions in **growth**, **repair**, and **reproduction**. Two types of cell division occur, **mitosis** and **meiosis**. **Mitosis** produces two genetically identical daughter cells referred to as **clones** and preserves the chromosome number ($2n$). **Meiosis** occurs in sexually reproducing organisms and results in cells that are **haploid**; they have half the chromosome number of the parent cell (n).

> **BE CAREFUL**
>
> Know the difference between mitosis and meiosis.

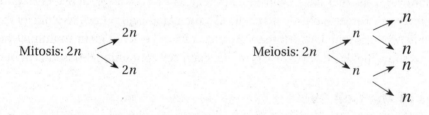

Any discussion about cell division must first consider the structure of the chromosome. A chromosome consists of a highly coiled and condensed strand of DNA. A replicated chromosome consists of two **sister chromatids**, where one is an exact copy of the other. The **centromere** is a specialized region that holds the two chromatids together. The **kinetochore** is a disc-shaped protein on the centromere that attaches the chromatid to the **mitotic spindle** during cell division. Figure 6.1 shows a sketch of a replicated chromosome.

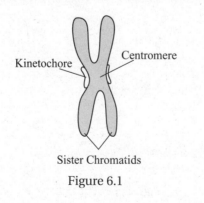

Figure 6.1

THE CELL CYCLE

Living and dividing cells pass through a regular sequence of growth and division called the **cell cycle**. The timing and rate of cell division are crucial to normal growth and development. In humans, the frequency of cell division varies with the cell type. Bone marrow cells are always dividing in order to produce a constant supply of red and white blood cells. Liver cells are arrested in G_0 but can be induced to divide or regenerate when liver tissue is damaged. Human intestine cells normally divide about twice per day to renew tissue destroyed during digestion. Specialized cells, like nerve cells, do not divide at all. In every case, however, the entire process is tightly regulated by a complex mechanism involving kinases and allosteric interactions. If something goes awry, the result can be uncontrolled cell division characteristic of cancer.

Two important factors limit cell size and promote cell division, the **ratio of the volume of a cell to the surface area** and the **capacity of the nucleus to control the entire cell**.

Ratio of the Cell Volume to Surface Area

As a cell grows, the area of the cell membrane *increases as the square* of the radius, while the volume of the cell *increases as the cube* of the radius. Therefore, as a cell grows larger, the volume inside the cell increases at a faster rate than does the cell membrane. Since a cell depends on the cell membrane for exchange of nutrients and waste products, the ratio of cell volume to membrane size is a major determinant of when the cell divides. See page 35.

Capacity of the Nucleus

The nucleus must be able to provide enough information to produce adequate quantities of all substances to meet the cell's needs. As a result, metabolically active cells are generally small. However, cells that have evolved a strategy to exist as large, active cells do exist in several kingdoms. Large, sophisticated cells like the paramecium have two nuclei that each control different cell functions. Human skeletal muscle cells are giant multinucleate cells. The fungus slime mold actually consists of one giant cell containing thousands of nuclei.

Phases of the Cell Cycle

The cell cycle consists of five major phases: G_1, **S**, and G_2 (which together make up **interphase**), **mitosis**, and **cytokinesis**.

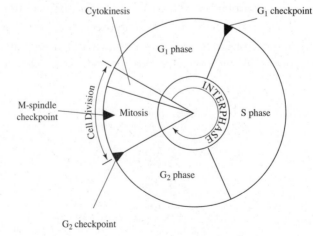

Figure 6.2 The cell cycle

INTERPHASE

Interphase consists of G_1, **S**, and G_2. The G_1 **phase** is a period of intense growth and biochemical activity. **S** stands for the synthesis or replication of DNA. G_2 is the phase when the cell continues to grow and to complete preparations for cell division. More than 90 percent of the life of a cell is spent in interphase. When a cell is in interphase and not dividing, the chromatin is threadlike, not condensed. Within the nucleus are one or more **nucleoli**. A single **centrosome**, consisting of two **centrioles**, may be seen in the cytoplasm of an animal cell. The centrosome is duplicated during S phase. At the G_2-M transition, the two centrosomes separate from one another and move to opposite poles. Plant cells lack centrosomes but have **microtubule organizing centers**, **MTOC**s. See Figure 6.3, which shows an animal cell during interphase.

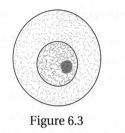

Figure 6.3

MITOSIS

Mitosis consists of the actual dividing of the nucleus. It is a continuous process. However, scientists have divided it into four arbitrary divisions: **prophase**, **metaphase**, **anaphase**, and **telophase**. Here are the characteristics of each phase. Figure 6.4 shows each of these phases.

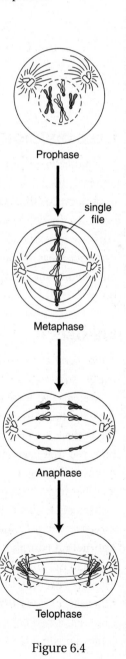

Prophase

Prophase

- The nuclear membrane begins to disintegrate.
- The strands of chromosomes begin to condense into discrete observable structures, like that in Figure 6.1 on page 89.
- The nucleolus disappears.
- In the cytoplasm, the mitotic spindle begins to form, extending from one centrosome to the other.

single file

Metaphase

- The *chromosomes line up in a single file* located on the equator or metaphase plate.
- Centrosomes are already positioned at opposite poles of the cell.
- Spindle fibers run from the **centrosomes** to the **kinetochores** in the **centromeres**.

Metaphase

Anaphase

- Centromeres of each chromosome separate, as spindle fibers pull apart the sister chromosomes.

Anaphase

Telophase

- Chromosomes cluster at opposite ends of the cell, and the nuclear membrane reforms.
- The supercoiled chromosomes begin to unravel and to return to their normal, pre-cell division condition as long, threadlike strands.
- Once two individual nucleoli form, mitosis is complete.

Telophase

Figure 6.4

CYTOKINESIS

Cytokinesis consists of the dividing of the cytoplasm. It begins during mitosis, often during anaphase. In animal cells, a **cleavage furrow** forms down the middle of the cell as **actin** and **myosin microfilaments** pinch in the cytoplasm.

In plant cells, a **cell plate** forms during telophase as vesicles from the Golgi coalesce down the middle of the cell. Daughter plant cells do not separate from each other. A new cell wall forms, and a sticky **middle lamella** cements adjacent cells together. See Figure 6.5.

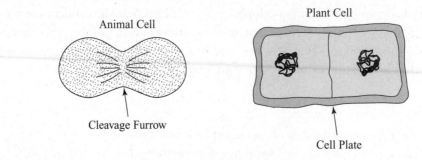

Figure 6.5 Cytokinesis

CELL DIVISION AND CANCEROUS CELLS

Normal cells grow and divide until they become too crowded; then they stop dividing and enter G_0 (G zero). This reaction to overcrowding is called **contact inhibition** or **density-dependent inhibition**. Another characteristic of normal animal cells is **anchorage dependence**. To divide, a cell must be attached or anchored to some surface, such as a Petri dish (in vitro) or an extracellular membrane (in vivo). Cancer cells show neither contact inhibition nor anchorage dependence. They divide uncontrollably and do not have to be anchored to any membrane. That is why cancer cells can migrate or metastasize to other regions of the body.

MEIOSIS

Meiosis generates the genetic diversity that is the raw material for natural selection and evolution.

Meiosis is a form of cell division that produces **gametes** (ova and sperm). These gametes have the **haploid** chromosome number (***n***), half the genetic material of the parent cell. During meiosis, the nucleus divides twice. Genetic material is randomly separated and recombined so that each gamete differs genetically from every other gamete. Sexual reproduction involves the fusion of two haploid gametes and restores the **diploid** chromosome number to its offspring. The two cell divisions in meiosis are called meiosis I and meiosis II.

Meiosis I, also called the **reduction division**, is the process by which *homologous chromosomes separate*. Each chromosome first pairs up precisely with its homologue into a **synaptonemal complex** by a process called **synapsis** and forms a structure known as a **tetrad** or **bivalent**. This pairing-up process is important. By aligning and binding the homologues together, accurate **crossing-over** is likely to occur. Crossing-over is the process by which *nonsister chromatids exchange genetic material*. It results in the *recombination* of genetic material. When crossing-over is occurring, what is visible under a microscope are **chiasmata** (chiasma, singular), Xs on the chromosome where homologous bits of DNA are switching places. *Crossing-over is a highly organized mechanism to ensure greater variation among gametes.*

Meiosis II is like mitosis. In this process, sister chromatids separate into different cells. The two stages of meiosis are further divided into phases. Each meiotic cell division consists of the same four stages as mitosis: prophase, metaphase, anaphase, and telophase.

Meiosis I

PROPHASE I

- **Synapsis**, the pairing of homologues, occurs.
- **Crossing-over**, the exchange of homologous bits of chromosomes, occurs.
- **Chiasmata**, the visible manifestations of the cross-over events, are visible.
- Sets the stage for separation (segregation) of DNA.

METAPHASE I

- The homologous pairs of chromosomes are lined up **double file** along the metaphase plate.
- **Spindle fibers** from the poles of the cell are attached to the centromeres of each pair of homologues.

ANAPHASE I

- Separation of homologous chromosomes as they are pulled by spindle fibers and migrate to opposite poles.

TELOPHASE I

- Homologous pairs continue to separate until they reach the poles of the cell. Each pole has the haploid number of chromosomes.

CYTOKINESIS I

- Cytokinesis usually occurs simultaneously with telophase I.

In some species, an interphase occurs between meiosis I and meiosis II. In other species, none occurs. In either case, chromosomes do not replicate between meiosis I and II because chromosomes already exist as double or replicated chromosomes.

Meiosis II

Meiosis II is functionally the same as mitosis and consists of the same phases: prophase, metaphase, anaphase, telophase, and cytokinesis. The chromosome number remains haploid, and daughter cells are genetically identical to the parent cell.

Figure 6.6 is a sketch comparing meiosis and mitosis. Each parent cell contains four chromosomes. The cell undergoing meiosis experiences one cross-over.

LOOK CLOSELY

Pay attention to
how metaphase
differs during
mitosis and
meiosis I.

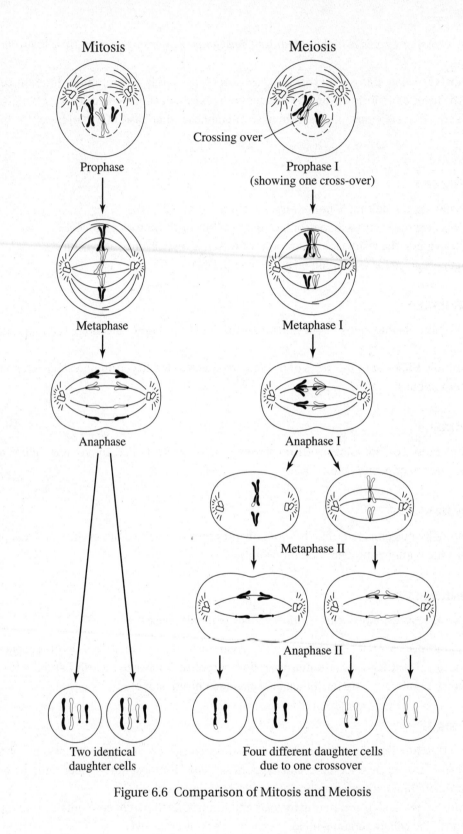

Figure 6.6 Comparison of Mitosis and Meiosis

MEIOSIS AND GENETIC VARIATION

Three types of genetic variation result from the processes of meiosis and fertilization. They are **independent assortment of chromosomes**, **crossing-over**, and **random fertilization of an ovum by a sperm**.

Independent Assortment of Chromosomes

During meiosis, homologous pairs of chromosomes separate depending on the random way in which they line up on the **metaphase plate** during metaphase I. Each pair of chromosomes can line up in two possible orientations. There is a 50 percent chance that a particular gamete will receive a maternal chromosome and a 50 percent chance it will receive a paternal chromosome. Given that there are 23 pairs of chromosomes in humans, the number of possible combinations of maternal and paternal chromosomes in each gamete is 2^{23}, or about 8 million.

Cross-over

Cross-over produces **recombinant chromosomes** that combine genes inherited from both parents. For humans, an average of two or three cross-over events occur in each chromosome pair. In addition, at metaphase II, these recombinant chromosomes line up on the metaphase plate in random fashion. This increases the possible types of gametes even more.

Random Fertilization

One human ovum represents one of approximately 8 million possible chromosome combinations. The same is true for the human sperm. Thus, when one sperm fertilizes one ovum, 8 million × 8 million recombinations are possible.

REGULATION AND TIMING OF THE CELL CYCLE

A **cell cycle control system** regulates the rate at which cells divide. Several **checkpoints** act as built-in stop signals that halt the cell unless they are overridden by go-ahead signals. Three checkpoints exist in G_1, G_2, and **M**. The G_1 checkpoint is known as the **restriction point** (**R**) and is the most important one in mammals. If it receives a go-ahead, the cell will most likely complete cell division. On the other hand, if it does not get the appropriate signal, the cell will exit the cycle and become a nondividing cell arrested in the G_0 (**G zero**) **phase**. Since the activity of a cell varies, the rate at which it needs to divide also varies. The timing of the cell cycle is initiated by growth factors and controlled by two kinds of molecules: **cyclins** and **protein kinases**. Protein kinases, common in *cell signal transduction*, catalyze the phosphorylation of target proteins that regulate the cell cycle. The class of protein kinases involved in this process is called **cyclin-dependent kinases** (**CDKs**). As the name implies, they are activated by binding to the protein **cyclin**, which exposes the active site of the **CDK** and activates the molecule. This is another example of **allosteric regulation**.

Several CDKs in different eukaryotic species regulate the cell cycle at specific stages called **cell cycle checkpoints**. Each CDK is activated by its own **cyclin**, and the cyclin is manufactured only at the right time. After the CDK acts, its cyclin is broken down by a protease. Cyclins are synthesized in response to various molecular signals, including growth factors.

Here is the summary of the chain reaction that controls the cell cycle:

Growth factor → Cyclin synthesis → CDK activation → Cell cycle events

The activity of a CDK, cyclin-dependent kinase, rises and falls with changes in the concentration of its cyclin partner. Figure 6.7 shows the fluctuation activity of MPF, the cyclin-CDK complex. Notice that the peaks of MPF activity correspond to a rise in cyclin concentration. The cyclin levels rise during the S and G_2 phases and then fall abruptly during the M phase. MPF triggers the cell's passage past the G_2 checkpoint into the M phase.

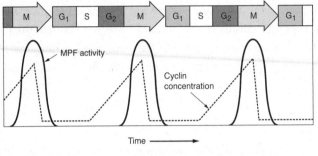

Figure 6.7

PROGRAMMED CELL DEATH—APOPTOSIS

Cells that are infected, damaged, or simply have come to the end of their life span die by **apoptosis**, a genetically programmed series of events that result in cell death. During this process, the DNA, organelles, and other cytoplasmic components are chopped up. The parts are packaged in vesicles that are engulfed by special scavenger cells.

Cells carry out apoptosis for several reasons.

1. During embryonic development when cells or tissues are no longer needed, they die and are engulfed by neighboring cells. A familiar example is provided by the cells in the tail of the tadpole, which undergo apoptosis during frog metamorphosis.

2. The cell has sustained too much genetic damage that could lead to cancer. This is common for epithelial cells on the surface of the skin, which have been exposed to extensive solar radiation. Skin cells normally die by apoptosis and are replaced every few days.

3. In plant cells, apoptosis is an important defense against infection by fungus and bacterium. By dying, cells at the site of infection leave no living tissue to spread infection inside the plant.

4. In mammals, including humans, several different pathways involving enzymes called **caspases** carry out apoptosis. Signals from different sources—inside or outside the cell—trigger the apoptosis pathway. The trigger can be from a neighboring cell, which makes use of a signal transduction pathway to begin the process. A signal may instead come from inside the cell itself, from irreparably damaged DNA in the nucleus, or from the endoplasmic reticulum when excessive protein misfolding has occurred.

The fact that the mechanism of apoptosis is basically similar in single-celled yeast and in mammalian cells indicates that *the mechanism for apoptosis evolved early in the evolution of eukaryotic cells.*

1. Which of the following does NOT occur by mitosis?

    (A) growth
    (B) production of gametes
    (C) repair
    (D) development in the embryo

The following two questions refer to the sketch below of a cell containing chromosomes.

2. How many chromosomes are in this cell?

    (A) 2
    (B) 4
    (C) 8
    (D) 16

3. How many chromatids are in this cell?

    (A) 2
    (B) 4
    (C) 8
    (D) 16

4. Which is a factor that limits cell size?

    (A) how active a cell is
    (B) what kind of activity a cell is engaged in
    (C) the ratio of volume to cell surface area
    (D) whether the cell is a plant cell or animal cell

5. In which stage of the life of a cell is the nucleolus always visible?

    (A) anaphase
    (B) telophase
    (C) cytokinesis
    (D) interphase

6. Which of the following cells is a giant multinucleate cell?

    (A) neuron
    (B) skeletal muscle
    (C) tracheids and vessels in plants
    (D) companion cell in plants

7. Which of the following is NOT found in plant cells?

    (A) cell plate
    (B) actin and myosin filaments
    (C) microtubule organizing center
    (D) cleavage furrow

8. If a cell has 24 chromosomes at the beginning of meiosis, how many chromosomes will it have at the end of meiosis?

    (A) 6
    (B) 12
    (C) 24
    (D) The number varies with the species.

9. If a cell has 24 chromosomes, how many will it have at the end of mitosis?

    (A) 6
    (B) 12
    (C) 24
    (D) 48

10. All of the following are true of meiosis EXCEPT

    (A) cross-over occurs during prophase I
    (B) there is no replication of chromosomes between meiosis I
        and meiosis II
    (C) in plants, spindle fibers are attached to the centriole
    (D) synapsis occurs during prophase I

Questions 11–13
Match the event of meiosis with the stages shown below.

11. Identify metaphase I of meiosis.

    (A) A
    (B) B
    (C) C
    (D) D

12. Identify metaphase II of meiosis.

    (A) A
    (B) B
    (C) C
    (D) D

13. Identify anaphase II.

    (A) A
    (B) B
    (C) C
    (D) D

    _____

14. The synaptonemal complex forms during

    (A) anaphase I of meiosis
    (B) prophase I of meiosis
    (C) anaphase II of meiosis
    (D) telophase I

15. Homologous chromosomes separate during

    (A) prophase I
    (B) prophase II
    (C) anaphase II
    (D) anaphase I

16. Chiasmata are most closely related to which of the following?

    (A) crossing-over
    (B) fertilization
    (C) cytokinesis
    (D) mitotic cell division

17. Which is NOT a source of genetic variation?

    (A) independent assortment of chromosomes
    (B) cross-over
    (C) random fertilization
    (D) mitosis

18. Which of the following cells are permanently arrested in the G_0 phase?

    (A) bone marrow
    (B) nerve cells
    (C) cancer cells
    (D) liver cells

19. Which is TRUE of the cell cycle?

    (A) The timing of cell division is controlled by cyclins and CDKs.
    (B) A characteristic of cancer cells is density-dependent inhibition.
    (C) The cell cycle is controlled solely by signals external to the cell.
    (D) The cell cycle is controlled solely by internal signals.

20. A cell that passes the restriction point will most likely

   (A) stop dividing
   (B) divide
   (C) show density-dependent inhibition
   (D) die

Answers to Multiple-Choice Questions

1. **(B)** The production of gametes, sperm, and eggs occurs by meiosis, where the chromosome number gets cut in half.

2. **(B)** There are 4 replicated chromosomes in this cell but 8 chromatids. The number of chromosomes is determined by the number of centromeres, 1 per chromosome.

3. **(C)** See #2.

4. **(C)** Two important factors limit cell size and promote cell division: ratio of the volume of a cell to the surface area and capacity of the nucleus.

5. **(D)** Most of the life of the cell is spent in interphase when the chromosomes are threadlike and not visible under a light microscope. When the cell divides, chromosomes must be condensed. When supercoiled or condensed, chromosomes appear like the Xs and Ys we commonly see them as. The nucleolus is not a real structure but threadlike chromosomes organized in a way that form a sphere.

6. **(B)** The skeletal muscle cell is a giant cell with many nuclei. At some time in its evolutionary history, it apparently underwent mitosis many times (which accounts for the many nuclei) without undergoing cytokinesis (which accounts for its large size).

7. **(D)** A cleavage furrow is a shallow groove in the cell surface in animal cells where cytokinesis is taking place. The cell plate is seen in dividing plant cells. The middle lamella is a layer between two adjacent plant cells.

8. **(B)** Meiosis cuts the chromosome number in half, from $2n$ to n. This occurs so that after fertilization, when two gametes fuse, the embryo will have the correct chromosome number, $2n$.

9. **(C)** The cells that result from mitotic cell division have the same number of chromosomes as the parent cell.

10. **(C)** Plants do not have centrioles, they have only microtubule organizing centers.

11. **(C)** During metaphase I of meiosis, homologous pairs line up on the metaphase plate in double file.

12. **(D)** During metaphase II of meiosis, single chromosomes line up on the metaphase plate in preparation for division. During meiosis II, sister chromatids separate.

13. **(A)** During anaphase II, sister chromatids are separating.

14. **(B)** The synaptonemal complex forms during prophase I and holds the two replicated chromosomes tightly together as a bivalent or tetrad so that crossing-over can occur without error.

15. **(D)** Homologous chromosomes separate during anaphase I of meiosis.

16. **(A)** Chiasmata are the microscopically visible regions of homologous chromatids where crossing-over has occurred.

17. **(D)** The daughter cells resulting from mitotic cell division are genetically identical to each other and to the mother cell. Sources of variation are independent assortment of chromosomes, crossover, random fertilization, and recombinant chromosomes.

18. **(B)** A cell arrested in the G_0 phase is not dividing. Most human body cells are not actively dividing. Highly specialized cells such as nerve and muscle cells never divide. Liver cells can be induced to divide when damaged, and human skin and bone marrow cells are always dividing. Also, cancer cells are always rapidly dividing.

19. **(A)** The timing of the cell cycle responds to external and internal cues and to fluctuations in levels of cyclins and cyclin-dependent kinases (CDKs). Normal cells stop dividing when crowded. This phenomenon is called contact inhibition or density-dependent inhibition. Cancer cells are characterized by uncontrolled growth.

20. **(B)** In mammalian cells, the G_1 checkpoint is known as the restriction point. If the cell receives the go-ahead signal at the G_1 checkpoint, it will usually complete the cycle and divide. In contrast, if the cell is not stimulated to pass the restriction point, it will switch into a nondividing mode known as G_0.

FREE-RESPONSE QUESTION

Directions: Answer all questions. You must answer the question in essay—**not** outline—form. You may use labeled diagrams to supplement your essay, but diagrams alone are *not* sufficient. Before you start to write, read each question carefully so that you understand what the question is asking.

An organism is heterozygous at two gene loci on different chromosomes.

a. Explain how these alleles are transmitted by the process of mitosis to daughter cells.

b. Explain how these alleles are distributed to gametes by the process of meiosis.

c. Explain how the behavior of these two pairs of chromosomes during meiosis provides the physical basis for two of Mendel's laws of heredity.

Typical Free-Response Answer

Author's note: These questions are about Mendelian inheritance and can be used as essays for that topic as well as this one. In this essay, the key words are in bold to show how many possible terms there are to discuss. You might consider underlining as you write to keep track of your key words.

a. Mitotic cell division produces daughter cells that are **genetically identical** to the parent cell. Mitosis is the division of the nucleus and consists of the following stages: **prophase**, **metaphase**, **anaphase**, and **telophase**. In preparation for mitosis, the **DNA replicates** itself. **Cytokinesis**, the actual division of the cytoplasm, almost always follows mitosis but is not part of mitosis. During prophase, the nuclear membrane begins to break apart and the **nucleolus** disappears as the chromosomes condense and become visible under a light microscope. In metaphase, chromosomes line up **single file** on the metaphase plate. **Sister chromatids** begin to separate during anaphase. They are pulled apart by **spindle fibers** connected at one end to the centrioles and at the other end to the **kinetochore** within the **centromere** of the chromosome. In telophase, the chromosomes reform into a circle as the nuclear membrane begins to reform. The two daughter cells that will result from this mitotic cell division each contain the same chromosomes, *T, t, Y,* and *y.*

b. Meiosis is a form of cell division that produces **gametes** (sex cells) with the **haploid chromosome number (*n*)**. Two stages occur in meiosis: **meiosis I** (**reduction division**) in which homologous chromosomes separate, and **meiosis II**, which is like mitosis, in which sister chromatids separate. The two stages of meiosis are further divided into phases. Each meiotic cell division consists of the same four stages as mitosis: prophase, metaphase, anaphase, and telophase.

   In meiosis I, the homologous pairs line up double file on the metaphase plate. One homologue comes from the mother, and one comes from the father. During anaphase I, the homologues are pulled apart by spindle fibers and migrate to the poles; one homologue goes to each daughter cell. How the homologues separate or **segregate** is determined by how they line up on the metaphase plate; and how they line up on the metaphase plate is random.

   If the four chromosomes line up like this:

the daughter cells will contain the alleles *TY* and *ty.*

If the four chromosomes line up like this:

the daughter cells will contain the alleles *Ty* and *tY*.

Meiosis II is like mitosis. After two meiotic cell divisions, there are **four haploid gametes**. Unlike cells that are formed by mitosis, these gametes are very different from each other.

c. Mendel's **law of segregation** states that during gamete formation, the two alleles for each trait are separated or segregated. This separation is random. In the example given, *T* separates from *t* and *Y* separates from *y* during anaphase of meiosis I.

Mendel's **law of independent assortment** states that during gamete formation, the alleles of a gene for one trait, such as *T* or *t*, segregate independently from the alleles of another gene, such as *Y* or *y*. This is shown in section B above. How they assort or segregate is random. Additionally, the law of independent assortment applies only if the two genes are on separate chromosomes. **Linked genes** will not assort independently.

Heredity

7

INTRODUCTION

The father of modern genetics is **Gregor Mendel**, an Austrian monk who, in the 1850s, bred garden peas in order to study patterns of inheritance. Mendel was successful because he brought an experimental and quantitative approach to the study of inheritance. First, he studied traits that were clear-cut, with no intermediates between varieties. Second, he collected data from a large sample, hundreds of plants from each of several generations. Mendel collected ten thousand plants in all. Third, he applied statistical analysis to his carefully collected data.

Until the nineteenth century, people thought that inheritance was blended, a mixture of fluids that passed from parents to children. In contrast, Mendel's theory of genetics is one of **particulate inheritance** in which inherited characteristics are carried by discrete units that he called *elementes*. These *elementes* eventually became known as genes.

BASICS OF PROBABILITY

Probability is the likelihood that a particular event will happen. If an event is an absolute certainty, its probability is 1. If the event cannot happen, its probability is 0. The probability of anything else happening is between 0 and 1. Probability cannot predict whether a particular event will actually occur. However, if the sample is large enough, probability can predict an average outcome.

Understanding probability is important to the study of genetics because predicting outcomes is what Punnett squares enable us to do. What is the chance that two brown-eyed people can give birth to a child with blue eyes? That is probability, and that is what this chapter is all about.

When Do You Multiply? Multiplication Rule

To find the probability of **two independent events** happening, **multiply** the chance of one happening by the chance that the other will happen. For example, the chance of a couple having two boys depends on two independent events. The chance of the first child being a boy is ½; and the chance of the next child being a boy is ½. Therefore, the chance that the couple will have two boys is ½ × ½ = ¼. The chance of having three boys is ½ × ½ × ½ = ⅛.

When Do You Add? Addition Rule

When more than one arrangement of events producing the specified outcome is possible, the probabilities for each outcome are added together. For example, if a couple is planning on having two children, what is the chance that they will have one boy and one girl (in either order)? Here is how you solve this problem. The probability of having a boy then a girl is ½ × ½ = ¼. The probability of having a girl and then a boy is ½ × ½ = ¼. Therefore, the probability of having one boy and one girl is ¼ + ¼ = ½.

LAW OF DOMINANCE

Mendel's first law is the **law of dominance**, which states that when two organisms, each **homozygous** (pure) for two opposing traits are crossed, the offspring will be **hybrid** (carry two different alleles) but will exhibit only the **dominant trait**. The trait that remains hidden is known as the **recessive trait**.

	T	T
t	Tt	Tt
t	Tt	Tt

Parent (P): TT × tt

Pure tall   Pure dwarf

Offspring (F_1): Tt

All hybrid tall

Law of dominance
All offspring are tall

LAW OF SEGREGATION

The **law of segregation** states that during the formation of gametes, the two traits carried by each parent separate. See Figure 7.1.

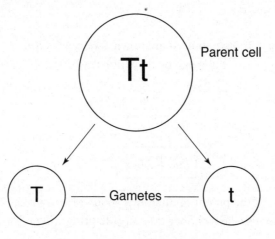

Figure 7.1 Law of Segregation

The cross that best exemplifies this law is the **monohybrid cross**, *Tt* × *Tt*. In the monohybrid cross, a trait that was not evident in either parent appears in the F$_1$ generation.

MONOHYBRID CROSS

The **monohybrid cross** (*Tt* × *Tt*) is a cross between two organisms that are each hybrid for one trait. The **phenotype** (appearance) ratio from this cross is 3 tall to 1 dwarf plant. The **genotype** (type of genes) **ratio**, 1 to 2 to 1, given as percentages: 25 percent homozygous dominant, 50 percent heterozygous, and 25 percent homozygous recessive. These results are always the same for any monohybrid cross.

	T	*t*
T	*TT*	*Tt*
t	*Tt*	*tt*

F$_1$: *Tt* × *Tt*

F$_2$: *TT, Tt,* or *tt*

Monohybrid cross

BACKCROSS OR TESTCROSS

The **testcross** or **backcross** is a way to determine the genotype of an individual plant or animal showing only the dominant trait. The individual in question (*B/__*) is crossed with a homozygous recessive individual (*b/b*). If the individual being tested is in fact homozygous dominant, all offspring of the testcross will be *B/b* and will show the dominant trait. There can be no offspring showing the recessive trait. If the individual being tested is hybrid (*B/b*), one-half of the offspring can be expected to show the recessive trait. Therefore, if any offspring show the recessive trait, the parent of unknown genotype must be hybrid.

	B	B
b	Bb	Bb
b	Bb	Bb

B = black

b = white

If the parent of unknown genotype is *BB*,
there can be no white offspring.

	B	b
b	Bb	bb
b	Bb	bb

B = black

b = white

If the parent of unknown genotype is hybrid,
there is a 50% chance that any offspring will be white.

LAW OF INDEPENDENT ASSORTMENT

The **law of independent assortment** applies when a cross is carried out between two individuals hybrid for two or more traits that are **not on the same chromosome**. This cross is called the **dihybrid cross**. This law states that during gamete formation, the alleles of a gene for one trait, such as height (*Tt*), segregate independently from the alleles of a gene for another trait such as seed color (*Yy*). Figure 7.2 represents a dihybrid individual (*TtYy*) where the traits will assort independently. In it, *T* = tall, *t* = short, *Y* = yellow seed, and *y* = green seed.

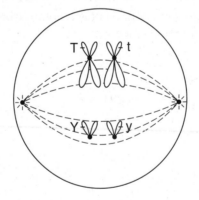

Figure 7.2

The genes for height and seed color are not on the same chromosome and will assort independently. The only factor that determines how these alleles **segregate** or assort is how the homologous pairs line up in metaphase of meiosis I, which is random.

During metaphase I if the homologous pairs happen to line up like this:

$$T \mid t$$

$$Y \mid y$$

they will produce these gametes: TY ty

If the homologous pairs happen to line up like this:

$$T \mid t$$

$$y \mid Y$$

they will produce these gametes: Ty tY

In contrast, if the gene for tall is **linked** to the gene for yellow seed color and the gene for short is linked to the gene for green seed color, the genes will **not assort independently**. If a plant is tall, it will have yellow seeds. If a plant is short, it will have green seeds. See Figure 7.3.

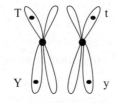

Figure 7.3 Linked Genes

The following describes a cross that adheres to the law of independent assortment. Two flowers are crossed that are homozygous for different traits and have the genes for height and seed color on different homologous chromosomes.

P: *T/T Y/Y*   ×   *t/t y/y*
        Homozygous tall plant   Homozygous short plant
           with yellow seeds   with green seeds

Gametes: *TY*   *ty*

F₁: *T/t Y/y*
                    Homozygous tall plant
                       with yellow seeds

The *T/T Y/Y* parent can produce gametes carrying only *TY* genes. The *t/t y/y* parent can produce gametes carrying only *ty* genes. A Punnett square is not needed because only one outcome is possible in the F₁: *T/t Y/y*. The phenotype of all members of the F₁ generation is tall plants with yellow seeds. Their genotype is known as **dihybrid**.

The Dihybrid Cross

A cross between two F_1 plants is called a **dihybrid cross** because it is a cross between individuals that are hybrid for two different traits, such as height and seed color. This cross can produce four different types of gametes: *TY, Ty, tY,* and *ty.* The figure below shows how to set up the Punnett square for this cross. There are 16 squares; therefore all the ratios are out of 16.

Dihybrid cross	*TY*	*Ty*	*tY*	*ty*
TY	*TTYY*	*TTYy*	*TtYY*	*TtYy*
Ty	*TTYy*	*TTyy*	*TtYy*	*Ttyy*
tY	*TtYY*	*TtYy*	*ttYY*	*ttYy*
ty	*TtYy*	*Ttyy*	*ttYy*	*ttyy*

Obviously, many different genotypes are possible in the resulting F_2 generation, but you need not pay attention to them. Just pay attention to, the **phenotype ratio** of the dihybrid cross. It is **9:3:3:1**—9 tall, yellow; 3 tall, green; 3 short, yellow; and 1 short, green. To show this as a probability:

F_2: 9/16 tall, yellow   3/16 tall, green   3/16 short, yellow   1/16 short, green

BEYOND MENDELIAN INHERITANCE

Mendelian principles apply to traits determined by a single gene for which there are only two alleles. An example is height in pea plants (one gene), which has only two alleles (tall and short). Now we will consider situations in which two or more genes are involved in determining a particular phenotype.

Incomplete Dominance

Incomplete dominance is characterized by **blending**. Here are two examples. A long watermelon (*LL*) crossed with a round watermelon (*RR*) produces all oval watermelons (*RL*). A black animal (*BB*) crossed with a white (*WW*) animal produces all grey (*BW*) animals. Since neither trait is dominant, the convention for writing the genes uses a different convention: all capital letters.

A red Japanese four o'clock flower (*RR*) crossed with a white Japanese four o'clock flower (*WW*) produces all pink offspring (*RW*).

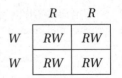

	R	*R*
W	*RW*	*RW*
W	*RW*	*RW*

If two pink four o'clocks are crossed, there is a 25 percent chance that the offspring will be red, a 25 percent chance the offspring will be white, and a 50 percent chance the offspring will be pink.

	R	W
R	RR	RW
W	RW	WW

Codominance

In **codominance**, *both traits show*. A good example is the *MN* blood groups in humans. (These are not related to *ABO* blood groups). There are three different blood groups: *M, N,* and *MN*. These groups are based on two distinct molecules located on the surface of the red blood cells. There is a **single gene locus** at which **two allelic variants** are possible. A person can be homozygous for one type of molecule (*MM*), be homozygous for the other molecule (*NN*), or be hybrid and have both molecules (*MN*) on their red blood cells. The *MN* genotype is not intermediate between *M* and *N* phenotypes. Both *M* and *N* traits are expressed because both molecules are present on the surface of the red blood cells; see Figure 7.4.

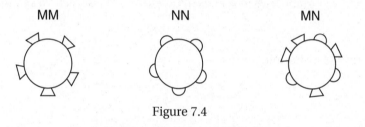

Figure 7.4

Multiple Alleles

Many genes in a population exist in only two allelic forms. For example, pea plants can be either tall (*T*) or short (*t*). When *there are more than two allelic forms of a gene*, that is referred to as **multiple alleles**. In humans there are four different blood types: **A**, **B**, **AB**, and **O** determined by the presence of specific molecules on the surface of the red blood cells. There are three alleles, *A, B,* and *O,* which determine the four different blood types. A and B are codominant and are often written as I^A and I^B. (*I* stands for immunoglobin.) When both alleles are present, they both manifest themselves, and the person has *AB* blood type. In addition, O is a recessive trait and is often written as i. A person can have any one of the six blood genotypes shown in Table 7.1.

Table 7.1

Human Blood Types and Genotypes		
Blood Type	**Genotype**	
A	Homozygous *A*:	*AA*
A	Hybrid *A*:	*Ai*
B	Homozygous *B*:	*BB*
B	Hybrid *B*:	*Bi*
AB	*AB*	
O	*ii*	

GENE INTERACTIONS

Pleiotropy

Pleiotropy is the ability of *one single gene to affect an organism in several or many ways.* An example of pleiotropy is the autosomal recessive disease **cystic fibrosis**, which is characterized by an abnormal thickening of mucus that coats certain cells. Instead of protecting the body, thick mucus builds up in the pancreas, lungs, digestive tract, and other organs leading to multiple, **pleiotropic** effects, including poor absorption of nutrients in the intestine and chronic bronchitis. Due to greater understanding of the disease process and better treatment and preventative measures, half the people with cystic fibrosis now survive into their 30s.

Epistasis

In **epistasis**, two separate genes control one trait, but *one gene masks the expression of the other gene.* The gene that masks the expression of the other gene is **epistatic** to the gene it masks.

The best way to explain this gene interaction is by an example involving agouti coat color in mice. Two genes control different enzymes in two different pathways that both contribute to coat color. Both genes *A* and *B* must be present in order to produce the agouti color. We can see how this is epistatic if we look at the dihybrid cross between two agouti-colored mice.

$$AaBb \times AaBb$$

Here are the results of this dihybrid cross:

A/_ B/_	→	agouti: 9/16
a/a B/_	→	black: 3/16
A/_ b/b	→	albino: 3/16
a/a b/b	→	albino: 1/16

If the two traits, *A* and *B*, exhibited simple dominance, we would expect a ratio of 9:3:3:1 as in the dihybrid cross we have seen before. In this case, though, we get only three different phenotypes in the ratio of 9 (agouti): 3 (black): 4 (albino). In the absence of *B*, even if *A* is present, the coat is colorless (albino). Therefore, we say that gene *B* is epistatic to gene *A*.

	AB	*aB*	*Ab*	*ab*
AB	*A/A B/B* Agouti	*A/a B/B* Agouti	*A/A B/b* Agouti	*A/a B/b* Agouti
aB	*A/a B/B* Agouti	*a/a B/B* Black	*A/a B/b* Agouti	*a/a B/b* Black
Ab	*A/A B/b* Agouti	*A/a B/b* Agouti	*A/A b/b* Albino	*A/a b/b* Albino
ab	*A/a B/b* Agouti	*a/a B/b* Black	*A/a b/b* Albino	*a/a b/b* Albino

POLYGENIC INHERITANCE

Many characteristics such as skin color, hair color, and height result from a blending of several separate genes that vary along a continuum. These traits are known as **polygenic**. Two parents who are short carry more genes for shortness than for tallness. However, they can have a child who inherits mostly genes for tallness from both parents and who will be taller than the parents. This wide variation in genotypes always results in a bell-shaped curve in an entire population. See Figure 7.5, which shows the distribution of skin pigmentation across a population.

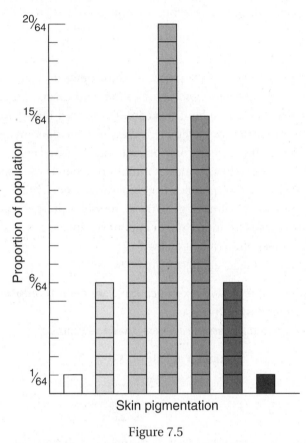

Figure 7.5

GENES AND THE ENVIRONMENT

The environment can alter the expression of genes. In fruit flies, the expression of the mutation for vestigial wings (short, shriveled wings) can be altered by temperature. When raised in a hot environment, fruit flies homozygous recessive for vestigial wings can grow wings almost as long as normal wild-type wings. Many human diseases have a **multifactorial basis**. There is an underlying genetic component with a significant environmental influence. Examples are heart disease, diabetes, cancer, alcoholism, schizophrenia, and bipolar disorder. Similarly, the development of intelligence is the result of an interaction of genetic predisposition and the environment, or **nurture and nature**.

Penetrance

Penetrance is the proportion or percentage of individuals in a group with a given genotype that actually show the expected phenotype. For example, many people who inherit a mutant allele for BRCA1 develop breast cancer in their lifetimes. For reasons that are unknown, some people with the allele do not develop breast cancer. Perhaps this is caused by an interaction with some other gene or some factor from the environment.

LINKED GENES

Genes on the same chromosome are called **linked genes**. Since there are many more genes than chromosomes, thousands of genes are linked. Humans have 46 chromosomes in every cell. Therefore, humans have 46 linkage groups. Linked genes tend to be inherited together and do not assort independently (unless they are separated by a cross-over event).

Sex-linkage

Of the 46 human chromosomes, 44 (22 pairs) are **autosomes** and 2 are **sex chromosomes**, X and Y. Traits *carried on the X chromosome* are called **sex-linked**. Few genes are carried on the Y chromosome. Females (XX) inherit two copies of the sex-linked genes. If a sex-linked trait is due to a **recessive mutation**, a female will express the phenotype only if she carries two mutated genes (X–X–). If she carries only one mutated X-linked gene, she will be a **carrier** (X–X). If a sex-linked trait is due to a **dominant mutation**, a female will express the phenotype with only one mutated gene (X–X). Males (XY) inherit only one X-linked gene. As a result, if the male inherits a mutated X-linked gene (X–Y), he will express the gene. Recessive sex-linked traits are much more common than dominant sex-linked traits; so males suffer with sex-linked conditions more often than females do.

Here are some important facts about sex-linked traits.

- Common examples of recessive sex-linked traits are **color blindness**, **hemophilia**, and Duchenne muscular dystrophy**.**
- *All daughters of affected fathers are carriers* (shaded squares).

Punnett square	X–	Y
X	X–X	XY
X	X–X	XY

- Sons cannot inherit a sex-linked trait from the father because the son inherits the Y chromosome from the father.
- A son has a 50 percent chance of inheriting a sex-linked trait from a carrier mother (shaded square).

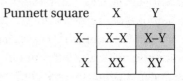

Punnett square	X	Y
X–	X–X	X–Y
X	XX	XY

- There is no carrier state for X-linked traits in males. If a male has the gene, he will express it.
- It is uncommon for a female to have a recessive sex-linked condition. In order to be affected, she must inherit a mutant gene from *both* parents.

CROSS-OVER AND LINKAGE MAPPING

The farther apart two genes are on one chromosome, the more likely they will be separated from each other during meiosis because a cross-over event will occur between them. At the site at which a cross-over and recombination occur, one can see a **chiasma**, a physical bridge built around the point of exchange. The result of a cross-over is a **recombination**. *Cross-over and recombination are a major source of variation in sexually reproducing organisms.*

Figure 7.6 shows one cross-over between homologous chromosomes. Without the cross-over, the resulting four gametes would contain the following genes: *AB, AB, ab,* and *ab*. There would be only two different types of gametes, *AB* and *ab*. With one cross-over (as shown), the four resulting gametes contain the following genes: *AB, Ab, aB* and *ab*. There would be four different types of gametes. That one cross-over results in twice the variation in the type of gametes possible.

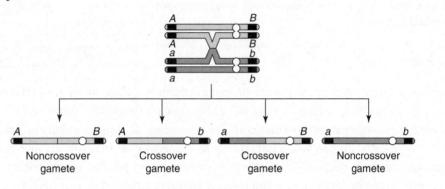

Figure 7.6

By convention, one **map unit** distance on a chromosome is *the distance within which recombination occurs 1 percent of the time.* The rate of cross-over gives no information about the actual distance between genes, but it tells us the order of the linked genes on the chromosome.

Here is one example. Genes *A, B,* and *D* are linked. The cross-over or **recombination frequencies** for *B* and *D* is 5 percent, *B* and *A* is 30 percent, and for *D* and *A* is 25 percent. The **linkage map** that can be constructed from this data is *BDA* or *ADB*. Whether you read it forward or backward does not matter.

A ⟵ 25 ⟶ D 5 ⟶ B
30

Recombination Frequencies

Here is a *testcross* between a dihybrid and a homozygous recessive for both traits

F$_1$ dihybrid	×	Pure black, vestigial-winged fly
G/g N/n	×	*g/g n/n*

Here are the ratios that were expected and the results:

Expected ratios from the testcross if all traits are located on different chromosomes, and therefore assort independently	Wild-type, gray normal	Black vestigial	Gray vestigial	Black normal
	G/g N/n (parental phenotype)	g/g n/n (parental phenotype)	G/g n/n (nonparental phenotype)	g/g N/n (nonparental phenotype)
	1	1	1	1
Expected ratios if genes are linked and located on the same chromosome. Alleles will always be inherited together.	G/g N/n (parental phenotype)	g/g n/n (parental phenotype)		
	1	1	0	0
Actual results:	965	944	206	185

In this experiment, the actual results do not conform to either prediction. The reason is that the genes for body color and wing size are located on the same chromosome. However, they are linked, although not permanently. The existence of small numbers of *nonparental phenotypes* (shaded in) can be explained only by an occasional break in the linkage—where **cross-over** occurred. Remember that *crossing-over accounts for the recombination of linked genes*. Here is how to determine recombination frequency given data such as this.

DETERMINING RECOMBINATION FREQUENCIES

Use the following formula to calculate recombination frequencies

$$\frac{\text{Number of recombinants}}{\text{Total number of offspring}} \times 100$$

Use the results from the cross above.
Number of recombinants: 206 + 185 = 391
Total number of offspring: 965 + 944 + 206 + 185 = 2300

$$\frac{391}{2300} \times 100 = 17\%$$

The recombination frequency for these linked genes, G/g and N/n, is 17%.

THE PEDIGREE

A **pedigree** is a family tree that indicates the phenotype of one trait being studied for every member of a family. *Geneticists use the pedigree to determine how a particular trait is inherited.* By convention, females are represented by a circle and males by a square. The carrier state is not always shown. If it is, though, it is sometimes represented by a half-shaded-in shape. A shape is completely shaded in if a person exhibits the trait.

The pedigree in Figure 7.7 shows three generations of deafness. Try to determine the pattern of inheritance. First, eliminate all possibilities. Dominance can be ruled out (either sex-linked or autosomal) because in order for a child to have the condition, she or he would have had to receive one mutant gene from one afflicted parent, and nowhere is that the case.

(All afflicted children have unaffected parents.) Also, you can rule out sex-linked recessive, because in order for F₃ generation daughter #1 to have the condition, she would have had to inherit two mutant traits (X–X–), one from each parent. However, her father does not have the condition. Therefore, the trait must be autosomal recessive.

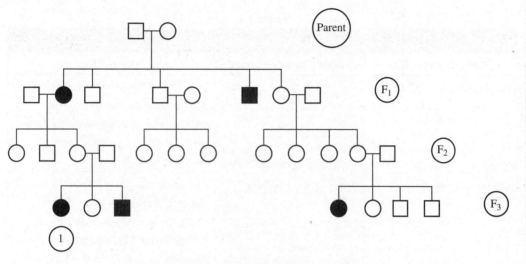

Figure 7.7 Three Generations of Deafness

X INACTIVATION—THE BARR BODY

Early in the development of the embryo of a female mammal, one of the X chromosomes is inactivated in every **somatic** (body) **cell**. This inactivation occurs randomly. The process results in an embryo that is a **genetic mosaic**; some cells have one X inactivated, some cells have the other X inactivated. Therefore, all the cells of female mammals are not identical. The inactivated chromosome condenses into a dark spot of chromatin and can be seen at the outer edge of the nucleus of all somatic cells in the female. This dark spot is called a **Barr body**.

Proof of X chromosome inactivation can be seen in the genetics of the female calico cat where the alleles for black and yellow fur are carried on the X chromosome. Male cats, having only a single X chromosome, can be either yellow (X^YY) **or** black (X^BY). Calico cats, which are almost always female, have coats with patches of both yellow and black (X^BX^Y). These patches of fur developed from embryonic cells with different deactivated X chromosomes. Some fur-producing cells contained the X^B active chromosome and produce black fur. Other fur-producing cells contain the X^Y active chromosome and produce yellow fur. The result is a cat with yellow and black patches of fur, the characteristic calico appearance.

Another example of X chromosome inactivation is evident in humans. A certain X-linked recessive mutation prevents the development of sweat glands. A woman who is heterozygous for this trait is not merely a carrier. Because of X inactivation, she has patches of normal skin and patches of skin lacking sweat glands.

MUTATIONS

Mutations are any changes in the genome. They can occur in the somatic (body) cells and be responsible for the spontaneous development of cancer, or they can occur during gametogenesis and affect future offspring. Even though radiation and certain chemicals cause mutations, when and where mutations occur is random.

There are two types of mutations, gene mutations and chromosome mutations. Gene mutations are caused by a change in the DNA sequence. Some human genetic disorders caused by gene and chromosome mutations are listed and described further in Table 7.2. The nature of gene mutations at the DNA level is discussed in the next chapter.

Table 7.2

Gene and Chromosome Mutations		
Genetic Disorder	**Pattern of Inheritance**	**Description**
Phenylketonuria (PKU)	Autosomal recessive	Inability to break down the amino acid phenylalanine. Requires elimination of phenylalanine from diet, otherwise serious mental retardation will result.
Cystic fibrosis	Autosomal recessive	The most common lethal genetic disease in the U.S. 1 out of 25 Caucasians is a carrier. Characterized by buildup of extra-cellular fluid in the lungs, digestive tract, etc.
Tay-Sachs disease	Autosomal recessive	Onset is early in life and is caused by lack of the enzyme necessary to break down lipids needed for normal brain function. It is common in Ashkenazi Jews and results in seizures, blindness, and early death.
Huntington's disease	Autosomal dominant	A degenerate disease of the nervous system resulting in certain and early death. Onset is usually in middle age.
Hemophilia	Sex-linked recessive	Caused by the absence of one or more proteins necessary for normal blood clotting.
Color blindness	Sex-linked recessive	Red-green color blindness is rarely more than an inconvenience.
Duchenne muscular dystrophy	Sex-linked recessive	Progressive weakening of muscle control and loss of coordination
Sickle cell disease	Autosomal recessive	A mutation in the gene for hemoglobin results in deformed red blood cells. Carriers of the sickle cell trait are resistant to malaria.

Chromosomal Disorder	Pattern of Inheritance	Description
Down syndrome	47 chromosomes due to trisomy 21	Characteristic facial features, mental retardation, prone to developing Alzheimer's and leukemia
Turner's syndrome	XO 45 chromosomes due to a missing sex chromosome	Small stature, female
Klinefelter's syndrome	XXY 47 chromosomes due to an extra X chromosome	Have male genitals, but the testes are abnormally small and the men are sterile

Although gene mutations cannot be seen under a microscope, **chromosome mutations** can. A procedure called a **karyotype** shows the size, number, and shape of chromosomes and can reveal the presence of certain abnormalities. Karyotypes can be used to scan for chromosomal abnormalities in developing fetuses. Figure 7.8 shows a karyotype of a male with Down syndrome due to an extra chromosome 21. The bottom of Table 7.2 "Chromosomal Disorder" describes three conditions that result from nondisjunction in the formation of the ovum or the sperm.

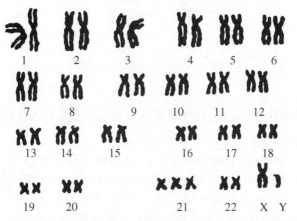

Figure 7.8 Karyotype of a male with trisomy 21

Chromosomal aberrations include:

- **Deletion**—when a fragment lacking a centromere is lost during cell division
- **Inversion**—when a chromosomal fragment reattaches to its original chromosome but in the reverse orientation
- **Translocation**—when a fragment of a chromosome becomes attached to a nonhomologous chromosome
- **Polyploidy**—when a cell or organism has extra sets of chromosomes.

NONDISJUNCTION

Nondisjunction is an error that sometimes occurs during meiosis in which homologous chromosomes fail to separate as they should. See Figure 7.9. When this happens, one gamete receives two of the same type of chromosome and another gamete receives no copy. The remaining chromosomes may be unaffected and normal. If either aberrant gamete unites with a normal gamete during fertilization, the resulting zygote will have an abnormal number of chromosomes. Any abnormal number of chromosomes is known as **aneuploidy**. If a chromosome is present in triplicate, the condition is known as **trisomy**. People with Down syndrome have an extra chromosome 21. The condition is referred to as **trisomy 21**. Cancer cells grown in culture almost always have extra chromosomes. An organism in which the cells have an extra set of chromosomes is referred to as **triploid** ($3n$). An organism with the $4n$ chromosome number is known as **tetraploid**. Strawberries are **octoploid**. An organism with extra sets of chromosomes is referred to as **polyploid**. Hugo DeVries, the scientist who coined the term **mutation**, was studying plants that were polyploidy. Polyploidy is common in plants and results in plants of abnormally large size. In some cases, it is responsible for the evolution of new species. As the word is used today, mutation refers to any **genetic** or **chromosomal abnormality**.

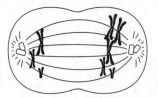

Figure 7.9 Nondisjunction

GENOMIC IMPRINTING AND EXTRANUCLEAR GENES

Two inheritance patterns are exceptions to Mendelian inheritance: genomic imprinting and extranuclear genes. **Genomic imprinting** is a variation in phenotype depending on whether a trait is inherited from the mother or from the father. It occurs during gamete formation and is caused by the silencing of a particular allele by methylation of DNA. Therefore, a zygote expresses only one allele of the imprinted gene. The imprint is carried to all body cells and passed down from generation to generation. Unlike sex-linked genes, which are located on the X chromosome, imprinted genes are located on autosomes.

Extranuclear genes are genes located in mitochondria and chloroplasts. The DNA in these organelles is small, circular, and carries only a small number of genes. Extranuclear genes have been linked to several rare and severe inherited diseases in humans. Since the products of most mitochondrial genes are involved with energy production, defects (mutations) in these genes cause weakness and deterioration in muscles. Mitochondrial DNA is inherited only from the mother because the father's mitochondria do not enter the egg during fertilization.

1. Which is TRUE about a testcross?

   (A) It is a mating between two hybrid individuals.
   (B) It is a mating between a hybrid individual and a homozygous recessive individual.
   (C) It is a mating between an individual of unknown genotype and a homozygous recessive individual.
   (D) It is a mating to determine which individual is homozygous recessive.

2. All are true of cross-over EXCEPT

   (A) it normally occurs between sister chromatids
   (B) it is the site of a chiasma
   (C) it is responsible for linked genes being inherited separately
   (D) it cannot occur between sex chromosomes in human males

3. A round watermelon is crossed with a long watermelon and all the offspring are oval. If two oval watermelons are crossed, what is the percent of watermelons that will be round?

   (A) 0
   (B) 25%
   (C) 50%
   (D) 75%

4. In peas, the trait for tall plants is dominant (*T*) and the trait for short plants is recessive (*t*). The trait for yellow seeds is dominant (*Y*) and the trait for green seeds is recessive (*y*). A cross between two plants results in 292 tall yellow plants and 103 short green plants. Which of the following are most likely to be the genotypes of the parents?

   (A) *TtYY* × *Ttyy*
   (B) *TTYy* × *TTYy*
   (C) *TTyy* × *TTYy*
   (D) *TtYy* × *TtYy*

5. A child is born with blood type O. All of the following could be the blood type of the parents EXCEPT

   (A) A and B
   (B) A and A
   (C) O and O
   (D) AB and O

6. *ABCDEF* → *ABEDCF*

   A rearrangement in the linear sequence of genes as shown in the diagram above is known as a/an

   (A) translocation
   (B) deletion
   (C) addition
   (D) inversion

7. A diploid organism has 36 chromosomes per cell. How many linkage groups does it have?

   (A) 9
   (B) 18
   (C) 36
   (D) 72

8. The expression of both alleles for a trait in a hybrid individual is

   (A) pleiotropy
   (B) epistasis
   (C) codominance
   (D) incomplete dominance

9. How many autosomes does the human male normally have?

   (A) 2
   (B) 22
   (C) 23
   (D) 44

10. A couple has 6 children, all girls. If the mother gives birth to a seventh child, what is the probability that the seventh child will be a girl?

    (A) $^6/_7$
    (B) $^1/_{128}$
    (C) $^1/_2$
    (D) 1

11. Assume that two genes, *A* and *B*, are not linked. If the probability of allele A being in a gamete is $^1/_2$ and the probability of allele *B* being in a gamete is $^1/_2$, then the probability of BOTH *A* and *B* being in the same gamete is

    (A) $^1/_2$
    (B) $^1/_4$
    (C) 1
    (D) $^1/_8$

12. Gene *R* controls the formation of feathers on a bird. In addition, it seems to be responsible for traits in several other body systems. What is the best explanation for this type of inheritance?

    (A) blending inheritance
    (B) codominance
    (C) pleiotropy
    (D) epistasis

13. In one strain of mice, fur color ranges from white to darkest brown with every shade of brown in between. This pattern of inheritance for fur color is most likely controlled by

   (A) multiple genes
   (B) a single gene with many alleles
   (C) pleiotropy
   (D) incomplete dominance

14. How is Huntington's disease inherited?

   (A) It is sex-linked recessive.
   (B) It is autosomal recessive.
   (C) It is sex-linked dominant.
   (D) It is autosomal dominant.

15. Two traits, A and B, are linked, but they are usually not inherited together. The most likely reason is

   (A) they are not on the same chromosome
   (B) they are not sex-linked
   (C) they are on the same chromosome but are far apart
   (D) they are close together on the same chromosome

16. A cross was made between two fruit flies, a white-eyed female and a wild male (red eyed). One hundred F_1 offspring were produced. All the males were white eyed and all the females were wild. When these F_1 flies were allowed to mate, the F_2 flies were observed and the following data was collected.

	Females		Males
P:	White eyed	×	Wild (red eyed)
F₁:	59 wild		51 white eyed
F₂:	24 wild		23 wild
	26 white eyed		27 white eyed

   What is the most likely pattern of inheritance for the white-eyed trait?

   (A) autosomal dominant
   (B) autosomal recessive
   (C) sex-linked dominant
   (D) sex-linked recessive

17. A man who has a sex-linked allele will pass it on to

   (A) all his daughters
   (B) all his sons
   (C) $\frac{1}{2}$ of his daughters
   (D) $\frac{1}{2}$ of his sons

18. The figure below shows a pedigree for a family that carries the gene for Huntington's disease. Individuals who express a particular trait are shown shaded in.

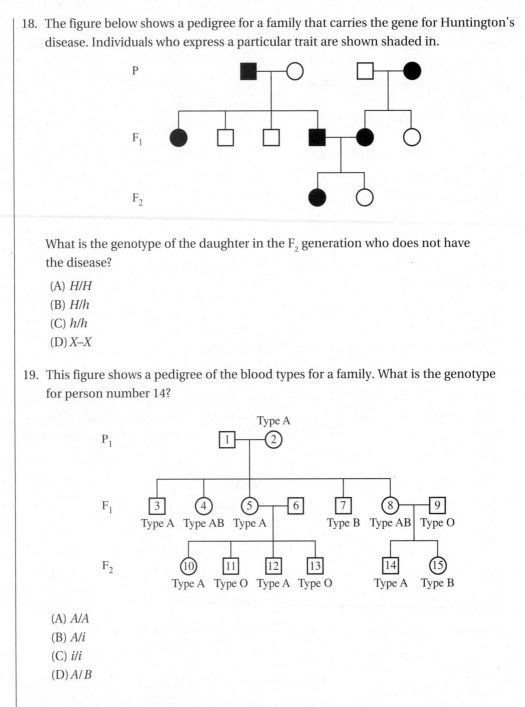

What is the genotype of the daughter in the F_2 generation who does not have the disease?

(A) *H/H*

(B) *H/h*

(C) *h/h*

(D) *X–X*

19. This figure shows a pedigree of the blood types for a family. What is the genotype for person number 14?

(A) *A/A*

(B) *A/i*

(C) *i/i*

(D) *A/B*

Answers to Multiple-Choice Questions

1. **(C)** A testcross is an actual mating between a pure recessive animal and an animal that shows the dominant phenotype but whose genotype is unknown.

2. **(A)** Cross-over occurs between homologous chromosomes, not sister chromatids. Sister chromatids are genetically identical. If cross-over did occur between sister chromosomes, it would be undetectable.

3. **(B)** Here is the cross.

	R	L
R	RR	RL
L	RL	LL

RR is round, *RL* is oval, and *LL* is long.
The inheritance is an example of incomplete dominance.

4. **(D)** Since there are four genes (*T, t, Y,* and *y*) but only two phenotypes in the offspring, the traits for height and seed color must be linked, that is, on the same chromosome. So solve the problem this way. Consider the traits separately at first. The phenotype ratio in the offspring for height is 3 : 1, tall to short. Therefore, the parents must be *Tt* and *Tt*. The phenotype ratio in the offspring for seed color is 3 : 1, yellow to green. Therefore, the parents must be *Yy* and *Yy*. Now, put both genotypes together. The parents must be *TtYy* and *TtYy*.

5. **(D)** Blood type O is homozygous recessive, *ii* or *I°I°*. The child must receive an *O* allele from each parent. Blood type AB has no *O* allele. Therefore, choice D is excluded as a parent.

6. **(D)** The section of the strand shows genes *CDE* inverted.

7. **(C)** Linked genes are located on one chromosome. Since there are 36 chromosomes, there are 36 linkage groups.

8. **(C)** In codominance, both traits show. An example is blood type in humans, where a person who has a gene for blood antigen A and another for blood antigen B has the AB blood type.

9. **(D)** Autosomes are the chromosomes other than sex chromosomes (X and Y).

10. **(C)** No matter how many children a couple has, the chance that the child will be a boy *or* a girl is always $1/2$. Although it is true that whether the sperm carries an X or a Y sex chromosome determines the sex of the child, that it is irrelevant to the question here.

11. **(B)** Since *A* and *B* are not linked, they assort independently. To find the probability of two independent events happening, multiply the chance of one happening by the chance of the other happening.

12. **(C)** When one gene seems to control the expression of several traits, the pattern of inheritance is pleiotropy. A classic example of pleiotropy in humans is cystic fibrosis.

13. **(A)** Examples of polygenic inheritance in humans are genes for skin color and height.

14. **(D)** Although the trait is autosomal dominant, symptoms do not usually appear until later in life, long after the person with the condition has passed the trait on to children and, perhaps, grandchildren.

15. **(C)** If genes are on the same chromosome but far apart, they will often be inherited separately because they will often be separated by cross-over.

16. **(D)** Here is the first cross.

	X	Y
X–	X–X	X–Y
X–	X–X	X–Y

All the female offspring are carriers (X–X), and all the male offspring have white eyes (X–Y).

Here is the second cross.

	X–	Y
X–	X–X–	X–Y
X	X–X	XY

There is a 50 percent chance that a male will be white eyed and a 50 percent chance he will be red eyed. There is a 50 percent chance a female will be white eyed and a 50 percent chance she will be red eyed (a carrier).

17. **(A)** A man gives his sons the Y chromosome and passes sex-linked traits to all his daughters.

18. **(C)** Females are represented as circles; males as squares unless otherwise stated. You should know that Huntington's disease is inherited as autosomal dominant. The F_2 daughter who does not have the condition must have inherited one healthy gene from each parent. She must be *h/h*, normal.

19. **(B)** Person 14 has blood type A. It can be *A/A* or *A/i*. Since his father has type O blood, person 14 must have inherited the A from his mother and the O from his father. His genotype therefore is *A/i*.

> **Directions:** Answer all questions. You must answer the question in essay—**not** outline—form. You may use labeled diagrams to supplement your essay, but diagrams alone are *not* sufficient. Before you start to write, read each question carefully so that you understand what the question is asking.

1. A person with Turner's syndrome has a genotype of XO, while a person with Klinefelter's syndrome has the genotype XXY. Explain how these two mutations come about.

2. Scientists understand that comparatively few genes are inherited in a simple Mendelian fashion. The expression of most genes is altered by many things, including other genes or the environment. Choose five of the terms below and explain what they mean.

   a. Pleiotropy
   b. Penetrance
   c. Epistasis
   d. Genomic imprinting
   e. Polygenic inheritance
   f. Incomplete dominance
   g. The effect of the environment on genes

The explanations of these terms can be found in this review chapter.

Typical Free-Response Answers

1. Both of these conditions, Turner's and Klinefelter's syndrome, are mutations that arise as a result of nondisjunction. Normally, during anaphase I of meiosis, homologous pairs separate (disjoin), with one homologue going into each of two daughter cells. Occasionally, one homologous pair does not separate as it should during anaphase I. As a result, both homologues go into one of the daughter cells, giving it an extra chromosome, while leaving the other daughter cell missing one chromosome. Klinefelter's syndrome results when an egg with two X chromosomes fuses with a sperm carrying a Y chromosome. The resulting zygote has the genotype XXY. Turner syndrome results when a gamete without any sex chromosome fuses with a normal gamete with one X chromosome. The resulting zygote has the genotype XO, where O means missing chromosome. A person with Turner's syndrome has 45 chromosomes, while a person with Klinefelter's syndrome has 47 chromosomes.

2. The explanations for these terms can be found within this chapter. (For further information, see pages 101 and 103.)

The Molecular Basis of Inheritance

8

- → THE SEARCH FOR INHERITABLE MATERIAL
- → STRUCTURE OF NUCLEIC ACIDS
- → DNA REPLICATION IN EUKARYOTES
- → FROM DNA TO PROTEIN
- → GENE MUTATION
- → THE GENETICS OF VIRUSES AND BACTERIA
- → PRIONS
- → THE HUMAN GENOME
- → RECOMBINANT DNA; CLONING GENES
- → TOOLS AND TECHNIQUES OF RECOMBINANT DNA
- → ETHICAL CONSIDERATIONS

INTRODUCTION

Today, everyone knows that DNA is the molecule of heredity. We know that DNA makes up chromosomes and that genes are located on the chromosomes. Today, we can even see the location of particular genes by tagging them with fluorescent dye.

However, until the 1940s, many scientists believed that proteins, not DNA, were the molecules that make up genes and constitute inherited material. Several factors contributed to that belief. First, proteins are a major component of all cells. Second, they are complex macromolecules that exist in seemingly limitless variety and have great specificity of function. Third, a great deal was known about the structure of proteins and very little was known about DNA. The work of many brilliant scientists has transformed our knowledge of the structure and function of the DNA molecule and led to the acceptance of DNA as the molecule responsible for heredity.

This chapter includes the history of the search for the heritable material, the structure of nucleic acids and how DNA makes proteins. It also includes an extensive review of genetic engineering and recombinant DNA techniques.

THE SEARCH FOR INHERITABLE MATERIAL

Griffith (1927) performed experiments with several different strains of the bacterium *Diplococcus pneumoniae*. Some strains are virulent and cause pneumonia in humans and mice, and some strains are harmless. Griffith discovered that *bacteria have the ability to transform harmless cells into virulent ones by transferring some genetic factor from one bacteria cell to* *another.* This phenomenon is known as **bacterial transformation**, and the experiment is known as the **transformation experiment**. See the information about bacterial transformation later in this chapter.

Avery, MacLeod, and McCarty (1944) published their classic findings that Griffith's **transformation factor** is, in fact, DNA. This research proved that DNA was the agent that carried the genetic characteristics from the virulent dead bacteria to the living nonvirulent bacteria. *This provided direct experimental evidence that DNA, not protein, was the genetic material.*

Hershey and Chase (1952) carried out experiments that lent strong support to *the theory that DNA is the genetic material.* They tagged bacteriophages with the radioactive isotopes ^{32}P and ^{35}S. Since proteins contain sulfur but not phosphorus and DNA contains phosphorus but not sulfur, the radioactive ^{32}P labeled the DNA of the phage viruses while ^{35}S labeled the protein coat of the phage viruses. Hershey and Chase found that when bacteria were infected with phage viruses, the radioactive phosphorus in the phage always entered the bacterium while the radioactive sulfur remained outside the cells. This proved that *DNA from the viral nucleus, not protein from the viral coat, was infecting bacteria and producing thousands of progeny.*

Rosalind Franklin (1950–53), while working in the lab of Maurice Wilkins, carried out the X-ray crystallography analysis of DNA that showed DNA to be a helix. Her work was critical to Watson and Crick. Although Maurice Wilkins shared the Nobel prize with Watson and Crick, Rosalind Franklin did not. She had died by the time the prize was awarded, and the prize is not awarded posthumously.

Watson and Crick (1953), while working at Cambridge University, *proposed the double helix structure of DNA* in a one-page paper in the British journal *Nature*. Throughout the 1940s, until 1953, many scientists worked to understand the structure of DNA. All the data that Watson and Crick used to build their model of DNA derived from other scientists who published earlier. Two major pieces of information they used were the biochemical analysis of DNA (from Erwin Chargaff) and the X-ray diffraction analysis of DNA (from Rosalind Franklin). However, the fact that much of the components of DNA were known before Watson and Crick began their model building does not detract from the brilliance of their achievement. Understanding the structure of DNA gives a foundation to understand how DNA could replicate itself. Watson and Crick received the Nobel prize in 1962 for correctly describing the structure of DNA.

Meselson and Stahl (1958) *proved that DNA replicates in a semiconservative fashion,* as Francis Crick predicted. They cultured bacteria in a medium containing heavy nitrogen (^{15}N), allowing the bacteria to incorporate this heavy nitrogen into their DNA as they replicated and divided. These bacteria were then transferred to a medium containing light nitrogen (^{14}N) and allowed to replicate and divide only once. The bacteria that resulted from this final replication were spun in a centrifuge and found to be midway in density between the bacteria grown in heavy nitrogen and those grown in light nitrogen. This demonstrated that the new bacteria contained DNA consisting of one heavy strand and one light strand. See Figure 8.1 of semiconservative replication.

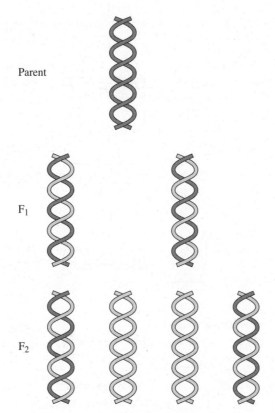

Parent

F₁

F₂

Figure 8.1 Semiconservative Replication

STRUCTURE OF NUCLEIC ACIDS
Deoxyribonucleic Acid (DNA)

The DNA molecule is a **double helix**, shaped like a twisted ladder, consisting of two strands running in opposite directions (antiparallel); see Figure 8.2. One strand runs **5′ to 3′** (right side up), the other **3′ to 5′** (upside down). DNA is a polymer consisting of repeating units of **nucleotides**. In DNA, these consist of a **5-carbon sugar** (**deoxyribose**), a **phosphate**, and a **nitrogen base**. The carbon atoms in deoxyribose are numbered 1 to 5. There are four nitrogenous bases in DNA: **adenine (A)**, **thymine (T)**, **cytosine (C)**, and **guanine (G)**. Of the four nitrogenous bases, adenine and guanine are **purines**, and thymine and cytosine are **pyrimidines**. The nitrogenous bases of opposite chains are paired to one another by **hydrogen bonds**: the adenine nucleotide bonds by a **double hydrogen bond** to the thymine nucleotide, and the cytosine nucleotide bonds by a **triple hydrogen bond** to the guanine nucleotide.

DNA gets packed and unpacked in the nucleus as needed. Eukaryotic DNA combines with a large amount of proteins called **histones** from which it separates only briefly during replication. This complex of DNA plus histones is called by the general name **chromatin**. The double helix of DNA wraps twice around a core of histones, forming structures called **nucleosomes** that look like beads on a string.

REMEMBER

The two strands of DNA run in opposite directions.

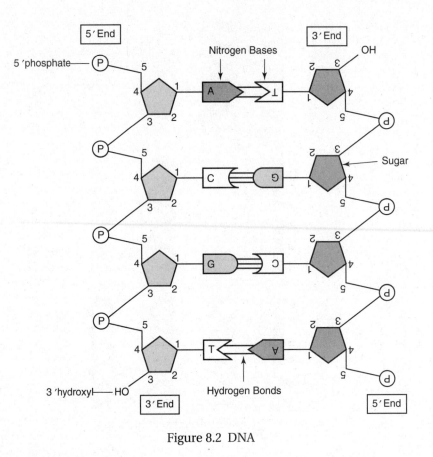

Figure 8.2 DNA

Ribonucleic Acid (RNA)

RNA is a single-stranded helix consisting of repeating nucleotides: adenine, cytosine, guanine, and **uracil (U)**, which replaces thymine. The 5-carbon sugar in RNA is **ribose**.

Figure 8.3 shows structural formulas for the purines—adenine and guanine—and for the pyrimidines—cytosine, thymine, and uracil.

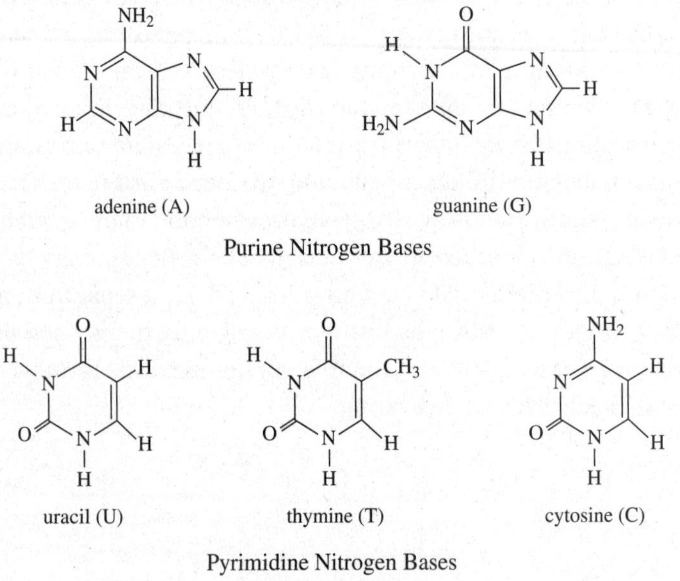

adenine (A)   guanine (G)

Purine Nitrogen Bases

uracil (U)   thymine (T)   cytosine (C)

Pyrimidine Nitrogen Bases

Figure 8.3

DNA REPLICATION IN EUKARYOTES

DNA replication, the making of an exact replica of the DNA molecule by **semiconservative replication**, was predicted by Watson and Crick and proven by Meselson and Stahl. The DNA double helix unzips, and each strand serves as a **template** for the formation of a new strand composed of complementary nucleotides: A with T and C with G. The two new molecules each consist of one old strand and one new strand. The following describes DNA replication in eukaryotes. See also Figures 8.1 and 8.4.

- Replication begins at special sites called **origins of replication**, where the two strands of DNA separate to form **replication bubbles**. Thousands of these bubbles can be seen along the DNA molecule by using electron microscopy. Replication bubbles speed up the process of replication along the giant DNA molecule that consists of *6 billion nucleotides*. A replication bubble expands as replication proceeds in *both directions at once*.

- At each end of the replication bubble is a **replication fork**, a Y-shaped region where the new strands of DNA are elongating. Eventually, all the replication bubbles fuse.

- The enzyme **DNA polymerase** catalyzes the antiparallel elongation of the new DNA strands. (At least 15 different types of DNA polymerase have been identified, but only one is involved in the elongation of the DNA strand.)

- DNA polymerase builds a new strand from the 5′ to the 3′ direction by moving along the template strand and pushing the replication fork ahead of it. In humans, the rate of elongation is about 50 nucleotides per second.

- *DNA polymerase cannot initiate synthesis*; it can only add nucleotides to the 3′ end of a preexisting chain. This preexisting chain actually consists of RNA and is called **RNA primer**. An enzyme called **primase** makes the primer by joining together RNA nucleotides.

- One of your cells can replicate its entire DNA in a few hours.

- DNA polymerase replicates the two original strands of DNA differently. Although it builds both new strands in the 5′ to 3′ direction, one strand is formed *toward the replication fork* in an unbroken, linear fashion. This is called the **leading strand**. The other strand, the **lagging strand**, forms in the direction *away from the replication fork* in a series of segments called **Okazaki fragments**. Okazaki fragments are about 100–200 nucleotides long and will be joined into one continuous strand by the enzyme **DNA ligase**.

- Other proteins and enzymes assist in replication of the DNA. **Helicases** are enzymes that untwist the double helix at the replication fork. They separate the two parental strands, making these strands available as templates. **Single-stranded binding proteins** act as scaffolding, holding the two DNA strands apart. **Topoisomerases** lessen the tension on the tightly wound helix by breaking, swiveling, and rejoining the DNA strands.

- DNA polymerases carry out **mismatch repair**, a kind of proofreading that corrects errors. Damaged regions of DNA are excised by **DNA nuclease**.

- Each time the DNA replicates, some nucleotides from the ends of the chromosomes are lost. To protect against the possible loss of genes at the ends of the chromosomes, eukaryotes have special nonsense nucleotide sequences (TTAGGG) at the ends of the chromosomes that repeat thousands of times. These protective ends are called **telomeres**. Telomeres are created and maintained by the enzyme **telomerase**. Normal body cells contain little telomerase, so every time the DNA replicates, the telomeres get shorter. This may serve as a clock that counts cell divisions and causes the cell to stop dividing as the cell ages.

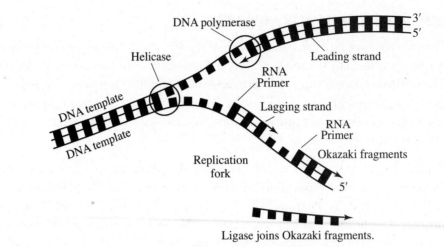

Figure 8.4 DNA Replication at Replication Fork

FROM DNA TO PROTEIN

The process whereby DNA makes proteins has been worked out in great detail. To summarize, the **triplet code** in DNA is **transcribed** into a **codon sequence** in messenger-RNA (mRNA) inside the nucleus. Next, this newly formed strand of **RNA**, known as pre-RNA, is **processed** or modified in the nucleus. Then the codon sequence leaves the nucleus and is **translated** into an amino acid sequence (a polypeptide) in the cytoplasm at the ribosome.

If the strand of DNA triplets to be transcribed is   5′-AAA TAA CCG GAC-3′

Then the strand of mRNA **codons** that forms is   3′-UUU AUU GGC CUG-5′

The transfer RNA (tRNA) **anticodon** strand
complementary to the mRNA strand is   AAA UAA CCG GAC

Figure 8.5 shows an overview of transcription, RNA processing, and translation.

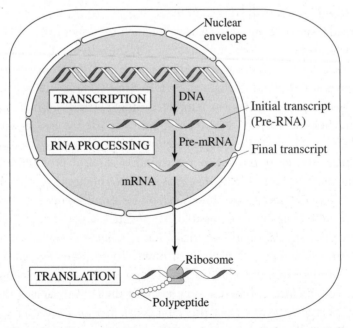

Figure 8.5 Transcription and Translation in a Eukaryotic Cell

Transcription

Transcription is the process by which the information in a DNA sequence is copied (transcribed) into a complementary RNA sequence. Although there are many kinds of RNA, three types are directly involved in protein synthesis. See Figure 8.6.

1. **MESSENGER RNA (mRNA) IS INVOLVED IN TRANSCRIPTION:** When a sequence of DNA is expressed, one of two strands of DNA is copied into mRNA according to the base-pairing rules, C with G and A with U (in RNA, uracil replaces the thymine in DNA).

2. **RIBOSOMAL RNA (rRNA) IS INVOLVED IN TRANSLATION:** rRNA is structural. Along with proteins, it makes up the ribosome, which consists of two subunits, one large and one small. The ribosome has one mRNA binding site and three tRNA binding sites, known as A, P, and E sites. A ribosome is a protein synthesis factory.

3. **TRANSCRIPTION RNA (tRNA) CARRIES AMINO ACIDS FROM THE CYTOPLASMIC POOL OF AMINO ACIDS TO mRNA AT THE RIBOSOME:** tRNA is shaped like a coverleaf and has a binding site for an amino acid at one end and another binding site for an **anticodon** sequence that binds to mRNA at the other.

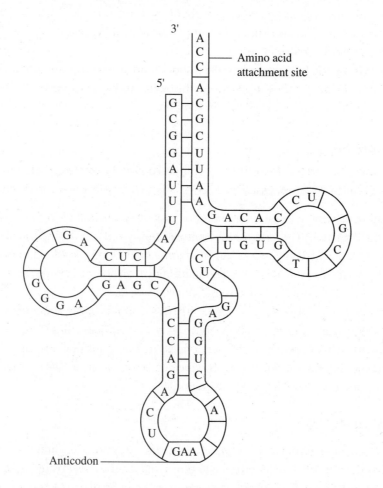

Figure 8.6 Transfer-RNA (tRNA)

- Transcription consists of three stages: **initiation**, **elongation**, and **termination**.
- **Initiation** begins when an enzyme, *RNA polymerase, recognizes and binds to DNA at the promoter region*. The promoter "tells" RNA polymerase where to begin transcription and which of the two strands to transcribe. A collection of proteins called **transcription factors** recognize a key area within the promoter, the **TATA box** (because of its many repeating thymine and adenine nucleotides), and mediate the binding of RNA polymerase to the DNA. The completed assembly of transcription factors and RNA polymerase bound to the promoter is called a **transcription initiation complex**. Once RNA polymerase is attached to the promoter, DNA transcription of the DNA **template** begins.
- **Elongation** of the strand continues as *RNA polymerase adds nucleotides to the 3′ end of a growing chain*. RNA polymerase pries the two strands of DNA apart and attaches RNA nucleotides according to the base pairing rules: C with G and A with U. The stretch of DNA that is transcribed into an mRNA molecule is called a **transcription unit**. Each unit consists of triplets of bases called **codons** (for example, AAU, CGA) that code for specific amino acids. A single gene can be transcribed into mRNA simultaneously by several molecules of RNA polymerase following each other in a caravan fashion. Like DNA polymerases, RNA polumerase has mechanisms for proofreading during transcription. Because mRNA is usually short-lived, any errors in mRNA are not as potentially harmful as errors in the DNA sequence.
- **Termination** is the final stage in transcription. Elongation continues for a short distance after the RNA polymerase transcribes the **termination sequence** (AAUAAA). At this point, mRNA is cut free from the DNA template.

RNA Processing

Before the newly formed pre-RNA strand is shipped out of the nucleus to the ribosome in the cytoplasm, it is altered or **processed** by a series of enzymes. Here are the details.

- A **5′ cap** consisting of a modified guanine nucleotide is added to the 5′ end. This cap helps the RNA strand bind to the ribosome in the cytoplasm during translation.
- A **poly (A) tail**, consisting of a string of adenine nucleotides, is added to the 3′ end. This tail protects the RNA strand from degradation by hydrolytic enzymes, and facilitates the release of mRNA from the nucleus into the cytoplasm.
- Noncoding regions of the mRNA called **introns** or **intervening sequences** are removed by **snRNPs**, small nuclear ribonucleoproteins, and **splicesomes**. This removal allows only **exons**, which are expressed sequences, to leave the nucleus. As a result of this processing, the mRNA that leaves the nucleus is a great deal shorter than the original transcription unit. See Figure 8.5 on page 134.

Alternative Splicing

Before the human genome was sequenced by the Human Genome Project, scientists expected that they would find about 100,000 genes. In fact, they discovered that humans have only about 24,000 genes. This surprised everyone but can be explained when you recognize that different mRNAs can be synthesized from the same primary transcript. In **alternative RNA splicing**, different RNA molecules are produced from the same primary transcript, depending on which RNA segments are treated as exons and which as introns. *Regulatory proteins* specific to a cell type control intron-exon choices by binding to regulatory sequences within the primary transcript. See Figure 8.7.

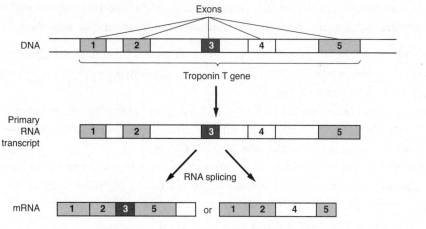

Figure 8.7 Alternative RNA Splicing

Translation of mRNA—Synthesis of a Polypeptide

Translation *is the process by which the codons of an mRNA sequence are changed into an amino acid sequence*; see Figure 8.8. Amino acids present in the cytoplasm are carried by tRNA molecules to the codons of the mRNA strand at the ribosome according to the base pairing rules (A with U and C with G). One end of the tRNA molecule bears a specific amino acid, and the other end bears a nucleotide triplet called an **anticodon**. Unlike mRNA which is broken down immediately after it is used, tRNA is used repeatedly. The energy for this process is provided by **GTP (guanosine triphosphate)**, a molecule closely related to ATP. Each amino acid is joined to the correct tRNA by a specific enzyme called **aminoacyl-tRNA synthetase**. There are only 20 different aminoacyl-tRNA synthetases, one for each amino acid. There are 64 codons; 61 of them code for amino acids. One codon, **AUG**, has two functions; it codes for methionine and is also a **start codon**. Three codons, **UAA**, **UGA**, and **UAG**, are **stop codons**, and terminate all sequences. Some tRNA molecules have anticodons that can recognize two or more different codons. This occurs because the pairing rules for the third base of a codon are not as strict as they are for the first two bases. This relaxation of base pairing rules is known as **wobble**. For example, the codons UCU, UCC, UCA, and UCG all code for the amino acid serine.

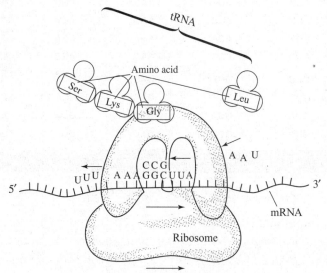

Figure 8.8 Translation

See Figure 8.6 for a sketch of tRNA and Figure 8.8 for a sketch of translation. The process of translation consists of three stages: **initiation**, **elongation**, and **termination**.

- **Initiation** begins when mRNA becomes attached to a subunit of the ribosome. This first codon is always **AUG**. It must be positioned correctly in order for transcription of an amino acid sequence to begin.
- **Elongation** continues as tRNA brings amino acids to the ribosome and a polypeptide chain is formed. One mRNA molecule is generally translated simultaneously by several ribosomes in clusters called **polyribosomes**.
- **Termination** of an mRNA strand is complete when a ribosome reaches one of three **termination** or **stop codons**. A **release factor** breaks the bond between the tRNA and the last amino acid of the polypeptide chain. The polypeptide is freed from the ribosome, and mRNA is broken down.

The Genetic Code

The complete genetic code is shown in Figure 8.9. There are 64 (4^3) possible combinations of the four nitrogenous bases. Notice that AUG, which codes for methionine, also codes for **start**, the initiation signal for translation. Three of the codons are **stop codons**, termination signals of translation.

FIRST BASE (5' end)	SECOND BASE: U	C	A	G	THIRD BASE (3' end)
U	UUU Phe	UCU Ser	UAU Tyr	UGU Cys	U
U	UUC Phe	UCC Ser	UAC Tyr	UGC Cys	C
U	UUA Leu	UCA Ser	UAA Stop	UGA Stop	A
U	UUG Leu	UCG Ser	UAG Stop	UGG Trp	G
C	CUU Leu	CCU Pro	CAU His	CGU Arg	U
C	CUC Leu	CCC Pro	CAC His	CGC Arg	C
C	CUA Leu	CCA Pro	CAA Gln	CGA Arg	A
C	CUG Leu	CCG Pro	CAG Gln	CGG Arg	G
A	AUU Ile	ACU Thr	AAU Asn	AGU Ser	U
A	AUC Ile	ACC Thr	AAC Asn	AGC Ser	C
A	AUA Ile	ACA Thr	AAA Lys	AGA Arg	A
A	AUG Met or start	ACG Thr	AAG Lys	AGG Arg	G
G	GUU Val	GCU Ala	GAU Asp	GGU Gly	U
G	GUC Val	GCC Ala	GAC Asp	GGC Gly	C
G	GUA Val	GCA Ala	GAA Glu	GGA Gly	A
G	GUG Val	GCG Ala	GAG Glu	GGG Gly	G

Figure 8.9 The Genetic Code

Remember this important statement about the genetic code: *There are redundancies in the code, but there is no ambiguity.* Examine the chart. The fact that there are four codons for leucine means there is redundancy. In every case, one codon codes for one particular amino acid. So there is no ambiguity.

With very few exceptions within bacteria, mitochondria and chloroplasts, this code is universal and unifies all life. It indicates that the code originated early in the evolution of life on Earth and that all living things descended from those first ancestral cells.

GENE MUTATION

Mutations are permanent changes in genetic material. They occur *spontaneously* and *randomly*. They can be caused by **mutagenic agents**, including toxic chemicals and radiation. A mutation in **somatic** (body) cells disrupts normal cell functions. Mutations that occur in gametes are transmitted to offspring and can change the **gene pool** of a population. *Mutations are the raw material for natural selection.*

Some regions of DNA are more vulnerable to mutations than others. For example, regions of As and Ts are subject to more breakages than regions of Cs and Gs because A and T are connected by a double hydrogen bond whereas Cs and Gs are connected by a triple hydrogen bond.

Point Mutation

The simplest mutation is a **point mutation**. This is a **base-pair substitution**, a chemical change in just one base pair in a single gene. Here is an example of a change in an English sentence analogous to a point mutation in DNA:

Point Mutation

THE FAT CAT SAW THE **DOG** → THE FAT CAT SAW THE **HOG**.

The inherited genetic disorder **sickle cell anemia** results from a single point mutation in a single base pair in the gene that codes for hemoglobin. This point mutation is responsible for the production of abnormal hemoglobin that can cause red blood cells to sickle when oxygen tension is low. When red blood cells sickle, a variety of tissues may be deprived of oxygen and suffer severe, permanent damage. The possibility exists, however, that a point mutation could result in a beneficial change for an organism or, because of **wobble** in the genetic code, result in no change in the proteins produced. (Wobble is the relaxation of the base-pairing rules for the third base in a codon.) Here is an example of wobble:

DNA	mRNA	Amino Acid Produced
AAA	UUU	Phenylalanine
AAG	UUC	Phenylalanine
↑ mutation		No change occurs in the amino acid.

Insertion or Deletion

A second type of gene mutation results from a single nucleotide **insertion** or **deletion**. To continue the three-letter word analogy, a deletion is the loss of one letter, and an insertion is the addition of a letter into the DNA sentence. Both mutations result in a **frameshift**, because the entire reading frame is altered.

Deletion of the Letter E shifts the reading frame.

$$\downarrow$$

THE FAT CAT SAW THE DOG → THF ATC ATS AWT HED OG

Insertion of the Letter T shifts the reading frame.

$$\downarrow$$

THE FAT CAT SAW THE DOG → THE FTA TCA TSA WTH EDO G

As a result of the frameshift, one of two things can happen. Either a mutated polypeptide is formed or no polypeptide is formed.

Missense Mutations

When point mutations or frameshifts change a codon *within a gene* into a stop codon, translation will be altered into a **missense** or **nonsense mutation**.

THE GENETICS OF VIRUSES AND BACTERIA

Since the early part of the twentieth century when Griffith discovered the transformation factor, knowledge of genetics has been based on work with the simplest biological systems—viruses and bacteria. Scientists' understanding of replication, transcription, and translation of DNA was worked out using bacteria as a model. Their understanding of how viruses and bacteria infect cells is the basis for how diseases are treated and how vaccines are developed. A worldwide industry of genetic engineering and recombinant DNA relies on bacteria like *Escherichia coli* and viruses like the phage viruses for research and therapeutic endeavors. Whereas Gregor Mendel depended on the garden pea and Thomas Hunt Morgan on the fruit fly, researchers now depend on bacteria and viruses.

The Genetics of Viruses

A virus is a parasite that can live only inside another cell. It commandeers the host cell machinery to transcribe and translate all the proteins it needs to fashion new viruses. In the process, thousands of new viruses are formed and the host cell is often destroyed. A virus consists of DNA or RNA enclosed in a protein coat called a **capsid**. Some viruses also have a viral **envelope** that is derived from membranes of host cells, cloaks the capsid, and aids the virus in infecting the host. Each type of virus can infect only one specific cell type because it gains entrance into a cell by binding to *specific receptors* on the cell surface. For example, the virus that causes colds in humans infects only the membranes of the respiratory system, and the virus that causes AIDS infects only one type of white blood cell. In addition, one virus can usually only infect one species. The range of organisms that a virus can attack is referred to as the **host range** of the virus. A sudden emergence of a new viral disease that affects humans, such as AIDS or H1N1, may result from a mutation in the virus that expands its host range.

- **BACTERIOPHAGES**—The most complex and best understood virus is the one that infects bacteria, the **bacteriophage**, or **phage** virus. The bacteriophage can reproduce in different ways.

1. In the **lytic cycle**, the phage enters a host cell, takes control of the cell machinery, replicates itself, and then causes the cell to burst, releasing a new generation of infectious phage viruses. These new viruses infect and kill thousands of cells in the same manner. A phage that replicates only by a lytic cycle is a virulent phage.

2. In the **lysogenic cycle**, viruses replicate without destroying the host cell. The phage virus becomes incorporated into a specific site in the host's DNA. It remains dormant within the host genome and is called a **prophage**. As the host cell divides, the phage is replicated along with it and a single infected cell gives rise to a population of infected cells. At some point, an environmental trigger causes the prophage to switch to the **lytic phase**. Viruses capable of both modes of reproducing, lytic and lysogenic, within a bacterium are called **temperate viruses**.

- **RETROVIRUSES** are viruses that contain RNA instead of DNA and replicate in an unusual way. Following infection of the host cell, the retrovirus RNA serves as a template for the synthesis of complementary DNA (cDNA) because it is complementary to the RNA from which it was copied. *Thus, these retroviruses reverse the usual flow of information from DNA to RNA.* This reverse transcription occurs under the direction of an enzyme called **reverse transcriptase**. A retrovirus usually inserts itself into the host genome, becomes a permanent resident, called a **prophage**, and is capable of making multiple copies of the viral genome for years. An example of a retrovirus is the HIV (human immunodeficiency virus), which causes AIDS.

- **TRANSDUCTION**—Phage viruses acquire bits of bacterial DNA as they infect one cell after another. This process, which leads to genetic recombination, is called **transduction**. Two types of transduction occur, **generalized** and **restricted (specialized)**. Generalized transduction moves random pieces of bacterial DNA as the phage lyses one cell and infects another during the lytic cycle. **Restricted transduction** involves the transfer of specific pieces of DNA. During the lysogenic cycle, a phage integrates into the host cell at a specific site. At a later time, when the phage ruptures out of the host DNA, it sometimes carries a piece of adjacent host DNA with it and inserts this host DNA into the next host it infects.

The Genetics of Bacteria

The bacterial chromosome is a circular, double-stranded DNA molecule, tightly condensed into a structure called a **nucleoid**, which has no nuclear membrane. Bacteria replicate their DNA in **both directions** from a **single point of origin**.

Although bacteria can reproduce by a primitive sexual method called **conjugation**, the main mode of reproduction is asexual, by **binary fission**. Binary fission results in a population with all identical genes, but mutations do occur spontaneously. Although mutations are rare, bacteria reproduce by the millions, and even one mutation in every 1,000 replications can amount to significant variation in the population as a whole.

- **Bacterial transformation** was discovered by **Frederick Griffith** in 1927 when he performed experiments with several different strains of the bacterium *Diplococcus pneumoniae*.

   Transformation is either a natural or an artificial process that provides a mechanism for the recombination of genetic information in some bacteria. Small pieces of extra-cellular DNA are taken up by a living bacterium, ultimately leading to a stable genetic change in the recipient cell. Bacterial transformation is very easy to carry out today.

- A **plasmid** is a foreign, small, circular, self-replicating DNA molecule that inhabits a bacterium. A bacterium can harbor many plasmids and will express the genes carried by the plasmid. The first plasmid discovered was the **F plasmid**. F stands for fertility. Bacteria that contain the F plasmid are called F+; those that do not carry the plasmid are called F-. The F plasmid contains genes for the production of **pili**, cytoplasmic bridges that connect to an adjacent cell and that allow DNA to move from one cell to another in a form of primitive sexual reproduction called **conjugation**. Another plasmid, the **R plasmid**, makes the cell in which it is carried resistant to specific antibiotics, such as ampicillin or tetracycline. In addition, the R plasmid can be transferred to other bacteria by conjugation. Bacteria that carry the R plasmid have a distinct evolutionary advantage over bacteria that are not resistant to antibiotics. Resistant bacteria will be selected for (survive) and their populations will increase while nonresistant bacteria die out. This is exactly what is happening today as an increasing number of populations of pathogenic bacteria, such as the one that causes tuberculosis, are becoming resistant to antibiotics. This is cause for serious concern in the health community.

The Operon

> **REMEMBER**
>
> **If an essay question on the AP exam is about regulation, the operon is a perfect example.**

The **operon** was discovered in the bacterium *E. coli* by **Jacob and Monod** in the 1940s. Although it is found only in bacteria, it is an important model of **gene regulation**. An operon is essentially a set of genes and the switches that control the expression of those genes. There are two types of operons: the **inducible (lac)** operon and the **repressible (tryptophan)** operon.

THE TRYPTOPHAN OPERON

The **tryptophan operon** consists of a **promoter** and five adjacent structural genes (A, B, C, D and E) that code for the five separate enzymes necessary to synthesize the amino acid tryptophan; see Figures 8.10 and 8.11. As long as **RNA polymerase** binds to the promoter, one long strand of mRNA containing start and stop codons is transcribed. If adequate tryptophan is present, tryptophan itself acts as a **corepressor** activating the **repressor**. The activated repressor binds to the **operator**, preventing RNA polymerase from binding to the promoter. Without RNA polymerase attached to DNA at the promoter, transcription ceases. The tryptophan operon is known as a repressible operon, meaning it is always switched on unless the repressor is activated.

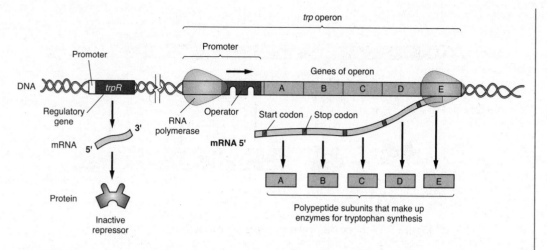

Figure 8.10 Tryptophan absent, repressor inactive, operon on, → tryptophan produced

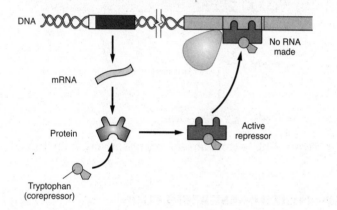

Figure 8.11 Tryptophan present, repressor active, operon off

THE LAC OPERON

In order for the *E. coli* in our intestines to utilize lactose as an energy source, three enzymes must be synthesized to break down lactose into glucose and galactose. These enzymes, β-galactosidase, permease, and transacetylase, are coded for by three genes in the *lac* operon (A, B, and C); see Figures 8.12 and 8.13. In order for these three genes to be transcribed, the **repressor** must be prevented from binding to the operator and RNA polymerase must bind to the promoter region. Allolactose, an isomer of lactose, is the **inducer** that facilitates this process by binding to the **active repressor** and inactivating it. When a person drinks milk, they ingest allolactose, the inducer, which deactivates the repressor, allowing RNA polymerase to bind to DNA. When RNA polymerase binds to DNA, transcription of the *lac* genes occurs and lactose can be utilized as an energy source.

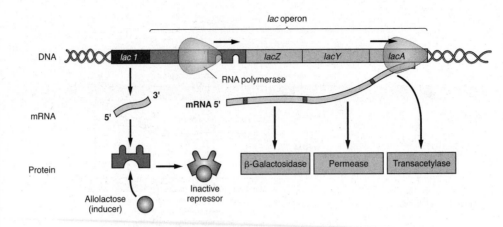

Figure 8.12 Lactose present, repressor inactive, operon on

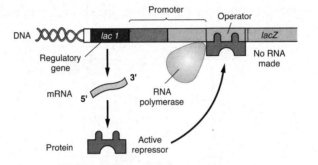

Figure 8.13 Lactose absent, repressor active, operon off

CAP AND cAMP—POSITIVE GENE REGULATION

When glucose and lactose are both present in the intestine, *E. coli* preferentially metabolize glucose and the enzymes for breaking down glucose are always present. However, when lactose is present and glucose is in short supply, *E. coli* switch to lactose as an energy source. This ability depends on the interaction of an allosteric regulatory protein, CAP (catabolite activator protein) and cAMP (cyclic AMP). Since the attachment of CAP to the promoter directly stimulates gene expression, this mechanism is an example of **positive gene regulation**.

VOCABULARY FOR THE OPERON

- **RNA polymerase**: Enzyme that transcribes a new RNA chain by linking ribonucleotides to nucleotides on a DNA template
- **Operator**: Sequence of nucleotides near the start of an operon to which the active repressor can attach. The binding of the repressor prevents RNA polymerase from attaching to the promoter and transcribing the operon's genes.
- **Promoter**: Nucleotide sequence in the DNA of a gene that is the binding site of RNA polymerase, positioning the RNA polymerase to begin to transcribe RNA at the appropriate position.
- **Repressor**: Protein that inhibits gene transcription. In the operon of prokaryotes, repressors bind to the operator.
- **Regulator gene**: Gene that codes for a repressor. It is located some distance from its operon and has its own promoter.

PRIONS

Prions are not cells and are not viruses. They are misfolded versions of a protein normally found in the brain. If prions get into a normal brain, they cause all the normal versions of the protein to misfold in the same way. Prions are infectious and cause several brain diseases: **scrapie** in sheep, **mad cow disease** in cattle, and **Creutzfeldt-Jakob disease** in humans. All known prion diseases are fatal.

THE HUMAN GENOME

The human genome consists of 3 billion base pairs of DNA and about 24,000 genes. Surprisingly, **98 percent** of human DNA does NOT code for protein product and has often been called **junk**. Of the **non-gene DNA**, some are **regulatory sequences** that control gene expression, some are **introns** that interrupt genes, and most are **repetitive sequences** that never get transcribed. A number of genetic disorders, including Huntington's disease, are caused by abnormally long stretches of **tandem repeats** (back-to-back repetitive sequences) *within* affected genes. Many of these tandem repeats make up the **telomeres**. Scientists have also identified certain noncoding regions of DNA, **polymorphic regions**, that are highly variable from one region to the next.

RECOMBINANT DNA; CLONING GENES

Recombinant DNA means taking DNA from two or more sources and combining them into one molecule. This occurs in nature during **viral transduction, bacterial transformation**, and **conjugation** and when **transposons** jump around the genome. Scientists can also manipulate and engineer genes in vitro (in the laboratory). The branch of science that uses **recombinant DNA techniques** for practical purposes is called **biotechnology** or **genetic engineering**.

Many tools and techniques have been developed to manipulate and engineer genes. The following sections contain a discussion of the uses for genetic engineering, an explanation of some techniques that are used, and a brief discussion of some ethical issues within the field.

The potential uses of **recombinant DNA** or **gene cloning** are many:

- To produce a **protein product**, such as human insulin, in large quantities as an inexpensive pharmaceutical.
- To **replace** a **nonfunctioning gene** in a person's cells with a functioning gene by **gene therapy**. Scientists are currently conducting clinical trials in this area with disappointing results. Sometimes the human subjects become ill from the viral **vector** used to carry the gene. Other times, the gene is inserted successfully and begins to produce the necessary protein but stops working in a short time. If scientists can master this technique, many lives will be improved.
- To prepare **multiple copies of a gene** itself **for analysis**. Since most genes exist in only one copy on a chromosome, the ability to make multiple copies is of great value as a research tool.
- To **engineer bacteria** to clean up the environment. Scientists have engineered many bacteria; one modified species can even eat **toxic waste**.

The Technique of Gene Cloning

- Isolate a gene of interest, for example, the gene for human insulin.
- Insert the gene into a **plasmid**.
- Insert the plasmid into a **vector**, a cell that will carry the plasmid, such as a bacterium. To accomplish this, a bacterium must be made **competent**, which means be able to take up a plasmid.
- **Clone** the gene. As the bacteria reproduce themselves by **fission**, the plasmid and the selected gene are also being **cloned**. Millions of copies of the gene are produced.
- Identify the bacteria that contain the selected gene and harvest it from the culture.

TOOLS AND TECHNIQUES OF RECOMBINANT DNA

Restriction Enzymes

Restriction enzymes were discovered in the late 1960s and are a basic biotechnology tool. They are extracted from bacteria, which use them to fend off attacks by invading bacteriophages. Restriction enzymes cut DNA at specific **recognition sequences** or **sites**, such as GAATTC. Often these cuts are staggered, leaving single-stranded **sticky ends** to form a temporary union with other sticky ends. The fragments that result from the cuts made by restriction enzymes are called **restriction fragments**.

Scientists have now isolated hundreds of different restriction enzymes. They are named for the bacteria in which they were found. Common examples are *Eco*RI (which was discovered in *E. coli*), *Bam*HI, and *Hind*III. Restriction enzymes have many uses, including **gene cloning**.

Gel Electrophoresis

Gel electrophoresis separates large molecules of DNA on the basis of their rate of movement through an **agarose gel** in an electric field. The smaller the molecule, the faster it runs. DNA, which is negative (due to the presence of phosphate groups, PO_4^{3-}), flows from **cathode** (-) to **anode** (+). The concentration of the gel can be altered to provide a greater impediment to the DNA, allowing for finer separation of smaller pieces.

Electrophoresis is also commonly used to separate proteins and amino acids. If DNA is going to be run through a gel, it must first be cut up by **restriction enzymes** into pieces small enough to migrate through the gel. Once separated on a gel, the DNA can be analyzed in many ways. The DNA strands can be **sequenced** to determine the sequence of bases A, C, T, and G. The gel can also be used in a comparison with other DNA samples. A **DNA probe** can also identify the location of a specific sequence within the DNA.

Figure 8.14 shows gel electrophoresis of DNA that was cut with restriction enzymes. Lane 1 has four bands of DNA, three larger pieces and one short piece. Lane 2 contains two pieces of DNA, one large and one tiny one. Lane 3 contains one very large, uncut piece of DNA. Lane 4 contains two pieces of DNA. *The smaller the piece of DNA, the farther it has traveled from the well.*

DNA Probe

A **DNA probe** is a **radioactively labeled single strand** of nucleic acid molecule used to tag a specific sequence in a DNA sample. The probe bonds to the complementary sequence wherever it occurs, and the radioactivity enables scientists to detect its location. The DNA probe is used to identify a person who carries an inherited genetic defect, such as sickle cell anemia, Tay-Sachs disease, Huntington's disease, and hundreds of others.

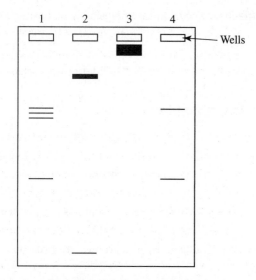

Figure 8.14 Gel Electrophoresis of DNA Fragments

Polymerase Chain Reaction (PCR)

Devised in 1985, **polymerase chain reaction** is a cell-free, automated technique by which a piece of DNA can be rapidly copied or amplified. Billions of copies of a fragment of DNA can be produced in a few hours. The DNA piece that is to be amplified is placed into a test tube with *Taq* **polymerase** (a heat-stable form of DNA polymerase extracted from extremophile bacteria), along with a supply of **nucleotides** (**A, C, T**, and **G**) and **primers** necessary for DNA synthesis. Once the DNA is amplified, these copies can be studied or used in a comparison with other DNA samples.

The PCR technique has limitations:

- Some information about the nucleotide sequence of the target DNA must be known in advance in order to make the necessary primers.
- The size of the piece that can be amplified must be very short.
- Contamination is a major problem. A few skin cells from the technician working with the sample could make obtaining accurate results difficult or impossible. Such an error could have dire consequences with a crime scene sample.

Restriction Fragment Length Polymorphisms (RFLPs)

A **restriction fragment** is a segment of DNA that results when DNA is treated with restriction enzymes. When scientists compared **noncoding regions** (junk DNA) of human DNA across a population, they discovered that the **restriction fragment pattern** is different in every individual. These differences have been named **restriction fragment length polymorphisms** or **RFLPs**, pronounced "riflips." A RFLP analysis of someone's DNA gives a human **DNA fingerprint** that looks like a bar code.

Each person's RFLPs are unique, except in identical twins, and are inherited in a Mendelian fashion. Because they are inherited in this way, they can be used very accurately in **paternity suits** to determine, with absolute certainty, if a particular man is the father of a particular child. In addition, RFLPs are routinely used to identify the perpetrator in rape and murder cases. DNA from the crime scene and the victim are compared against DNA from the suspect. Because of the accuracy of RFLP analysis, these cases can be solved with a high degree of certainty, and some suspects have been convicted on DNA evidence alone. By contrast, several incidents have occurred where prisoners who have been jailed for many years for violent crimes were proven innocent by DNA evidence and released.

Complementary DNA (cDNA)

When scientists try to clone a human gene in a bacterium, the introns (long intervening, noncoding sequences) present a problem. Bacteria lack introns and have no way to edit them out after transcription. Therefore, in order to clone a human gene in a bacterium, scientists must insert a gene with no introns. To do this, scientists extract fully processed mRNA from cells and then use the enzyme **reverse transcriptase** (obtained from **retroviruses**) to make DNA transcripts of this RNA. The resulting DNA molecule carries the complete coding sequence of interest but without introns. The DNA produced by retroviruses in this way is called **complementary DNA** or **cDNA**.

ETHICAL CONSIDERATIONS

Many people are concerned about potential problems arising from genetic engineering. This section discusses some of those concerns.

Safety

Much of the milk available in stores comes from cows that have been given a genetically engineered bovine growth hormone (BGH) to increase the quantity of milk they produce. Many people are concerned that this hormone will find its way into the milk and cause problems for the people who drink it.

Vegetable seeds have been genetically engineered to produce special characteristics in the vegetables people eat. Again, individuals are concerned that the genes that have been inserted into the vegetables may be dangerous to those who eat them.

Privacy

DNA probes are being coupled with the technology of the semiconductor industry to produce **DNA chips** that are about 1/2 inch square and can hold personal information about someone's genetic makeup. The chips scan a person for mutations in over 7,000 genes, including mutations in the immune system or the breast cancer genes (BRCA I and II) and for a predisposition to other cancers or heart attacks.

Although many people might like to know if they carry these mutations, the possibility that the personal information on a DNA chip might not remain private has caused much controversy. For example, if your health insurance company learned that you carry a harmful gene, it might not insure you or might charge you a higher premium. Similarly, if a company where you have applied for employment learns about a defect in your personal genetic makeup, based on the possibility that you might be disabled with a serious illness in the future, it might refuse to hire you.

Questions 1–4

Questions 1–4 refer to the following four choices.

(A) Translation
(B) Replication
(C) Transcription
(D) Transformation

1. The process in which DNA makes messenger RNA

2. The process in which DNA is synthesized from a template strand

3. The process in which foreign DNA is taken up by a bacterial cell

4. The process in which a polypeptide strand is synthesized using mRNA as a template

5. All of the following are true about electrophoresis EXCEPT

(A) it can be used only to analyze DNA
(B) the heavier the fragment, the slower it moves
(C) the fragments of DNA are negatively charged and migrate to the positive pole
(D) a buffer must cover the gel to allow a current to pass through the system

6. *Eco*RI is

(A) a bacterium
(B) a bacteriophage
(C) a type of DNA used extensively in research
(D) a restriction enzyme

7. In DNA replication, the role of DNA polymerase is to

(A) bring two separate strands back together after new ones are formed
(B) join the RNA nucleotides together to make the primer
(C) build a new strand of DNA from 5′ to 3′
(D) unwind the tightly wound helix

8. Which is NOT used in the normal replication of DNA?

(A) RNA primer
(B) ligase
(C) restriction enzymes
(D) polymerase

9. DNA replication can best be described as

(A) semiconservative
(B) conservative
(C) degenerate
(D) comparative

Questions 10–13

Questions 10–13 refer to these scientists famous for their work with DNA.

(A) Hershey and Chase
(B) Griffith
(C) Meselson and Stahl
(D) Rosalind Franklin

10. Discovered transformation in bacteria

11. Proved that DNA replicates by semiconservative replication

12. Proved that the nuclear material in a bacteriophage, not the protein coat, infects a bacterium

13. The first to analyze DNA by X-ray crystallography

---

14. If a segment of DNA is 5'-TGA AGA CCG-3', the RNA that results from the transcription of this segment will be

(A) 5'-TGA AGA CCG-3'
(B) 3'-ACU UCU GGC-5'
(C) 3'-ACT TCT GGC-5'
(D) 3'-CGG UCU UCA-5'

15. Once transcribed, eukaryotic RNA normally undergoes substantial alteration that results primarily from

(A) removal of exons
(B) removal of introns
(C) addition of introns
(D) combining of RNA strands by a ligase

16. Which of the following contain a pyrimidine and a purine?

(A) adenine and guanine
(B) uracil and thymine
(C) cytosine and uracil
(D) adenine and cytosine

17. What happens when T7 bacteriophages are grown in radioactive phosphorus?

(A) They can no longer infect bacteria.
(B) They die.
(C) Their DNA becomes radioactive.
(D) Their protein coat becomes radioactive.

18. Which of the following acts as a primer that initiates the synthesis of a new strand of DNA?

(A) single-strand binding protein

(B) RNA

(C) DNA

(D) topoisomerases

19. If guanine makes up 28% of the nucleotides in a sample of DNA from an organism, then thymine would make up _____ % of the nucleotides.

(A) 28

(B) 56

(C) 22

(D) 44

20. If AUU is the codon, what is the anticodon?

(A) AUU

(B) TAA

(C) UUA

(D) UAA

21. Which of the following is an example of wobble?

(A) amino acids carried to the ribosome to form a polypeptide chain

(B) the excision of introns from mRNA

(C) the binding of a primer to DNA

(D) four codons can all code for the same amino acid

22. Which of the following is TRUE about sickle cell anemia?

(A) It is caused by a chromosome mutation that resulted from nondisjunction.

(B) It is common in people from the Middle East.

(C) It is caused by a point mutation.

(D) A person with sickle cell anemia is resistant to many other genetic disorders.

23. A particular triplet code on DNA is AAA. What is the anticodon for it?

(A) AAA

(B) TTT

(C) UUU

(D) CCC

24. What are the regions of DNA called that code for proteins?

(A) introns

(B) codons

(C) anticodons

(D) exons

25. Prions are

   (A) bacteriophages that cause disease

   (B) infectious proteins

   (C) a bacterium that infects viruses

   (D) the cause of sickle cell anemia

26. Which word would **best** describe the operon?

   (A) respiration

   (B) transport

   (C) regulation

   (D) nutrition

Questions 27–30

Questions 27–30 refer to the operon.

   (A) Lactose

   (B) Repressor

   (C) Regulator

   (D) Promoter

27. Acts as an inducer in the *lac* operon

28. Binding site for RNA polymerase

29. Codes for the repressor

30. Binds at the operator

---

31. Which is TRUE of biotechnology techniques?

   (A) PCR is used to cut DNA molecules.

   (B) A DNA probe consists of a radioactive single strand of DNA.

   (C) Restriction enzymes were first discovered in bacteriophage viruses.

   (D) *Eco*RI is a name for a DNA probe.

32. Gel electrophoresis is used to

   (A) amplify small pieces of DNA

   (B) make bacterial cells competent

   (C) separate DNA that has already been cut up by restriction enzymes

   (D) cause DNA to twist back into a helix after amplification

33. Mad cow disease is caused by a

   (A) virus

   (B) prion

   (C) bacterium

   (D) genetic mutation

34. Which enzyme permanently seals together DNA fragments that have complementary sticky ends?

   (A) DNA polymerase
   (B) single-stranded binding protein
   (C) reverse transcriptase
   (D) DNA ligase

Answers to Multiple-Choice Questions

1. **(C)** Transcription is the process by which DNA makes RNA. There are three types of RNA; mRNA, tRNA and rRNA. Transcription occurs in three stages: initiation, elongation, and termination.

2. **(B)** DNA makes an exact copy of itself during replication. This process occurs in a semiconservative fashion, as proven by Meselson and Stahl.

3. **(D)** Griffith was the first to recognize the phenomenon of transformation while working with pneumococcus bacteria.

4. **(A)** Translation is the process by which the codons of mRNA sequence are changed into an amino acid sequence. Amino acids present in the cytoplasm are carried by tRNA molecules to the codons of the mRNA strand at the ribosome according to the base-pairing rules (A with U and C with G).

5. **(A)** Electrophoresis is commonly used in the separation of proteins as well as in the separation of DNA.

6. **(D)** *Eco*RI stands for *E. coli* restriction enzyme #1. It was the first restriction enzyme discovered.

7. **(C)** DNA polymerase builds a new strand of DNA from the 5′ end to the 3′ end (of the new strand). DNA polymerase can only add nucleotides to an existing strand of DNA.

8. **(C)** Restriction enzymes are a laboratory tool for cutting pieces of DNA at specific restriction sites.

9. **(A)** Replication of DNA is semiconservative. This means that when one double helix makes a copy of itself, the two new DNA molecules each consist of one new strand and one old strand. This was hypothesized by Watson and Crick and confirmed experimentally by Meselson and Stahl.

10. **(B)** Griffith discovered bacterial transformation in 1927.

11. **(C)** Semiconservative replication was hypothesized by Watson and Crick and confirmed experimentally by Meselson and Stahl.

12. **(A)** Hershey and Chase carried out experiments where they tagged bacteriophages with ^{32}P and ^{35}S. They proved that the DNA from the viral (phage) nucleus, not protein from the viral coat, was infecting bacteria and producing thousands of progeny.

13. **(D)** Rosalind Franklin, while working in the lab of Maurice Wilkins, carried out the X-ray crystallography analysis of DNA that showed DNA to be a helix.

14. **(B)** You are given a strand of DNA, which makes a strand of mRNA. Follow the base-pairing rules: T with A, C with G, C with G, and A with U. Remember, RNA contains uracil instead of thymine. If the DNA segment is 5'-TGA AGA CCG-3', then the mRNA strand complementary to that is 3'-ACU UCU GGC-5'.

15. **(B)** Once transcription has occurred, the new RNA molecule undergoes RNA processing. During this process, introns (intervening sequences) are removed with the help of snRNPs and a 5' cap and poly(A) tail are added.

16. **(D)** Pyrimidines often have the letter *y* in them. They are thymine, cytosine, and uracil, which replaces thymine in RNA. Adenine is the purine; thymine is the pyrimidine.

17. **(C)** A bacteriophage virus consists of nuclear material surrounded by a protein coat. Proteins contain sulfur, and DNA contains phosphorus (in the phosphates). When a phage virus is grown in radioactive phosphorus, the phosphorus gets incorporated into the DNA, not into the protein coat.

18. **(B)** DNA polymerase can only add nucleotides to an existing strand of nucleotides. RNA primer binds to the DNA, and DNA polymerase attaches nucleotides to the RNA primer.

19. **(C)** If guanine makes up 28% of the DNA, then there must be an equal amount of cytosine (28%), for a total of 56%. That leaves 44% for adenine and thymine. Divide 44 by 2 = 22%, which is the percentage of thymine in the DNA.

20. **(D)** The codon is the nucleotide triplet associated with mRNA; the anticodon is the nucleotide sequence associated with tRNA. Codons and anticodons are complementary to each other.

21. **(D)** The pairing rules are not as strict for the third codon in mRNA as they are for the first two. One example is that UUU and UUA both code for phenylalanine.

22. **(C)** Sickle cell anemia is caused by a gene mutation in the gene that codes for hemoglobin. Sickle cell disease is common where malaria is endemic, in West Africa and southeast Asia. People who are carriers for the sickle cell trait are resistant to malaria. Sickle cell disease does occur in the Middle East but is not common there. Sickle cell does occur in Caucasians, but rarely.

23. **(A)** DNA (triplet code) makes RNA (codon) makes protein (anticodon). If the triplet in DNA is AAA, then the codon on mRNA is UUU, and the anticodon on tRNA is AAA.

24. **(D)** The regions of DNA that code for proteins are called exons or expressed sequences.

25. **(B)** A prion is a misfolded version of a protein normally found in the brain. Prions are infectious. If they get into the brain, they will convert the normal proteins to abnormal ones. They have been identified as the infectious agent in several diseases, including scrapie in sheep, Creutzfeldt-Jakob in humans, and mad cow disease.

26. **(C)** The operon is the means by which prokaryotes regulate gene expression. An operon consists of a cluster of related genes and the DNA that controls them, such as, a promoter and operator.

27. **(A)**

28. **(D)**

29. **(C)**

30. **(B)**

31. **(B)** A DNA probe is a single radioactive strand of DNA used to tag and identify a specific sequence in a strand of DNA. PCR is a cell-free system that amplifies small pieces of DNA rapidly. Restriction enzymes are found in bacteria. *Eco*RI was the first restriction enzyme discovered. Every person has a unique set of RFLPs.

32. **(C)** Restriction enzymes cut DNA at specific recognition sites. Gel electrophoresis separates DNA that has already been cut up by restriction enzymes. Single-stranded binding proteins, helicases, and topoisomerases help DNA twist during replication.

33. **(B)** A prion is a misfolded and infectious protein. See question 25.

34. **(D)** DNA ligase seals together DNA fragments that have complementary sticky ends.

FREE-RESPONSE QUESTIONS

Directions: Answer all questions. You must answer the question in essay—not outline—form. You may use labeled diagrams to supplement your essay, but diagrams alone are *not* sufficient. Before you start to write, read each question carefully so that you understand what the question is asking.

1. Explain the process by which DNA makes proteins.

2. By using techniques of genetic engineering, scientists are able to learn more about the human genome and to use these techniques for the betterment of humankind. Describe these techniques or procedures below, and explain how each contributes to our understanding of the human genome or for what practical purpose it can be used.
   a. Polymerase chain reaction
   b. Restriction fragment length polymorphism (RFLP) analysis
   c. Gene cloning

3. All humans are almost genetically identical. However, every person has a unique DNA fingerprint. Explain this contradiction.

Typical Free-Response Answers

> *Note: This is an essay that has appeared on the AP Exam before and could appear again in some form. You must understand molecular biology thoroughly because it is the basis of so much of modern biological theory. Once again, key words are in boldface to remind you that you must include correct scientific terminology and clear definitions to get full credit.*

1. The process whereby DNA makes proteins has been worked out in great detail. To summarize, the **triplet code** in DNA is transcribed into a **codon sequence** in **messenger RNA (mRNA)** inside the nucleus. Next, this newly formed strand of **RNA is processed** in the nucleus. Then the codon sequence is **translated into an amino acid sequence** in the cytoplasm at the ribosome by **tRNA**.

    The first stage of the process is **transcription,** where DNA makes mRNA that carries the message directly to the ribosome in the cytoplasm. Transcription consists of three stages: **initiation, elongation**, and **termination**. During initiation, the enzyme **RNA polymerase** recognizes and binds to DNA at the **promoter**. Once RNA polymerase is attached to the promoter, DNA transcription of the DNA template begins. The next step, elongation, continues as RNA polymerase adds nucleotides to the 3′ end of a growing chain. RNA polymerase pries the two strands of DNA apart and attaches RNA nucleotides according to the base-pairing rules: **C (cytosine)** with **G (guanine)** and **A (adenine)** with **U (uracil)**. The stretch of DNA that is transcribed into an mRNA molecule is called a **transcription unit** and consists of triplets of bases called **codons** that code for specific amino acids. When RNA transcribes a **termination sequence**, the process stops.

    Before the newly formed **transcription unit** is shipped out to the ribosome from the nucleus, it is altered or processed by a series of enzymes. A **5′ cap** is added to the 5′ end, which helps protect the RNA strand from degradation by hydrolytic enzymes and which also helps the RNA strand bind to the ribosome in the cytoplasm. In addition, a **poly(A) tail**, is added to the 3′ end to protect the RNA strand from degradation by hydrolytic enzymes and to facilitate the release of the RNA into the cytoplasm. A major part of the processing is the removal of **introns** (noncoding regions) from the transcription unit by **SnRNPs** (small nuclear ribonucleoproteins) and spliceosomes. With processing complete, only **exons**, expressed sequences, move out to the ribosome for translation.

    Translation is the process by which the codons of an mRNA sequence are changed into an amino acid sequence. Amino acids present in the cytoplasm are carried by **tRNA** molecules to the codons of the mRNA strand at the ribosome according to the base-pairing rules (**A** with **U** and **C** with **G**). One end of the tRNA molecule bears a specific amino acid, and the other end bears a nucleotide triplet called an anticodon. Some tRNA molecules have **anticodons** that can recognize two or more different codons. This is because the pairing rules for the third base of a codon are not as strict as they are for the first two bases. This relaxation of base-pairing rules is known as **wobble**. For example, codons UCU, UCC, UCA, and UCG all code for the amino acid serine. The process of translation consists of three stages: initiation, elongation, and termination. Translation of an mRNA strand is complete when a ribosome reaches one of the **termination or stop codons**. The mRNA is broken down, the tRNA is reused, and the polypeptide is freed from the **ribosome**.

2a. **Polymerase chain reaction**

Devised in 1985, polymerase chain reaction is a cell-free, automated technique by which a piece of DNA can be rapidly copied or amplified. Billions of copies of a fragment of DNA can be produced in a few hours. The DNA piece that is to be amplified is placed into a test tube with **Taq polymerase** (a heat-stable form of DNA polymerase extracted from bacteria that live in very hot places, such as hot springs), along with a supply of nucleotides (A, C, T, and G) and primers necessary for DNA synthesis. Once the DNA is amplified, these copies can be studied or used in a comparison with other DNA samples.

The PCR technique has limitations. First, some information about the nucleotide sequence of the target DNA must be known in advance in order to make the necessary primers. Second, the size of the piece that can be amplified must be very short. Contamination is a major problem. Third, a few skin cells from the technician working with the sample could make obtaining accurate results difficult or impossible. This could have dire consequences with a crime scene sample.

2b. **Restriction length polymorphisms (RFLP) analysis**

A restriction fragment is a segment of DNA resulting from treatment of DNA with restriction enzymes. When scientists compared noncoding regions of human DNA across a population, they discovered that the restriction fragment pattern is different in every individual. These differences have been named restriction fragment length polymorphisms or RFLPs, pronounced "riflips." A RFLP analysis of someone's DNA gives a human DNA fingerprint that looks like a bar code.

Each person's RFLPs are unique, except in identical twins, and are inherited in a Mendelian fashion. Because they are inherited in this way, they can be used very accurately in paternity suits to determine, with absolute certainty, if a particular man is the father of a particular child. In addition, RFLPs are routinely used to identify the perpetrator in rape and murder cases. DNA from the crime scene and the victim are compared against DNA from the suspect. Because of the accuracy of RFLP analysis, these cases can be solved with a high degree of certainty, and some suspects have been convicted on DNA evidence alone. By contrast, several incidents have occurred where men who have been jailed for many years for violent crimes were proven innocent by DNA evidence and released.

2c. **Gene cloning**

The potential uses of recombinant DNA or gene cloning are many. Here are several.

- To produce a protein product such as insulin or human growth hormone, in large quantities as an inexpensive pharmaceutical.
- To replace a nonfunctioning gene in a person's cells with a functioning gene by gene therapy. Scientists are currently conducting clinical trials in this area with disappointing results. If scientists can master this technique, many lives will be improved.

- To prepare multiple copies of a gene itself for analysis. Since most genes exist in only one copy in a cell, the ability to make multiple copies is of great value.
- To engineer bacteria to clean up the environment. Scientists have engineered many bacteria; one can even eat toxic waste.

In order to clone a gene, you must first isolate a gene of interest, for example, the gene for human insulin. Next insert the gene into a **plasmid.** Then insert the plasmid into a **vector**, a cell that will carry the plasmid, such as a bacterium like *E. coli.* To accomplish this, a bacterium must be made **competent**, be able to take up a plasmid. As the bacteria reproduce themselves by fission, the plasmid and the selected gene are also being cloned.

3. Genetically, all humans are almost identical. We can say this because only about 3 percent of all DNA contains genes that code for proteins. Furthermore, all the genes for a particular trait are identical. For example, everyone's gene for brown eyes is identical, and all genes for normal hemoglobin are identical.

    If all genes for a particular trait are identical, then how can each person have a unique DNA fingerprint? The answer lies not in the DNA that codes for genes but in the DNA that does not code for proteins—the **introns**, junk and **repetitive sequences**. Scientists have identified certain noncoding regions of DNA that are highly variable from one individual to the next. These regions are referred to as **polymorphic regions**. By analyzing many polymorphic regions, scientists have identified and agreed upon certain standard regions to analyze. This standardized analysis produces the DNA fingerprint that looks like a bar code and that is unique for every individual.

Classification

9

> *Much of this chapter contains background information, but focus on:*
>
> **The Three-Domain Classification System**
>
> **Characteristics of Mammals**
>
> **Characteristics of Primates**
>
> **Phylogenetic Trees**

INTRODUCTION

Taxonomy or **classification** is the naming and classification of species. It began in the eighteenth century when Linnaeus developed the system used today, the **system of binomial nomenclature**. This system has two main characteristics: a two-part name for every organism (for example, human is *Homo sapiens* and lion is *Panthera leo*); and a hierarchical classification of species into broader groups of organisms. These broader groups or **taxa**, in order from the general to the specific are **kingdom**, **phylum**, **class**, **order**, **family**, **genus**, and **species**.

THE THREE-DOMAIN CLASSIFICATION SYSTEM

Over the years, the classification system that scientists use has changed several times. Currently, most scientists use a system based on DNA analysis that accurately reflects evolutionary history and the relationships among organisms. This system is called the **three-domain system**. In it, all life is organized into three domains: **Bacteria**, **Archaea**, and **Eukarya**.

Figure 9.1 shows the organization of the current three-domain system of classification. (*The term **Monera** is no longer used in this system. Instead, prokaryotes are spread across two different domains, Archaea and Bacteria.*)

Table 9.1 shows a comparison of Bacteria, Archaea, and Eukaryotes. Notice, for some characteristics, the archaea resemble eukaryotes more than they resemble prokaryotes.

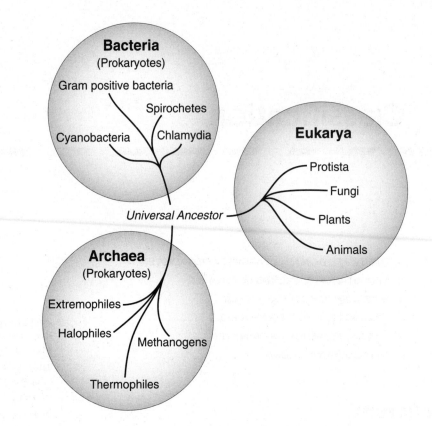

Figure 9.1 The Three-Domain Classification System

Table 9.1

Bacteria, Archaea, and Eukaryotes			
Feature	Bacteria	Archaea	Eukaryotes
Membrane-enclosed organelles	Absent	Absent	Present
Peptidoglycan in cell wall	Present	Absent	Absent
RNA Polymerase	One type	Several kinds	Several kinds
Introns (noncoding regions of genes)	Absent	Present in some genes	Present
Antibiotic sensitivity to streptomycin, chloramphenicol	Inhibited	Not inhibited	Not inhibited

Domain Bacteria

- All are single-celled **prokaryotes** with no internal membranes (no nucleus, mitochondria, or chloroplasts).
- Some are anaerobes; some are aerobes.
- Bacteria play a vital role in the ecosystem as **decomposers** that recycle dead organic matter.
- Many are **pathogens**, causing disease.
- Bacteria play a vital role in **genetic engineering**. The bacteria from the human intestine, *Escherichia coli*, are transformed to manufacture human insulin.
- Some bacteria carry out **conjugation**, a primitive form of sexual reproduction where individuals exchange genetic material.
- They have a thick, rigid cell wall containing a substance known as **peptidoglycan**.
- Some carry out photosynthesis, but others do not.
- No introns (noncoding regions within the DNA).
- Corresponds roughly to the old grouping Eubacteria and includes blue-green algae, bacteria like *E. coli* that live in the human intestine, those that cause disease like *Clostridium botulinum* and *Streptococcus*, and those necessary in the nitrogen cycle, like nitrogen-fixing bacteria and nitrifying bacteria.
- Viruses are placed here because we do not know where else to place them.

Domain Archaea

- Unicellular
- Prokaryotic—no internal membranes such as a nucleus
- Includes **extremophiles**, organisms that live in extreme environments, like

  1. **METHANOGENS**—obtain energy in a unique way by producing methane from hydrogen
  2. **HALOPHILES**—thrive in environments with high salt concentrations like Utah's Great Salt Lake
  3. **THERMOPHILES**—thrive in very high temperatures, like in the hot springs in Yellowstone Park or in deep-sea hydrothermal vents

- Introns present in some genes
- No peptidoglycan

Domain Eukarya

Eukarya is a domain that includes four of the original kingdoms, protista, fungi, plants, and animals.

- All organisms have a nucleus and internal organelles
- No peptidoglycan in cells
- Includes the four remaining kingdoms: Protista, Fungi, Plantae, and Animalia (see Table 9.2)

Today, taxonomy as a separate discipline is being replaced by the study of **systematics**, which includes taxonomy but considers biological diversity in an evolutionary context. Systematics focuses on tracing the ancestry of organisms. This is particularly important in light of current advances in DNA techniques that allow scientists to compare two species at the molecular level.

REMEMBER

The term "monera" is no longer used as the name of a kingdom.

Table 9.2

The Four Kingdoms of Eukarya	
Kingdom	**Characteristics**
Protista	Includes the widest variety of organisms, but all are eukaryotes
	Includes organisms that do not fit into the fungi or plant kingdoms, such as seaweeds and slime molds
	Consists of single and primitive multicelled organisms
	Includes heterotrophs and autotrophs
	Amoeba and paramecium are heterotrophs
	Euglena are primarily autotrophic with red eyespot and chlorophyll to carry out photosynthesis
	Protozoans like amoeba and paramecium are classified by how they move
	Mobility by varied methods: amoeba—pseudopods; paramecium—cilia; euglena—flagella
	Some carry out conjugation, a primitive form of sexual reproduction
	Some cause serious diseases like amoebic dysentery and malaria
Fungi	All are heterotrophs and eukaryotes
	Secrete hydrolytic enzymes outside the body where extracellular digestion occurs, then the building blocks of the nutrients are absorbed into the body of the fungus by diffusion
	Are important in the ecosystem as decomposers
	Cell walls are composed of chitin, not cellulose
	Examples: yeast, mold, mushrooms, the fungus that causes athlete's foot
Plantae	All are autotrophic eukaryotes
	Some plants have vascular tissue (Tracheophytes), some do not have any vascular tissue (Bryophytes)
	Examples: mosses, ferns, cone-bearing and flowering plants
Animalia	All are heterotrophic, multicellular eukaryotes
	Are grouped in 35 phyla; but this book discusses 9 main phyla: Porifera, Cnidaria, Platyhelminthes, Nematoda, Annelida, Mollusca, Arthropoda, Echinodermata, and Chordata
	Most animals reproduce sexually with a dominant diploid stage
	In most species, a small, flagellated sperm fertilizes a larger, nonmotile egg
	This category is monophyletic, meaning all animal lineages can be traced back to one common ancestor
	Traditionally classified based primarily on anatomical features (homologous structures) and embryonic development

EVOLUTIONARY TRENDS IN ANIMALS

Organisms began as tiny, primitive, single-celled organisms that lived in the oceans. The first **multicellular** eukaryotic **organisms** evolved about 1.5 billion years ago. The appearance of each phylum of animal represents the evolution of a new and successful body plan. These important trends include: specialization of tissues, germ layers, body symmetry, cephalization, and body cavity formation. Specifics of these trends are summarized below. Table 9.3 summarizes this information.

Table 9.3

Trends in Animal Development from the Primitive to the Complex	
From the Primitive	**To the Complex**
No symmetry or radial symmetry with little or no sensory apparatus	Bilateral symmetry with a head end and complex sensory apparatus
No cephalization	Cephalization
Two cell layers: ectoderm and endoderm (diploblastic)	Three cell layers: ectoderm, mesoderm, and endoderm (triploblastic)
No coelom	Pseudocoelom to coelom
No true tissues	True tissues, organs, and organ systems
Life in water	Life on land and all the modification it requires
Sessile	Motile
Few organs, but no organ systems	Many organ systems and much specialization

Specialized Cells, Tissues, and Organs

We need to begin with some definitions.

- The **cell** is the basic unit of all forms of life. A neuron is a cell.
- A **tissue** is a group of similar cells that perform a particular function. The sciatic nerve is a tissue.
- An **organ** is a group of tissues that work together to perform related functions. The brain is an organ.

Sponges (Porifera) consist of a loose federation of cells, which are not considered tissue because the cells are relatively unspecialized. They possess cells that can sense and react to the environment but do not have real nerve or muscular tissue.

Cnidarians like the hydra and jellyfish possess only the most primitive and simplest forms of tissue.

As larger and more complex animals evolved, specialized cells joined to form real tissues, organs, and organ systems. Flatworms have organs, but no organ systems.

More complex animals, like annelids (segmented worms) and arthropods, have organ systems.

Germ Layers

Germ layers are the main layers that form various tissues and organs of the body. They are formed early in embryonic development as a result of gastrulation. Complex animals are **triploblastic**. They consist of the ectoderm, endoderm, and mesoderm.

- The **ectoderm**, or outermost layer, becomes the skin and nervous system, including the nerve cord and brain.
- The **endoderm**, the innermost layer, becomes the viscera (guts) or the digestive system.
- The **mesoderm**, middle layer, becomes the blood and bones.

Primitive animals, like the **Porifera** and **Cnidarians**, have only two cell layers and are called **diploblastic**. Their bodies consist of ectoderm, endoderm, and **mesoglea** (middle glue), which connects the two layers together.

Bilateral Symmetry

Whereas primitive animals exhibit radial symmetry (see Figure 9.3), most sophisticated animals exhibit **bilateral symmetry** (see Figure 9.2). The echinoderms seem to be an exception because they exhibit bilateral symmetry only as larvae and revert to radial symmetry as adults. In bilateral symmetry, the body is organized along a **longitudinal axis** with right and left sides that mirror each other.

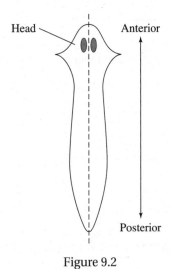

Figure 9.2
Bilateral Symmetry
in Flatworm-Planaria

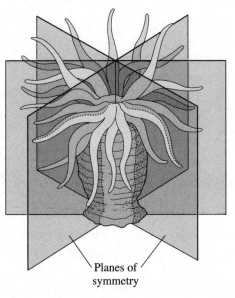

Figure 9.3
Radial Symmetry

Cephalization—Development of a Head End

Along with bilateral symmetry comes a front end, **anterior**, and a rear end, **posterior**. Sensory apparatus and a brain are clustered at the anterior, while digestive, excretory, and reproductive are located posterior. This enables animals to move faster to flee or to capture prey more effectively. Bilateral animals are all **triploblastic**, with an **ectoderm**, **mesoderm**, and **endoderm**.

The Coelom

The **coelom** is a fluid-filled body cavity (see Figure 9.4). A true coelom arises from within the **mesoderm** and is completely surrounded by mesoderm tissue. It is a significant advance in the course of animal evolution because it provides space for elaborate body systems like a transport or respiratory system. Consider how much space lungs need to expand in the chest cavity or where 30 feet (9 m) of intestines would coil without space in the abdomen. These major organs could not have evolved without the coelom.

Primitive animals (**Porifera, Cnidaria**, and **Platyhelminthes**) have no coelom at all and are called the **acoelomates**. Their three germ layers are packed together with no body cavity except the digestive cavity. **Nematodes** or roundworms are **pseudocoelomates** with a fluid-filled tube between the **endoderm** and the **mesoderm**. The pseudocoelom acts as a **hydrostatic skeleton**, increasing the effectiveness of the animal's muscular contractions in movement. See Figure 9.4.

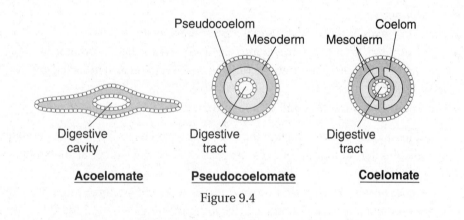

Figure 9.4

NINE COMMON ANIMAL PHYLA

Zoologists recognize about 35 phyla of animals. You must be familiar with **nine common phyla** with representative animals in each. Since much of animal classification is based on embryonic development, in order to understand the classification of animals, you must also understand the basics of embryonic development. See Figure 9.5. As you study the individual animal phyla, think in terms of strategies animals have evolved to adapt to a particular environment. Also, notice the trends in development in animals from the primitive to the complex, Table 9.3. For more information, see the review chapter entitled "Animal Reproduction and Development."

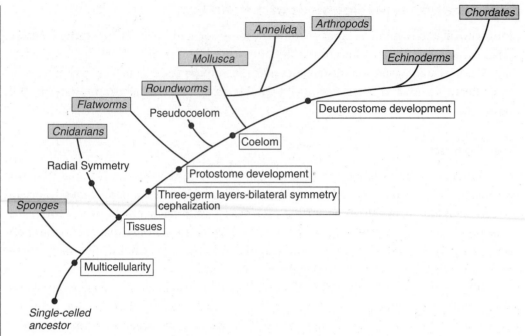

Figure 9.5 Trends in Nine Common Animal Phyla

Porifera—The Sponges—Invertebrates

- No symmetry
- Have no nerves or muscles; are **sessile**, meaning they do not move
- Filter nutrients from water drawn into a central cavity called a **spongocoel**
- Consist of two cell layers only: **ectoderm** and **endoderm** connected by noncellular **mesoglea**
- Have **no true tissues or organs** although they have different types of cells

  - ✔ **Choanocytes**, collar cells, line the body cavity and have flagella that circulate water
  - ✔ **Spicules** for support—sponges are classified by the material that makes up the spicules
  - ✔ **Amoebocytes** are cells that move on their own and perform numerous functions: reproduction, carrying food particles to nonfeeding cells, and secretion of material that forms the **spicules**

- Evolved from colonial organisms; if a sponge is squeezed through fine cheesecloth, it will separate into individual cells that will spontaneously reaggregate into a sponge
- Reproduce asexually by **fragmentation** as well as sexually; are **hermaphrodites**

Cnidarians—Hydra and Jellyfish

- Invertebrate
- Radial symmetry
- Body plan is the **polyp** (vase shaped) or the **medusa** (upside down bowl shaped)
- Life cycle—some go through a **planula larva** (free-swimming) stage then go through two reproductive stages: asexually reproducing (polyp) and sexually reproducing (medusa)
- Two cell layers: **ectoderm** and **endoderm** connected by noncellular **mesoglea**
- Have a **gastrovascular cavity** where **extracellular digestion** occurs

- Also carry out **intracellular digestion** inside body cells, carried out in lysosomes
- Have no transport system because every cell is in contact with environment.
- All members have **stinging cells**: **cnidocytes**

The Hydra

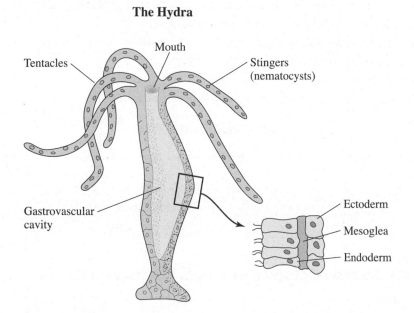

Platyhelminthes—Flatworms Including Tapeworms

- Invertebrate
- Simplest animals with bilateral symmetry, an anterior end, three distinct cell layers, and cephalization
- Have true tissues and organs
- Digestive cavity has only one opening for both ingestion and egestion so food cannot be processed continuously
- Flatworms are **acoelomate**, they have no coelom; they have a solid body with no room for true digestive or respiratory systems to circulate food molecules or oxygen; have solved this problem in two ways

   ✔ The body is very flat, which keeps the body cells in direct contact with oxygen in the environment
   ✔ The digestive cavity is branched so that food can be spread to all regions of the body

Nematoda—Roundworms

- Invertebrate
- Unsegmented worms with bilateral symmetry but little sensory apparatus
- Protostome pseudocoelomate
- **Pseudocoelom** transports nutrients, but there is inadequate room for a circulatory system
- Many are parasitic, *Trichinella* causes **trichinosis** acquired from uncooked pork
- One species, *Caenorhabditis elegans,* is widely used as a model in studying the link between genes and development

Annelida—Segmented Worms: Earthworms, Leeches

- Invertebrate
- **Protostome coelomates** with bilateral symmetry but little sensory apparatus
- Digestive tract is a **tube within a tube** consisting of **crop**, **gizzard**, and **intestine**
- **Nephridia** for excretion of nitrogenous waste, urea
- **Closed circulatory system**—heart consists of five pairs of **aortic arches**
- Blood contains **hemoglobin** and carries oxygen
- Diffusion of oxygen and carbon dioxide through moist skin
- Are **hermaphroditic**, but the animal does not self-fertilize

Earthworm-Digestive Tract

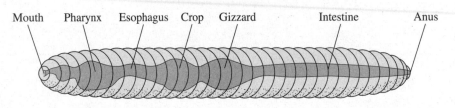

Mouth   Pharynx   Esophagus   Crop   Gizzard   Intestine   Anus

Mollusca—Squids, Octopuses, Slugs, Clams, and Snails

- Invertebrate
- **Protostome coelomates**
- Have a **soft body** often protected by a hard **calcium-containing shell**
- Have bilateral symmetry with three distinct body zones

  - ✔ **Head-foot**, which contains both sensory and motor organs
  - ✔ **Visceral mass**, which contains the organs of digestion, excretion, and reproduction
  - ✔ **Mantle**, a specialized tissue that surrounds the visceral mass and secretes the shell

- **Radula**, a movable, tooth-bearing structure, acts like a tongue
- **Open circulatory system** with blood-filled spaces called **hemocoels**; lack capillaries
- Most have **gills and nephridia**

Arthropoda—Insects (Grasshoppers), Crustaceans (Shrimp, Crabs), and Arachnids (Spiders)

- Invertebrate
- **Protostome coelomates**
- Jointed appendages
- Segmented: head, thorax, abdomen
- Having more sensory apparatus than the annelids gives them more speed and freedom of movement
- **Chitinous exoskeleton** protects the animal and aids in movement.
- **Open circulatory system** with a **tubular heart** and **hemocoels**; lack capillaries
- **Malpighian tubules** for removal of nitrogenous wastes, uric acid
- Air ducts called **trachea** bring air from the environment into **hemocoels**
- Some have **book lungs** or book gills

Echinodermata—Sea Stars (Starfish) and Sea Urchins

- Invertebrate
- **Deuterostome coelomates**
- Most are sessile or slow moving
- Bilateral symmetry as an embryo but reverts to the primitive radial symmetry as an adult; the radial anatomy is an adaptation to a sedentary lifestyle
- **Water vascular system**, which is a modified coelom, creates hydrostatic support for **tube feet**, the locomotive structures
- Reproduces by sexual reproduction where fertilization is external
- Also reproduces by fragmentation and **regeneration**; any piece of a sea star that contains part of the central canal will form an entirely new organism
- Sea stars have an **endoskeleton** consisting of calcium plates, which grow with the body

Chordata—Fishes, Amphibians, Reptiles, Birds, and Mammals

- Vertebrate
- **Deuterostome coelomates**
- Have a **notochord**—a rod that extends the length of the body and serves as a flexible axis
- Dorsal, hollow nerve cord
- **Tail** aids in movement and balance—the coccyx bone in humans is a vestige of a tail
- Birds and mammals are endotherms and homeotherms—maintain a consistent body temperature; all others are ectotherms, although some reptiles can raise their body temperature to a limited extent

CHARACTERISTICS OF MAMMALS

- Mothers nourish their babies with milk
- Have hair or fur, both made of **keratin**
- Are **homeotherms**
- **Placental mammals** (eutherians) are born and the embryo develops internally in a uterus connected to the mother by a **placenta**, where nutrients diffuse from mother to embryo
- Some, the **marsupials**, including kangaroos, are born very early in embryonic development and the joey completes its development while nursing in the mother's pouch
- **Monotremes**, egg-laying mammals, like the duck-billed platypus and the spiny anteater, derive nutrients from a shelled egg

Table 9.4 shows a classification of three mammals using the current system of taxonomy. The domain Eukarya is the new addition.

Table 9.4

Taxon	Human	Lion	Dog
Domain	Eukarya	Eukarya	Eukarya
Kingdom	Animalia	Animalia	Animalia
Phylum	Chordata	Chordata	Chordata
Class	Mammalia	Mammalia	Mammalia
Order	Primates	Carnivora	Carnivora
Family	Hominidae	Felidae	Canidae
Genus	*Homo*	*Panthera*	*Canus*
Species	*sapiens*	*leo*	*familiaris*

Taxonomic Classification of Three Mammals

CHARACTERISTICS OF PRIMATES

Primates descended from insectivores, probably from small, tree-dwelling mammals. Primates have **dexterous hands** and **opposable thumbs**, which allow them to do fine motor tasks. Claws have been replaced by **nails**, and hands and fingers contain many nerve endings and are sensitive. The **eyes** of a primate are **front facing** and set close together. Front-facing eyes enhances face-to-face communication, depth perception, and hand-eye coordination. Although mammals devote much energy to **parenting** young, primates engage in the most intense parenting of any mammal. Primates usually have single births and **nurture their young for a long time**. The primates include humans, gorillas, chimpanzees, orangutans, gibbons, and the old-world and new-world monkeys.

PHYLOGENETIC TREES

All living things evolved from a common ancestor almost 4 billion years ago, which is why the principles of biology apply to all organisms. The evolutionary history of these relationships is known as **phylogeny**. A **phylogenetic tree** or **cladogram** is a diagrammatic reconstruction of that history. Phylogenetic trees used to be based on morphology and physical behaviors. Since more genomes have been sequenced, biologists are now constructing phylogenetic trees based on DNA and evolutionary relationships. Phylogenetic analysis of human mitochondrial DNA has allowed scientists to construct a history of human migration out of Africa. The growing database of DNA sequences enables researchers to study more species, but it also makes building phylogenetic trees more complex. A phylogenetic tree can usually be built in several different ways, and *every phylogenetic tree is a hypothesis*. Scientists narrow down the possibilities by using the principle of **maximum parsimony**, which states that one should follow the simplest explanation that coincides with the facts.

Table 9.5 shows shared and derived traits for building a phylogenetic tree. A " — " indicates that a trait is absent; a " + " indicates that a trait is present.

Table 9.5

Derived Traits for Building a Phylogenetic Tree						
	Lancelet (Outgroup)*	Lamprey	Bass	Frog	Turtle	Leopard
Vertebral column (backbone)	−	+	+	+	+	+
Hinged jaws	−	−	+	+	+	+
Four walking legs	−	−	+	+	+	+
Amnion	−	−	−	−	+	+
Hair	−	−	−	−	−	+

Figure 9.6 shows a **phylogenetic tree** for the animals listed in Table 9.5. It includes *ingroups*, the organisms of interest—lamprey, bass, frog, turtle, and leopard—and as a point of reference, the *outgroup*, lancelet. The outgroup is the group that diverged before the lineage evolved, in this case, vertebrates. When two lineages diverge, the split is depicted as a *node*. In this example, the nodes are development of a vertebral column, hinged jaws, four walking legs, amnion, and the development of hair. All animals share characteristics with their ancestors and also differ from them. In this tree, all the animals (except the lancelets) share a vertebral column. That trait is known as a **shared ancestral trait** or **character**. In contrast, each animal in one **clade** or lineage has a trait that is not shared with their ancestors. That new trait is known as a **shared derived trait** or **character**. For example, hair in the leopard (and in all mammals) is a derived trait.

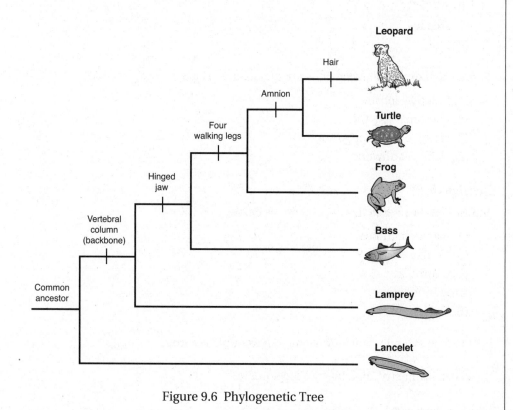

Figure 9.6 Phylogenetic Tree

1. Which of the following is the least inclusive?

   (A) kingdom
   (B) species
   (C) family
   (D) domain

2. Which is not true of Protista?

   (A) Some are autotrophs.
   (B) Some are heterotrophs.
   (C) An example is euglena.
   (D) They do not have internal membranes.

3. Which of the following contains prokaryote organisms capable of surviving extreme conditions of heat and salt concentration?

   (A) archaea
   (B) viruses
   (C) protists
   (D) fungi

4. Which of the following is best characterized as being eukaryotic, heterotrophic, and having cell walls made of chitin?

   (A) plants
   (B) animals
   (C) archaea
   (D) fungi

5. In which of the following pairs are the organisms most closely related?

   (A) fruit fly—lobster
   (B) sea stars—jellyfish
   (C) earthworm—tape worm
   (D) clam—earthworm

Questions 6–9*

Questions 6–11 refer to this list of animals below:

   (A) Platyhelminthes
   (B) Nematoda
   (C) Chordata
   (D) Echinodermata
   (E) Cnidaria

6. Bilateral symmetry, deuterostome. An example is a frog.

7. Radial symmetry, diploblastic, acoelomate

*These questions have 5 answer choices.

8. Bilateral symmetry in the larval stage, radial symmetry as an adult, deuterostome, endoskeleton

9. All organisms in the kingdom contain stinging cells

---

10. A randomly selected group of organisms from a family would show more genetic variation than a randomly selected group from a

    (A) genus
    (B) kingdom
    (C) class
    (D) domain

11. The only taxon that actually exists as a natural unit is

    (A) class
    (B) phylum
    (C) order
    (D) species

12. Which of the following is NOT generally a characteristic of complex, advanced organisms?

    (A) mesoderm
    (B) bilateral symmetry
    (C) true tissues and organs
    (D) sessile

Answers to Multiple-Choice Questions

1. **(B)** The least inclusive grouping includes the one where the organisms are the most similar, the species. The order of taxa from the most general to the most specific is domain, kingdom, phylum, class, order, family, genus, and species.

2. **(D)** Protista are all eukaryotes and have cells with internal membranes. They include paramecium and amoeba which are heterotrophs, and euglena, which is an autotroph. The prokaryotes are organisms without internal membranes.

3. **(A)** Archaea is the group that includes the extremophiles. Most biologists believe that the prokaryotes and the archaea diverged from each other in very ancient times. One basic way in which the two differ is in their nucleic acids.

4. **(D)** Fungi are all heterotrophs. They secrete digestive enzymes and digest food outside the organism. This is called extracellular digestion. Once the food is digested and broken down into building blocks, it is absorbed into the body of the organism by diffusion. The cell walls of mushrooms consist of chitin.

5. **(A)** Both fruit flies and lobsters are in the phylum Arthropoda—animals with jointed appendages. Arthropoda includes arachnids, crustaceans, and insects. Sea stars are Echinodermata, jellyfish are Cnidaria, earthworms are Annelida, tapeworms are Platyhelminthes, and clams are Mollusca.

6. **(C)**

7. **(E)**

8. **(D)**

9. **(E)**

10. **(A)** The genus is a narrower grouping than the family, so it contains animals with fewer differences.

11. **(D)** Whether you call it a species or something else; it still only contains one type of organism. The other taxa contain many different types of organisms.

12. **(D)** It is generally true that bilateral symmetry and motility are characteristics of advanced organisms. Echinoderms are an exception. Motility, the ability to move from place to place, is an important characteristic of an advanced organism because it enables an organism to search for food, a mate, or safety.

FREE-RESPONSE QUESTIONS

> **Directions:** Answer all questions. You must answer the question in essay—**not** outline—form. You may use labeled diagrams to supplement your essay, but diagrams alone are *not* sufficient. Before you start to write, read each question carefully so that you understand what the question is asking.

1. Describe the differences between the terms in each of the following pairs:

   a. Acoelomate—Coelomate
   b. Radial Symmetry—Bilateral Symmetry

2. Explain how each of the features listed in question 1 can be used to construct an evolutionary history of these common animal phyla:

   Porifera
   Cnidaria
   Platyhelminthes
   Nematoda
   Annelida
   Mollusca
   Arthropoda
   Echinodermata
   Chordata
   (See Table 9.4 on page 170 for answers.)

Typical Free-Response Answers

1a. A coelom is a fluid-filled body cavity and arises from within **mesoderm** tissue early in embryonic development. It is a significant advance in the course of animal evolution because it provides space for elaborate body systems like a transport system. More advanced phyla, such as echinodermata and chordata have a coelom. An animal that does not have a coelom (acoelomate) lacks internal cavities and complex organs. Primitive acoelomate phyla are Cnidaria, Porifera, and Platyhelminthes. Some animals have a pseudocoelom, an internal cavity that is only partly lined with mesoderm tissue. This fluid-filled pseudocoelom functions as a **hydrostatic skeleton**, providing support and making movement easier. An animal phylum with a pseudocoelom is the Nematoda.

1b. If symmetry is **radial**, several planes can pass through the long axis and divide the animal into similar parts. An example of an animal with radial symmetry is the hydra. If only one plane can bisect the animal into left and right halves, the symmetry is **bilateral**. Primitive organisms show radial symmetry; advanced organisms show bilateral symmetry. Chordates all have bilateral symmetry. The embryo of the echinoderm demonstrates bilateral symmetry but reverts to the primitive radial symmetry as an adult.

2. There is great diversity among all the animals, and they can be grouped and distinguished by several characteristics. These characteristics include the number of cell layers, whether they have a true coelom, symmetry at early cleavage, body plan, and whether they are protostomes or deuterostomes. The phyla listed above begin with the most primitive and oldest on top and end with the most advanced and most recently evolved at the bottom.

   The most primitive animals, the **Porifera**, or sponges, and **Cnidaria** are diploblastic; they lack a mesoderm. Instead, they have a mesoglea, a noncellular, gluey layer between the ectoderm and endoderm that helps keep them together. All of the other phyla are triploblastic with three germ layers. The ectoderm will become the skin and nervous system. The endoderm will become the internal organs. The mesoderm will become the blood, bones, and muscle. The most primitive phylum, the Porifera, includes the sponges. They have no symmetry. The slightly more advanced Cnidaria have radial symmetry. All triploblastic animals have bilateral symmetry. Animals with no true coelom, for example, Porifera, Cnidaria, and **Platyhelminthes**, are primitive. The **Nematoda** lack a true coelom, but they have a pseudocoelom. All the others have a true coelom. The two most advanced phyla are the **Echinodermata** and **Chordata**; they are both deuterostomes. All the remaining phyla are more primitive and are protostomes.

Evolution

10

INTRODUCTION

Evolution is the change in allelic frequencies in a population. Sometimes populations evolve rapidly, as in the case of the development and spread of antibiotic-resistant bacteria. However, evolution mostly takes place over hundreds, thousands, or millions of years. The theory of evolution is supported by scientific evidence from many disciplines.

HISTORY OF EARTH

Events in Earth's history can be dated by several methods. Studying undisturbed *sedimentary rock layers* and the fossils within them reveals the *relative age* of rocks and fossils. The *absolute age* of rocks can be accurately measured by *radiometric dating*, which is based on the decay of radioactive isotopes and half-life. Although radiometric dating is a powerful tool, sometimes there is inadequate rock or fossils available to measure, so scientists must use other techniques. *Paleomagnetic dating* uses the fact that Earth's magnetic poles shift and sometimes even reverse. These changes are recorded in rock layers. Other scientific methods of dating Earth use changes in sea levels, molecular clocks, and measurements of continental drift.

By employing all these methods, scientists have learned that Earth has radically changed since it formed 4½ billion years ago. Continents have shifted. The climate has gone through periods of extreme cooling and warming. The oceans have risen and lowered repeatedly. Volcanic activity has drastically changed Earth and killed off untold numbers of species. Meteorites bombarding Earth from outer space were responsible for the dinosaurs going extinct and giving mammals a chance to expand around the globe. Five major extinctions have occurred. Additionally, 4 billion years ago, the atmosphere contained no free oxygen. However, cyanobacteria, oxygen-generating organisms that form rocklike structures called *stromatolites*, provided free oxygen for the oceans and atmosphere. This high oxygen level made possible a rapid diversification of life on land and in the seas. A relatively short period known as the *Cambrian explosion* (535–525 million years ago) was characterized by the sudden appearance of many present-day animal phyla.

EVIDENCE FOR EVOLUTION

Many areas of scientific study provide evidence for evolution. Here are six.

1. Fossil Record

The fossil record reveals the existence of species that have become extinct or have evolved into other species. **Radiometric dating** and **half-life** accurately measure the age of fossils. Prokaryotes were the first organisms to develop on earth, and they are the oldest fossils. Paleontologists have discovered many transitional forms that link older fossils to modern species, such as the transition from *Eohippus* to the modern horse, *Equus*. *Archaeopteryx* is a fossil that links reptiles and birds. The fossil record also indicates that all the organisms alive today are only a tiny fraction of all the organisms that ever lived. In other words, most life that existed on Earth went extinct.

2. Comparative Anatomy

The study of different structures contributes to scientists' understanding of the evolution of anatomical structures and of evolutionary relationships.

- The wing of a bat, the lateral fin of a whale, and the human arm all have the same internal bone structure, although the function of each varies. These structures, known as **homologous structures**, have a common origin and reflect a common ancestry. See Figure 10.1.
- **Analogous structures**, such as a bat's wing and a fly's wing, have the same function. However, the similarity is superficial and reflects an adaptation to similar environments, not descent from a recent common ancestor.
- **Vestigial structures**, such as the appendix, are evidence that structures have evolved. The appendix is a vestige of a structure needed when human ancestors ate a very different diet.

3. Comparative Biochemistry

Organisms that have a common ancestor will have common biochemical pathways. The more closely related the organisms are to each other, the more similar their biochemistry is. Humans and mice are both mammals. This close relationship is the reason that medical researchers can test new medicines on mice and extrapolate the results to humans.

4. Comparative Embryology

Closely related organisms go through similar stages in their embryonic development. For example, all vertebrate embryos go through a stage in which they have gill pouches on the sides of their throats. In fish, the gill pouches develop into gills. In mammals, they develop into eustachian tubes in the ears.

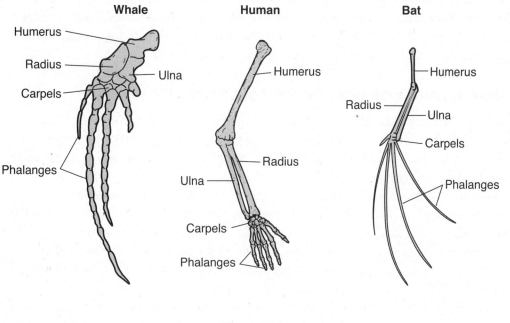

Figure 10.1

5. Molecular Biology

Since all aerobic organisms contain cells that carry out aerobic cell respiration, they all contain the polypeptide **cytochrome c**. A comparison of the amino acid sequence of cytochrome c among different organisms shows which organisms are most closely related. The cytochrome c in human cells is almost identical to that of our closest relatives, the chimpanzee and gorilla, but differs from that of a pig.

6. Biogeography

According to the theory of **plate tectonics**, continents and oceans rest on giant plates of the Earth's crust that float on top of the hot mantle. Convection currents in the mantle are responsible for the slow, continuous movement of the plates known as **continental drift**. For example, the North American plate drifts northwestward while it pulls away from the Eurasian plate at a rate of about 2 cm per year. Continental drift causes mountains to form as plates collide. About 45 million years ago, the Indian plate collided with and sank beneath (subducted) the Eurasian plate, forcing it upwards and forming the Himalayan Mountains.

The cumulative effect of plate movement over millions of years has changed the flora and fauna of the Earth. About 250 million years ago, plate movement brought all the previously separated land masses together into one supercontinent called **Pangaea**. Ocean basins became deeper, which lowered sea level and drained shallow waters. A multitude of species went extinct.

Continental drift has changed the distributions of life on Earth. One example is the fact that marsupials are located almost exclusively in Australia, while other continents are home to eutherians, the true placental mammals. Fossil evidence suggests that marsupials originated in what is now Asia and reached Australia via South America and Antarctica while the continents were still joined. When the continents broke apart and moved to different climates, Australia was set afloat like a giant raft, carrying both marsupials and eutherians. In

Australia the marsupials diversified, filling every available niche (*adaptive radiation*), while the true placental mammals, which were not adapted to the Australian environment and climate, went extinct. On other continents the opposite occurred; only the true placental mammals survived and diversified, while marsupials went extinct.

HISTORICAL CONTEXT FOR EVOLUTIONARY THEORY

Aristotle spoke for the ancient world with his theory of *Scala Natura*. According to this theory, all life-forms can be arranged on a ladder of increasing complexity, each with its own allotted rung. The species are permanent and do not evolve. Humans are at the pinnacle of this ladder of increasing complexity.

Carolus Linnaeus or **Carl von Linné** (1707–1778) specialized in **taxonomy**, the branch of biology concerned with naming and classifying the diverse forms of life. He believed that scientists should study life and that a classification system would reveal a divine plan. He developed the naming system used today: **binomial nomenclature**. In this system, every organism has a unique name consisting of two parts: a **genus** name and a **species** name. For example, the scientific name of humans is *Homo sapiens*.

Cuvier, who died in 1832 before Darwin published his thesis, studied fossils and realized that each stratum of earth is characterized by different fossils. He believed that a series of **catastrophes** was responsible for the changes in the organisms on earth and was a strong opponent of evolution. Cuvier's detailed study of fossils, however, was very important in the development of Darwin's theory.

James Hutton, one of the most influential geologists of his day, published his theory of **gradualism** in 1795. He stated that the earth had been molded, not by sudden, violent events, but by slow, gradual change. The effects of wind, weather, and the flow of water that he saw in his lifetime were the same forces that formed the various geologic features on earth, such as mountain ranges and canyons. His theories were important because they were based on the idea that the earth had a very long history and that change is the normal course of events.

Lyell was a leading geologist of Darwin's era. He stated that geological change results from slow, continuous actions. He believed that the earth was much older than the 6,000 years thought by early theologians. His text, *Principles of Geology*, was a great influence on Darwin.

Lamarck was a contemporary of Darwin who also developed a theory of evolution. He published his theory in 1809, the year Darwin was born. His theory relies on the ideas of **inheritance of acquired characteristics** and **use and disuse**. He stated that individual organisms change in response to their environment. According to Lamarck, the giraffe developed a long neck because it ate leaves of the tall acacia tree for nourishment and had to stretch to reach them. The animals stretched their necks and passed the acquired trait of an elongated neck onto their offspring. Although this theory may seem funny today, it was widely accepted in the early nineteenth century.

Wallace, a naturalist and author, published an essay discussing the process of natural selection identical to Darwin's, which had not yet been published. Many people credit Wallace, along with Darwin, for the theory of natural selection.

Darwin was a naturalist and author who, when he was 22, left England aboard the HMS Beagle to visit the Galapagos Islands, South America, Africa, and Australia. By the early 1840s, Darwin had worked out his **theory of natural selection** or **descent with modification** as the mechanism for how populations evolve, but he did not publish them. Perhaps he was afraid of the furor his theories would cause. He finally published "**On the Origin of the Species**" in 1859 when he was spurred on to publish by the appearance of a similar treatise by Wallace. Darwin's theory challenged the traditional view of a young Earth (about 6,000 years old) inhabited by unchanging species.

DARWIN'S THEORY OF NATURAL SELECTION

Natural selection is a major mechanism of evolution; it acts on phenotypic variation in populations. Here are the tenets of **Darwin's theory of natural selection**:

- **Populations tend to** grow exponentially, **overpopulate**, and exceed their resources. Darwin developed this idea after reading Malthus's work, a treatise on population growth, disease, and famine published in 1798.
- **Overpopulation results in competition and a struggle for existence**.
- **In any population, there is variation and an unequal ability of individuals** to survive and reproduce. Darwin, however, could not explain the origin of variation in a population. (Mendel's theory of genetics, published in 1865, would have given Darwin an understanding of why variation occurs in a population. However, the ramifications of Mendel's theories were not understood until many years after he presented them.)
- **Only the best-fit individuals survive and get to pass on their traits to offspring**.
- **Evolution occurs as advantageous traits accumulate in a population.** No individual organism changes in response to pressure from the environment. Rather, the frequency of an allele within a population changes.

How the Giraffe Got Its Long Neck

According to **Darwin's theory**, ancestral giraffes were short-necked animals although neck length varied from individual to individual. As the population of animals competing for the limited food supply increased, the taller individuals had a better chance of surviving than those with shorter necks. Over time, the proportion of giraffes in the population with longer necks increased until only long-necked giraffes existed.

How Insects "Become" Resistant to Pesticides

Insects do not actually become resistant to pesticides. Instead, some insects are naturally resistant to a particular chemical insecticide. When the environment is sprayed with that insecticide, the resistant insects have the **selective advantage**. All the insects not resistant to the insecticide die, and the remaining resistant ones breed quickly with no competition. The entirely new population is resistant to the insecticide. Insecticide resistance is an example of **directional selection**.

TYPES OF SELECTION

Natural selection can alter the frequency of inherited traits in a population in five different ways, depending on which phenotypes in a population are favored. Five types of selection are **stabilizing**, **diversifying**, **directional**, **sexual**, and **artificial**; see Figure 10.2.

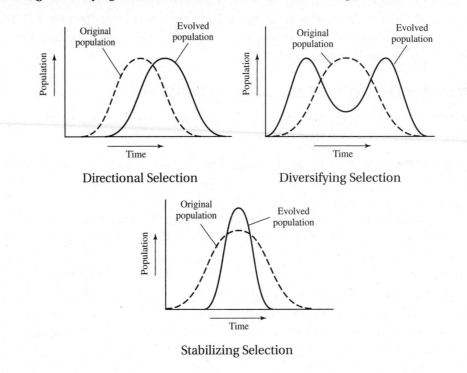

Directional Selection

Diversifying Selection

Stabilizing Selection

Figure 10.2

1. Stabilizing Selection

Stabilizing selection, sometimes called purifying selection, eliminates the extremes and favors the more common intermediate forms. Many mutant forms are weeded out in this way.

- In humans, stabilizing selection keeps the majority of birth weights in the 6–8 pound (2.7–3.6 kg) range. For babies much smaller and much larger, infant mortality is greater.
- In Swiss starlings, genotypes that lead to a clutch size (the number of eggs a bird lays) of up to five will have more surviving young than birds of the same species that lay a larger or smaller number of eggs.

2. Disruptive or Diversifying Selection

Disruptive selection increases the extreme types in a population at the expense of intermediate forms. What may result is called **balanced polymorphism**, one population divided into two distinct types. Over great lengths of time, disruptive selection may result in the formation of two new species.

Imagine that an environment with very light rocks and very dark soil is colonized by light, intermediate-colored, and dark mice. The frequency of very light and very dark mice, which are both camouflaged, would increase, while the intermediate-colored mice would die out because of predation. Pressure from the environment selects for two extreme characteristics.

3. Directional Selection

Changing environmental conditions give rise to **directional selection**, where one phenotype replaces another in the gene pool. Here are two examples of directional selection.

REMEMBER

No single organism changes in response to a change in the environment. Rather, the frequency of an allele in the population may change.

- One example of **directional selection** is **industrial melanism** in **peppered moths**, *Biston betularia*. Until 1845 in England, most peppered moths were light; a few individuals were found to be dark. With increasing industrialization, smoke and soot polluted the environment, making all the plants and rocks dark. By 1900, all moths in the industrialized regions were dark; only a few light-colored individuals could be found. Before the industrial revolution, white moths were camouflaged in their environment and dark moths were easy prey for predators. After the environment was darkened by heavy pollution, dark moths were camouflaged and had the selective advantage. Within a relatively short time, dark moths replaced the light moths in the population.

- **Directional selection** can produce rapid shifts in allelic frequencies. For example, soon after the discovery of **antibiotics**, bacteria appeared that were resistant to these drugs. Scientists now know that the genes for antibiotic resistance are carried on **plasmids**, small DNA molecules, which can be transferred from one bacterial cell to another and which can spread the mutation for **antibiotic resistance** very rapidly within the bacterial population. The appearance of antibiotics themselves does not induce mutations for resistance; it merely selects against susceptible bacteria by killing them. Since only resistant individuals survive to reproduce, the next generation will all be resistant. Joshua Lederberg carried out the experiment that proved that some bacteria are resistant to antibiotics prior to any exposure.

4. Sexual Selection

Sexual selection is selection based on variation in secondary sexual characteristics related to competing for and attracting mates. In males, the evolution of horns, antlers, large stature, and great strength are the result of sexual selection. Male elephant seals fight for supremacy of a harem that may consist of as many as fifty females. In baboons, long canines are important for male-male competition. Differences in appearance between males and females are known as **sexual dimorphism**. In many species of birds, the females are colored in a way to blend in with their surroundings, thus protecting them and their young. The males, on the other hand, have bright, conspicuous plumage because they must compete for the attention of the females.

5. Artificial Selection

Humans breed plants and animals by seeking individuals with desired traits as breeding stock. This is known as **artificial selection**. Racehorses are bred for speed, and laying hens are bred to produce more and larger eggs. Humans have bred cabbage, brussels sprouts, kale, kohlrabi, cauliflower, and broccoli all from the wild mustard plant by selecting for different traits.

PRESERVING VARIATION IN A POPULATION

Variation in a population is necessary in order for a population to evolve as the environment changes. Although Darwin could not explain the origin of variation, he knew that variation exists in every population. A good example of this is the existence of hundreds of breeds of dogs. All dogs belong to one species, *Canus familiaris.*

While it would seem that natural selection would tend to reduce genetic variation by removing unfavorable genotypes from a population, nature has many mechanisms to preserve it. Here are eight mechanisms that *preserve* diversity or variation in a gene pool or population: balanced polymorphism, geographic variation, sexual reproduction, outbreeding, diploidy, heterozygote superiority, frequency-dependent selection, and evolutionary neutral traits.

1. Balanced Polymorphism

Balanced polymorphism is the presence of two or more phenotypically distinct forms of a trait in a single population of a species. The shells of one genus of land snail exhibit a wide range of colors and banding patterns. Banded snails living on dark, mottled ground are less visible than unbanded ones and therefore are preyed upon less frequently. In areas where the background is fairly uniform, unbanded snails have the selective advantage. Each **morph** is better adapted in a different area, but both varieties continue to exist.

2. Geographic Variation

Two different varieties of rabbit continue to exist in two different regions in North America. Rabbits in the cold, snowy northern regions are camouflaged with white fur and have short ears to conserve body heat. Rabbits living in warmer, southern regions have mottled fur to blend in with surrounding woodsy areas and long ears to radiate off excess body heat. Such a graded variation in the phenotype of an organism is known as a **cline**. Because the variation in rabbit appearance is due to differences in northern and southern environments, this is an example of a **north-south cline**.

3. Sexual Reproduction

Sexual reproduction provides variation due to the shuffling and recombination of alleles during meiosis and fertilization.

- **Independent assortment of chromosomes** during metaphase I results in the recombination of unlinked genes.
- **Crossing-over** is the exchange of genetic material of homologous chromosomes and occurs during meiosis I. It produces individual chromosomes that combine genes inherited from two parents. In humans, two or three crossover events occur per homologous pair.
- The **random fertilization** of one ovum by one sperm out of millions results in enormous variety among the offspring.

4. Outbreeding

Outbreeding is the mating of organisms within one species that are not closely related. It is the opposite of inbreeding, the mating of closely related individuals. Outbreeding maintains both variation within a species and a strong gene pool. Inbreeding weakens the gene pool because if organisms that are closely related interbreed, detrimental recessive traits tend to appear in homozygous recessive individuals. Many mechanisms have evolved to promote outbreeding.

In lions, the dominant male of a pride chases away the young maturing males before they become sexually mature. This ensures that these young males will not inbreed with their female siblings. These young males roam the land, often over great distances, looking for another pride to join. If one of these young male lions can successfully overthrow the king of another pride, he will inseminate all the females of that new pride and develop his own lineage.

5. Diploidy

Diploidy, the $2n$ condition, maintains and hides a huge pool of alleles that may be harmful in the present environment but that could be advantageous when conditions change in the future.

6. Heterozygote Advantage

Heterozygote advantage preserves multiple alleles in a population. It is a phenomenon in which the hybrid individual is selected for because it has greater reproductive success. The hybrids are sometimes better adapted than the homozygotes. Notice that heterozygote advantage is defined in terms of genotype, not phenotype.

In the case of **sickle cell anemia** in West Africa, people who are hybrid (*Ss*) for the sickle cell trait have the selective advantage over other individuals. Those who are hybrid have normal hemoglobin and do not suffer from sickle cell disease. However, they are resistant to malaria, which is endemic in West Africa. Those individuals who are homozygous for the sickle cell trait (*ss*) are at a great disadvantage because they have abnormal hemoglobin and suffer from and may die of sickle cell disease. People who are homozygous for normal hemoglobin (*SS*) do not have sickle cell disease but are susceptible to and may die of **malaria**. Thus, the mutation for sickling is retained in the gene pool.

7. Frequency-Dependent Selection

Another mechanism that preserves variety in a population is known as **frequency-dependent selection** or the **minority advantage**. This acts to decrease the frequency of the more common phenotypes and increase the frequency of the less common ones. In predator-prey relationships, predators develop a **search image**, or standard representation of prey, that enables them to hunt a particular kind of prey effectively. If the prey individuals differ, the most common type will be preyed upon disproportionately while the less common individuals will be preyed upon to a lesser extent. Since these rare individuals have the selective advantage, they will become more common for a time, will lose their selective advantage, and will eventually be selected against.

8. Evolutionary Neutral Traits

Evolutionary neutral traits are traits that seem to have no selective advantage. One example is the different **blood types** in humans. Scientists do not understand where they evolved from or why they have remained (been conserved) in the human population. Perhaps they actually influence survival and reproductive success in ways that are difficult to perceive or measure.

CAUSES OF EVOLUTION OF A POPULATION

The agents of change for a population, that is, those things that cause **evolution** are **genetic drift**, **gene flow**, **mutations**, **nonrandom mating**, and **natural selection**.

Genetic Drift

Genetic drift is change in the gene pool due to chance. It is a fluctuation in frequency of alleles from one generation to another and is unpredictable. It tends to limit diversity. There are two examples: the **bottleneck effect** and the **founder effect**. Here are two examples.

- **Bottleneck effect**: Natural disasters such as fire, earthquake, and flood reduce the size of a population *unselectively*, resulting in a loss of genetic variation. The resulting population is much smaller and not representative of the original one. Certain alleles may be under or overrepresented compared with the original population. This is known as the **bottleneck effect**. Here are two examples.

  The high rate of Tay-Sachs disease among Eastern European Jews is attributed to a population bottleneck experienced by Jews in the Middle Ages. During that period, many Jews were persecuted and killed, and the population was reduced to a small fraction of its original size. Of the individuals who remained alive, there happen to have been a disproportionate percentage of people who carried the Tay-Sachs gene. Since Jews in Europe remained isolated and did not intermarry with other Europeans to any great extent, the incidence of the trait remained unusually high in that population.

  From the 1820s to the 1880s along the California coast, the northern elephant seal was hunted almost to extinction. Since 1884, when the seal was placed under government protection, the population has increased to about 35,000; all are descendants from that original group and have little genetic variation.

- **The founder effect**: When a small population breaks away from a larger one to colonize a new area, it is most likely not genetically representative of the original larger population. Rare alleles may be overrepresented. This is known as the **founder effect** and occurred in the Old Order of Amish of Lancaster, Pennsylvania. All of the colonists descended from a small group of settlers who came to the United States from Germany in the 1770s. Apparently one or more of the settlers carried the rare but dominant gene for **polydactyly**, having extra fingers and toes. Due to the extreme isolation and intermarriage of the close community, this population now has a high incidence of polydactyly.

Notice that whether a trait is dominant or recessive merely determines if it is expressed or remains hidden. It does not determine how common the trait is in a population. The increase or decrease in allelic frequency of a trait results from genetic drift or from the trait being advantageous or disadvantageous.

Gene Flow

Gene flow is the movement of alleles into or out of a population. It can occur as a result of the migration of fertile individuals or gametes between populations. For example, pollen from one valley can be carried by the wind across a mountain to another valley. Gene flow tends to increase diversity.

Mutations

Mutations are changes in genetic material and are the raw material for evolutionary change. They increase diversity. A single point mutation can introduce a new allele into a population. *Although mutations at one locus are rare, the cumulative effect of mutations at all loci in a population can be significant.*

Nonrandom Mating

Individuals choose their mates for a specific reason. The selection of a mate *serves to eliminate the less-fit individuals.* Snow geese exist in two phenotypically distinct forms, white and blue. Blue snow geese tend to mate with blue geese, and white geese tend to mate with white geese. If, for some reason, the blue geese became more attractive and both blue and white geese began mating with only blue geese, the population would evolve quickly, favoring blue geese.

Natural Selection

Natural selection is the major mechanism of evolution in any population. Those individuals who are better adapted in a particular environment exhibit *better reproductive success.* They have more offspring that survive and pass their genes on to more offspring.

HARDY-WEINBERG EQUILIBRIUM—CHARACTERISTICS OF STABLE POPULATIONS

Hardy and **Weinberg**, two scientists, described the characteristics of a **stable, nonevolving population**, that is, one in which allelic frequencies do not change. For example, if the frequency of an allele for a particular trait is 0.5 and the population is not evolving, in 1,000 years the frequency of that allele will still be 0.5.

According to Hardy-Weinberg, if the population is stable, the following must be true:

1. **THE POPULATION MUST BE VERY LARGE.** In a large population, a small change in the gene pool will be diluted by the sheer number of individuals and no change in the frequency of alleles will occur. (In a small population, the smallest change in the gene pool will have a major effect in allelic frequencies.)

2. **THE POPULATION MUST BE ISOLATED FROM OTHER POPULATIONS.** There must be no migration of organisms into or out of the gene pool because that could alter allelic frequencies.

3. **THERE MUST BE NO MUTATIONS IN THE POPULATION.** A mutation in the gene pool could cause a change in allelic frequency by introducing a new allele.

4. **MATING MUST BE RANDOM.** If individuals select mates, then those individuals that are better adapted will have a reproductive advantage and the population will evolve.

5. **NO NATURAL SELECTION.** Natural selection causes changes in relative frequencies of alleles in a gene pool.

The Hardy-Weinberg Equation

The Hardy-Weinberg equation enables us *to calculate frequencies of alleles in a population.* Although it can be applied to complex situations of inheritance, for the purpose of explanation here, we will discuss a simple case—a gene locus with only two alleles. Scientists use the letter p to stand for the **dominant allele** and the letter q to stand for the **recessive allele**.

The Hardy-Weinberg equation is

$$p^2 + 2pq + q^2 = 1 \quad \text{or} \quad p + q = 1$$

The monohybrid cross is the basis for this equation.

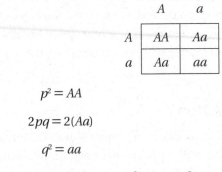

$$p^2 = AA$$

$$2pq = 2(Aa)$$

$$q^2 = aa$$

As of 2013, you will be allowed to use your calculator on the exam. So expect more difficult Hardy-Weinberg problems. However, the principle will be the same. Here are three sample problems.

PROBLEM 1

If 9% of the population has blue eyes, what percent of the population is hybrid for brown eyes? Homozygous for brown eyes?

To solve this problem, follow these steps:

1. The trait for blue eyes is homozygous recessive, *b/b*, and is represented by q^2.

    $q^2 = 9\%$ Converting to a decimal: $q^2 = .09$

2. To solve for q, the square root of $.09 = 0.3$

3. Since $p + q = 1$, if $q = 0.3$, then $p = 0.7$

4. The hybrid brown condition is represented by $2pq$. So—

5. To solve for the percent of the population that is hybrid, substitute values for $2(p)(q)$.

    The percentage of the population that is hybrid brown is $2(.7)(.3) = 42\%$.

6. Homozygous dominant is represented by p^2.

    The percentage of the population that is homozygous brown is $p^2 = (.7)^2 = 49\%$

Determine the percent of the population that is homozygous dominant if the percent of the population that is homozygous recessive is 16%.

1. Homozygous recessive $= q^2 = 16$. Therefore, $q^2 = .16$ and $q = 0.4$

2. If $q = 0.4$, then $p = 0.6$

3. Therefore, the percentage of the population that is homozygous dominant $= p^2 = .36 = 36\%$.

Determine the percent of the population that is hybrid if the allelic frequency of the recessive trait is 0.5.

1. In this example, you are given the value of q (not q^2). You only need to subtract from 1 to get the value of p.

2. If $p + q = 1$ and $q = 0.5$; then $p + 0.5 = 1$

3. Since both $p = 0.5$ and $p = 0.5$. The percentage of the population that is hybrid is $2pq = 0.5 \times 0.5 \times 2 = 50\%$.

SPECIATION AND REPRODUCTIVE ISOLATION

The definition of a **species** is a population whose members have the potential to interbreed in nature and produce viable, fertile offspring. Lions and tigers can be induced to interbreed in captivity but would not do so naturally. Therefore, they are considered separate species. Horses and donkeys can interbreed in nature and produce a mule that is not fertile. Therefore, the horse and donkey belong to different species.

A **species** is defined in terms of **reproductive isolation**, meaning that one group of genes becomes isolated from another to begin a separate evolutionary history. Once separated, the two isolated populations may begin to diverge genetically under the pressure of different selective forces in different environments. If enough time elapses and differing selective forces are sufficiently great, the two populations may become so different that, even if they were brought back together, interbreeding would not naturally occur. At that point, **speciation** is said to have taken place. *Anything that fragments a population and isolates small groups of individuals may cause speciation.* The following describes different modes of speciation due to different modes of isolation, and Figure 10.3 shows diagrams of **allopatric** and **sympatric speciation**.

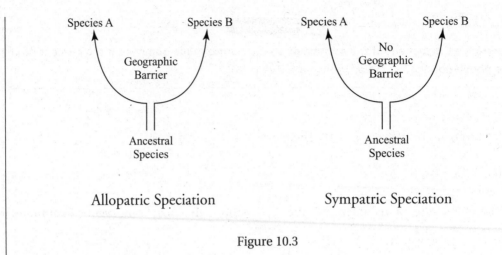

Allopatric Speciation   Sympatric Speciation

Figure 10.3

Allopatric Speciation

Allopatric speciation is caused by **geographic isolation**: separation by mountain ranges, canyons, rivers, lakes, glaciers, altitude, or longitude.

Sympatric Speciation

Under certain circumstances, speciation may occur without geographic isolation, in which case the cause of the speciation is **sympatric**. Examples of **sympatric speciation** are **polyploidy**, **habitat isolation**, **behavioral isolation**, **temporal isolation**, and **reproductive isolation**.

- **Polyploidy** is the condition where a cell has more than two complete sets of chromosomes ($4n$, $8n$, etc.). It is common in plants and can occur naturally or through breeding. It results from nondisjunction during meiosis when gametes with the $2n$ chromosome number are fertilized by another abnormal ($2n$) gamete, resulting in a daughter cell with $4n$ chromosomes. Plants that are polyploid cannot breed with others of the same species that are not polyploid and are functionally isolated from them. See Figure 10.4.

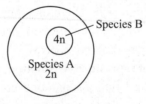

Figure 10.4 Polyploidy

- **Habitat isolation:** Two organisms live in the same area but encounter each other rarely. Two species of one genus of snake can be found in the same geographic area, but one inhabits the water while the other is mainly terrestrial.

- **Behavioral isolation**: Sticklebacks, small saltwater fish that have been studied extensively, have elaborate mating behavior. At breeding time, in response to increased sunlight, the males change in color and develop a red underbelly. The male builds a nest and courts the female with a dance that triggers a complex set of movements between the partners. If either partner fails in any step of the mating dance, no mating occurs and no young are produced.

  Male fireflies of various species signal to females of their kind by blinking the lights on their tails in a particular pattern. Females respond only to characteristics of their own species, flashing back to attract males. If, for any reason, the female does not respond with the correct blinking pattern, no mating occurs. The two animals become isolated from each other.

- **Temporal isolation**: Temporal refers to time. A flowering plant colonizes a region with areas that are warm and sunny and areas that are cool and shady. Flowers in the regions that are warmer become sexually mature sooner than flowers in the cooler areas. This separates flowers in the two different environments into two separate populations.

- **Reproductive isolation**: Closely related species may be unable to mate because of a variety of reasons. Differences in the structure of genitalia may prevent insemination. Difference in flower shape may prevent pollination. Things that prevent mating are called **prezygotic barriers**. For example, a small male dog and a large female dog cannot mate because of the enormous size differences between the two animals. Things that prevent the production of fertile offspring, once mating has occurred, are called **postzygotic barriers**. One example might be that a particular zygote is not viable. Both prezygotic and postzygotic barriers result in **reproductive isolation**.

PATTERNS OF EVOLUTION

The evolution of different species is classified into five patterns: divergent, convergent, parallel, coevolution, and adaptation radiation; see Figure 10.5.

Divergent Evolution

Divergent evolution occurs when a population becomes isolated (for any reason) from the rest of the species, becomes exposed to new selective pressures, and evolves into a new species. All the examples of allopatric and sympatric speciation on the previous pages are examples of divergent evolution.

Convergent Evolution

When unrelated species occupy the same environment, they are subjected to similar selective pressures and show similar adaptations. The classic example of **convergent evolution** is the whale, which has the streamlined appearance of a large fish because the two evolved in the same environment. The underlying bone structure of the whale, however, reveals an ancestry common to mammals, not to fish.

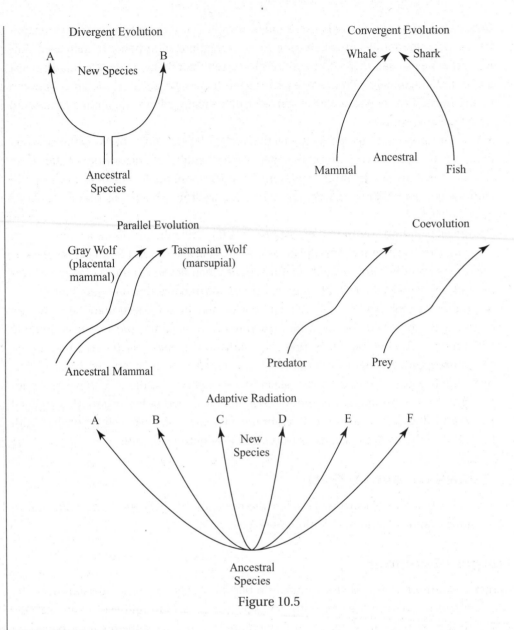

Figure 10.5

Parallel Evolution

Parallel evolution describes two related species that have made similar evolutionary adaptations after their divergence from a common ancestor. The classic example of this includes the marsupial mammals of Australia and the placental mammals of North America. (The only mammals in Australia are the ones that have been introduced from abroad, like rabbits.) There are striking similarities between some placental mammals like the gray wolf of North America and the marsupial Tasmanian wolf of Australia because they share a common ancestor and evolved in similar environments.

Coevolution

Coevolution is the reciprocal evolutionary set of adaptations of two interacting species. All predator-prey relationships are examples and the relationship between the **monarch butterfly** and **milkweed plant** is another. The **milkweed plant** contains poisons that deter herbivores from eating them. The butterfly lays its eggs in the milkweed plant and when the larvae (caterpillars) hatch, they feed on the milkweed and absorb the poisonous chemicals from the plant. They store the poison in their tissues. This poison, which is present in the adult butterfly, makes the butterfly toxic to any animal who tries to eat it. (The butterfly exhibits bright conspicuous warning colors that deter predators.)

Adaptive Radiation

Adaptive radiation is the emergence of numerous species from a common ancestor introduced into an environment. Each newly emerging form specializes to fill an ecological niche. All 14 species of **Darwin's finches** that live on the **Galapagos Islands** today diverged from a single ancestral species perhaps 10,000 years ago. There are currently six ground finches, six tree finches, one warbler finch, and one bud eater.

MODERN THEORY OF EVOLUTION
Gradualism

Gradualism is the theory that organisms descend from a common ancestor gradually, over a long period of time, in a linear or branching fashion. Big changes occur by an accumulation of many small ones. According to this theory, fossils should exist as evidence of every stage in the evolution of every species with no missing links. However, the fossil record is at odds with this theory because scientists rarely find **transitional forms** or **missing links**.

Punctuated Equilibrium

The favored theory of evolution today is called **punctuated equilibrium** and was developed by **Stephen J. Gould** and **Niles Eldridge** after they observed that the gradualism theory was not supported by fossil record. The theory proposes that new species appear suddenly after long periods of stasis. A new species changes most as it buds from a parent species and then changes little for the rest of its existence. The sudden appearance of the new species can be explained by the **allopatric model** of speciation. A new species arises in a different place and expands its range, outcompeting and replacing the ancestral species. See Figure 10.6, which has sketches showing gradualism and punctuated equilibrium.

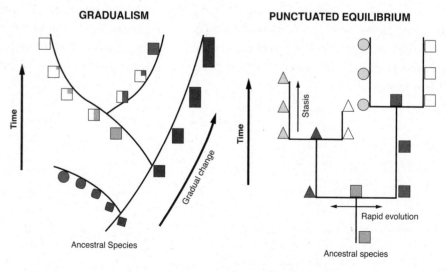

Figure 10.6 Punctuated Equilibrium

THE ORIGIN OF LIFE

The ancient atmosphere consisted of CH_4, NH_3, CO, CO_2, N_2 and H_2O, but lacked free O_2. There was probably intense lightning and ultraviolet (U.V.) radiation that penetrated the primitive atmosphere, providing energy for chemical reactions. Scientists have tried to mimic this early atmosphere to determine how the first organic molecules and earliest life developed. Here is a synopsis of those classic experiments.

A. I. Oparin and **J. B. S. Haldane**, in the 1920s, hypothesized separately that under the conditions of early earth, organic molecules could form. Without corrosively reactive molecular oxygen present to react with and degrade them, organic molecules could form and remain.

Stanley Miller and **Harold Urey**, in the 1950s, tested the Oparin-Haldane hypothesis and proved that almost any energy source would have converted the molecules in the early atmosphere into a variety of organic molecules, including **amino acids**. They used electricity to mimic lightning and U.V. light that must have been present in great amounts in the early atmosphere.

Sidney Fox, in more recent years, carried out similar experiments. However, he began with organic molecules (not the original inorganic ones) and was able to produce membrane-bound, cell-like structures he called proteinoid microspheres that would last for several hours.

The Heterotroph Hypothesis and the Theory of Endosymbiosis

The first cells on earth were **anaerobic heterotrophic prokaryotes**. They simply absorbed organic molecules from the surrounding primordial soup to use as a nutrient source. They probably began to evolve about 3.5 billion years ago. Eukaryotes did not evolve until another 2 billion years after the evolution of prokaryotes (about 1.5 billion years ago). They arose as a result of **endosymbiosis** according to Lynn Margulis, who developed the **theory**. She states that mitochondria and chloroplasts (and perhaps nuclei) were once free-living prokaryotes that took up residence inside larger prokaryotic cells. The mutually beneficial symbiotic relationship worked out so well that it became permanent. There are several points to prove that mitochondria and chloroplasts are **endosymbionts**.

- Chloroplasts and mitochondria have their own DNA.
- DNA is more like prokaryotic DNA than eukaryotic DNA. It is not wrapped with **histones**.
- These organelles have double membranes. The inner one belongs to the symbiont; the outer one belongs to the host plasma membrane. The theory states that the chloroplast and mitochondria were taken up by the host cell by some sort of endosymbiosis process, such as phagocytosis.

Questions 1–5

Questions 1–5 refer to the fields of study listed below. Choose the one that has provided each of the following pieces of evidence that biological evolution has occurred.

(A) Comparative biochemistry
(B) Comparative anatomy
(C) Comparative embryology
(D) Geographic distribution

1. Giraffes have the same number of vertebrae in the neck as do humans.

2. Kangaroos are found only in Australia.

3. Human embryos have tails.

4. Humans and sea stars both have radial cleavage in early embryonic development.

5. Humans can be made temporarily immune to various human diseases by receiving antibodies against those diseases from horses.

_____

6. The condition in which there are barriers to successful interbreeding between individuals of different species in the same community is referred to as

(A) sexual dependency
(B) reproductive isolation
(C) geographic isolation
(D) adaptive radiation

7. The wing of the bat and a human's arm have different functions and appear very different. Yet, the underlying anatomy is basically the same. Therefore, these structures are examples of

(A) geographic isolation
(B) analogous structures
(C) homology
(D) reproductive isolation

8. In a population that is in Hardy-Weinberg equilibrium, the frequency of a particular recessive allele a is 0.4. What is the percentage of the population heterozygous for this allele?

(A) 4%
(B) 16%
(C) 32%
(D) 48%

9. According to the Hardy-Weinberg equation, the dominant trait is represented by

(A) p

(B) q

(C) q^2

(D) p^2

Questions 10–13

Matching Column

(A) Stabilizing selection

(B) Disruptive selection

(C) Directional selection

(D) Sexual selection

10. The population of peppered moths in England changed from white to black in fifty years.

11. Human newborns usually weigh between 6–8 pounds (2.7–3.6 kg).

12. In one region of New Jersey there exist two distinct types of one species of snake.

13. Large horns and giant antlers are characteristic of the male.

_____

14. Miller's classic experiment demonstrated that a discharge of sparks through a mixture of gases could result in the formation of a large variety of organic compounds. Miller used all of the following gases in his experiment EXCEPT

(A) methane

(B) ammonia

(C) water

(D) oxygen

15. Which is an example of a cline?

(A) Males of a species have long antlers to fight other males of that species.

(B) In many species of birds, males have bright plumage to attract the female.

(C) In one species of rabbit, the ones that evolved in the cold, snowy north are white, while the ones that evolved in the south are brown.

(D) The hybrid tomato plant is stronger and produces better fruit than the pure genotype.

16. Who synthesized proteinoid microspheres in the laboratory using an apparatus that mimicked the early earth?

(A) Haldane

(B) Fox

(C) Urey

(D) Miller

17. The differences in sparrow songs among sympatric species of sparrows are examples of

 (A) geographic isolation
 (B) convergent evolution
 (C) behavioral isolation
 (D) physiological isolation

18. Which part of the theory of evolution did Darwin develop after reading Thomas Malthus?

 (A) Evolution occurs as advantageous traits accumulate in a population.
 (B) In any population, there is variation and an unequal ability of individuals to survive and reproduce.
 (C) Only the best-fit individuals survive and get to pass on their traits to offspring.
 (D) Populations tend to grow exponentially, overpopulate, and exceed their resources.

19. In a population of 1,000 people, 90 have blue eyes. What percent of the population has hybrid brown eyes?

 (A) 3%
 (B) 9%
 (C) 21%
 (D) 42%

20. The average length of a rabbit's ears decreases the farther north the rabbits live. This variation is an example of a

 (A) genetic drift
 (B) a cline
 (C) geographic isolation
 (D) founder effect

Questions 21–24
Matching Column

 (A) Founder effect
 (B) Parallel evolution
 (C) Adaptive radiation
 (D) Convergent evolution

21. Darwin's finches

22. The establishment of a genetically unique population through genetic drift

23. The independent development of similarities between unrelated groups resulting from adaptation to similar environments

24. The Tasmanian wolf in Australia is a marsupial but looks very similar to the gray wolf, a placental mammal of North America

Answers to Multiple-Choice Questions

1. **(B)**

2. **(D)**

3. **(C)**

4. **(C)**

5. **(A)**

6. **(B)** Any barrier that isolates organisms fosters evolution. Barriers to interbreeding are caused by reproductive isolation. Geographic isolation refers to organisms being isolated by geography, such as mountains or rivers. Balanced polymorphism refers to two different versions of the same species living in one area, such as a speckled snail and a plain snail. Both are camouflaged in different environments. Sexual dependency is not related to the topic in any way.

7. **(C)** Homologous structures demonstrate a common ancestry. They may not look alike, but they have an underlying common structure. Analogous structures may have the same function and look alike, but they do not have a common structure, nor do they have a common ancestry.

8. **(D)** The question provides you with the frequency of the allele. It provides you with q. Since the frequency of the recessive allele is 0.4, the frequency of the dominant allele is 0.6. The formula for the hybrid $= 2pq$. Therefore, substituting, $2 \times 0.4 \times 0.6 = 48\%$.

9. **(A)** According to Hardy-Weinberg equilibrium, p is the dominant allele and q is the recessive allele.

10. **(C)** The black peppered moths replaced the white peppered moths. Since one characteristic replaced another, this is directional selection.

11. **(A)** Stabilizing selection tends to eliminate the extremes in a population.

12. **(B)** Disruptive selection tends to select for the extremes. Originally, there was probably a range of coloration of snakes in the area in question. Over time, pressure from the environment selected against different colorations until only two remained.

13. **(D)** Sexual selection has to do with the selection for traits that attract a mate.

14. **(D)** Free oxygen was not available in the early earth's atmosphere. Scientists believe that since oxygen is very reactive, had it been present in the ancient atmosphere, it would have reacted with and degraded all the other chemicals in the atmosphere. The consequence would be that evolution of the early earth would not have occurred as it did.

15. **(C)** A cline is a change in some trait along some geographic axis, such as a north-south cline. In this example, the animal is camouflaged by its colorings.

16. **(B)** Sidney Fox was able to produce these cell-like structures, which he called proteinoid microspheres, when he began with amino acids in his experiment.

17. **(C)** A bird's song is a behavior; therefore, this is an example of behavioral isolation.

18. **(D)** Malthus was a mathematician studying populations. He stated that populations tend to overpopulation and exceed their resources. This leads to starvation, disease, and death.

19. **(D)** Of the total population, 9% have blue eyes (90 out of 1000), so $q^2 = 0.09$ and $q = 0.3$. Therefore, $p = 0.7$ and the frequency of hybrid brown $= 2pq = 42\%$.

20. **(B)** The rabbit's ears get shorter and grow closer to the head to retain heat. This is an example of a north-south cline.

21. **(C)** Adaptive radiation is the emergence of numerous species from one common ancestor introduced into a new environment. Today, 13 different species of finches are on the Galapagos Islands where originally there was only 1 species. Each species fills a different niche.

22. **(A)** Genetic drift is evolution through chance. The founder effect is one example of genetic drift. Another is the bottleneck effect.

23. **(D)** The classic example of this can be seen in the whale and the shark. The two animals are unrelated; the whale is a mammal and the shark is a fish. However, they look alike because they experience the same environmental pressures. They both have a streamlined appearance with fins because that design is best for living in the ocean, not because they are related or have a recent common ancestor.

24. **(B)** Eutherians (placental mammals) and marsupials are closely related although they diverged several million years ago. Although they live thousands of miles apart, these two animals live in similar environments and are under the same selective pressures from their respective environments. As a result, they have evolved along similar parallel lines.

FREE-RESPONSE QUESTIONS

> **Directions:** Answer all questions. You must answer the question in essay—not outline—form. You may use labeled diagrams to supplement your essay, but diagrams alone are *not* sufficient. Before you start to write, read each question carefully so that you understand what the question is asking.

1. Explain Charles Darwin's theory of evolution by natural selection.

2. Each of the following refers to one aspect of evolution. Explain each in terms of natural selection.
   a. Convergent evolution and the similarities among species in a particular biome.
   b. Insecticide resistance
   c. Speciation and isolation
   d. Heterozygote advantage

Typical Free-Response Answers

Note: This essay has two main sections (a and b), and section b is divided into four parts. Since the question demands much more from part b than part a, assume b is worth more and answer it accordingly.

If you run into trouble because you find that you understand the concept but cannot remember the name of the organism you are writing about, that is not very important. Call it organism "X." Showing that you understand the concept is the important thing.

1. Darwin's theory of evolution is known as the theory of natural selection. According to Darwin, populations tend to grow exponentially, to overpopulate, and to exceed the carrying capacity of their environment. This overpopulation results in a struggle for existence, where only the best-adapted organisms survive and gain a greater share of limited resources. Those who survive long enough pass their traits to the next generation. The less-fit individuals will not survive and will not reproduce. Therefore, the genes that pass to the next generation will be best adapted for that particular environment.

2a. Convergent evolution

The existence of convergent evolution demonstrates the power of the environment to select the direction of evolution. In convergent evolution, unrelated species often come to resemble one another because they are subject to the same environmental pressures. A perfect example can be seen with two unrelated families of plants, the cactus and the euphorbs, which are found in deserts in different parts of the world, the southwest of North America, and Central Asia. Both families of plants developed fleshy photosynthetic stems adapted for water storage, protective spines, and greatly reduced leaves. It is likely that there was once an array of different kinds of plants in each region, but the climate changed and became very dry. The only plants that could survive were the ones that bore adaptations that enabled them to survive in the dry environment. Traits that were well adapted for the environment were selected for, while the ones that were disadvantageous died out.

2b. Insecticide resistance

Insecticide resistance is an example of directional selection. Some insects are resistant to a particular chemical insecticide. The origin of this resistance is unknown, but it probably derives from mutation. When the environment is sprayed with that insecticide, the resistant insects have the selective advantage. All the nonresistant insects die and the remaining ones breed quickly, without any competition for resources. The new population is entirely resistant to the insecticide.

2c. Speciation and isolation

Anytime a population becomes isolated from another, the two isolated populations may begin to diverge genetically under the pressure of different selective forces in different environments. If enough time elapses and differing selective forces are sufficiently great, the two populations may become so different that, even if they were brought back together, interbreeding would not occur. The two populations would have become two different species.

There are two types of isolation, **allopatric** and **sympatric**. Allopatric isolation is caused by geographic separation, such as mountain ranges, rivers, glaciers, or canyons. One population of wildebeest might become separated from the larger group during a migration and remain isolated forever. Sympatric isolation occurs without geographic isolation. Examples are polyploidy, **habitat isolation**, **temporal** or **behavioral isolation**, and **reproductive isolation**.

An example of reproductive isolation can be seen in the stickleback fish, which has an elaborate premating behavior. If either partner fails to respond correctly in any step of the behavior, no mating occurs and the two individuals are effectively isolated from one another. Another cause of isolation is polyploidy, which is common in plants. It results from **nondisjunction** during gametogenesis and can produce gametes that are diploid. When these diploid gametes fuse with normal gametes, the resulting plants are triploid and unable to mate with any diploid individuals. This is an explanation of how a small population has become isolated.

2d.   Heterozygote advantage

There are many instances where the hybrid individual is more fit than the homozygous condition. A perfect example can be seen in West Africa where sickle cell anemia is endemic and so is malaria. People who are homozygous for the sickle cell trait (*ss*) have abnormal hemoglobin and suffer with sickle cell disease. People who are homozygous for normal hemoglobin (*SS*) are susceptible to malaria. However, people who are hybrid for the sickle cell trait do not have sickle cell disease and are resistant to malaria. Without serious medical intervention, the population homozygous for the sickle cell trait may die of sickle cell disease and the population homozygous for normal hemoglobin may die of malaria. The hybrid condition (*Ss*) has the selective advantage.

Plants

11

→ CLASSIFICATION OF PLANTS

→ BRYOPHYTES

→ TRACHEOPHYTES

→ STRATEGIES THAT ENABLED PLANTS TO MOVE TO LAND

→ PRIMARY AND SECONDARY GROWTH

→ PLANT TISSUE

→ ROOTS

→ STEMS

→ THE LEAF

→ TRANSPORT IN PLANTS

→ PLANT REPRODUCTION

→ ALTERNATION OF GENERATIONS

→ PLANT RESPONSES TO STIMULI

Much of this chapter contains background information, but focus on:

Transport in Plants

Plant Responses to Stimuli

INTRODUCTION

Plants are defined as multicelled, eukaryotic, photosynthetic **autotrophs**. Their cell walls are made of cellulose, and their surplus carbohydrate is stored as starch. The life cycle of plants is characterized by **alternation of generations**. One generation is the gametophyte generation, where all the cells of the plant body are haploid (n). The other, alternate generation is the sporophyte generation, where the cells of the plant body are diploid ($2n$). See Figure 11.11.

Plants evolved from aquatic green algae about 500 million years ago. Along with fungi and animals, they colonized the land during the Paleozoic era. This gradual move from the ancestral aquatic environment to the land occurred as organisms evolved adaptations to a dry environment.

Today, most plants live on land. They have diversified into almost 300,000 different species inhabiting all but the harshest environment. Plants stabilize the soil they live in and provide a home for billions of insects and larger animals. They release oxygen into the atmosphere and absorb carbon dioxide. Most of the world depends on the following plants for survival: rice, beans, soy, corn, and wheat.

Plants are organized into two groups. Those plants with no transport vessels (xylem and phloem) are called **bryophytes**. Those with transport vessels are called **tracheophytes**.

CLASSIFICATION OF PLANTS

1. **BRYOPHYTES**—Non-vascular plants

   Ex: mosses, liverworts, hornworts

2. **TRACHEOPHYTES**—Vascular plants

   A. **Seedless plants**, like ferns, that reproduce by **spores**

   B. **Seed plants**

   1. **Gymnosperms**—cone-bearing

      Ex: cedars, sequoias, redwoods, pines, yews, and junipers

   2. **Angiosperms** or **Anthophyta**—the **Flowering plants**

      Ex: roses, daisies, apples, and lemons

      • **Monocotyledon** (monocots)

        Ex: grasses such as corn, wheat, rye, and oats

      • **Dicotyledon** (dicots)

        Ex: peanuts

   Figure 11.1 is a cladogram showing the evolution of land plants.

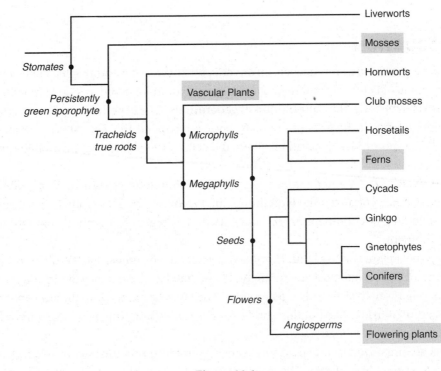

Figure 11.1

BRYOPHYTES

Bryophytes are primitive plants that lack transport vessels (xylem and phloem) and must therefore absorb water by diffusion from the air. In addition, their flagellated sperm must swim through water to fertilize an egg. They also lack any **lignin**-fortified tissue that is necessary to support a tall plant. As a result, bryophytes are restricted to moist habitats and are tiny. Bryophytes play a significant role in diverse terrestrial ecosystems. They grow on rocks, soil, and trees. Sphagnum or peat moss is used as fuel in much of the world.

Like all plants they exhibit alternation of generations. (See page 215.)

TRACHEOPHYTES

Tracheophytes are plants with vascular tissue. Prior to the evolution of tracheophytes, bryophytes were the dominant form of plant life on Earth. However, having an efficient transport system enabled the tracheophytes to outcompete bryophytes. The characteristics of tracheophytes include:

- Xylem and phloem for transport
- Lignified transport vessels to support the plant
- Roots to absorb water while also anchoring and supporting the plant
- Leaves that increase the photosynthetic surface
- Life cycle with a dominant sporophyte generation

Tracheophytes are divided into two groups, those with seeds and those without. Those with seeds, the seed plants, are more advanced and far more numerous than the seedless plants. An example of seedless plants is ferns. The seed plants can be further divided into **gymnosperms**—those bearing cones, and **angiosperms**—those bearing flowers and fruits.

Ferns—Seedless Plants

The ferns are the most widespread **seedless tracheophytes**. They are primitive plants and reproduce by spores instead of by seeds. They are **homosporous**, which means that they produce only one type of spore which then develops into a bisexual gametophyte. Although they have transport tissues and can grow several feet tall, ferns are still restricted to moist habitats. This is true because their sperm are flagellated and must swim from the antheridium to the archegonium to fertilize the egg.

Seed Plants

In contrast to seedless tracheophytes like ferns, seed plants are **heterosporous**. That is, they produce two kinds of spores, megaspores and microspores. **Megaspores** develop into female gametophytes. **Microspores** develop into male gametophytes. In addition, the sperm of seed plants have no flagella and therefore do not require a watery environment in order for fertilization to occur. There are two types of seed plants, gymnosperms and angiosperms.

GYMNOSPERMS: THE CONIFERS

Gymnosperms were the first seed plants to appear on Earth. The seeds of gymnosperm are said to be *naked* because they are not enclosed inside a fruit as are seeds in angiosperms. Instead, they are exposed on modified leaves that form cones, which are better adapted for a dry environment. These modifications for a dry environment include needle-shaped leaves, which have a thick, protective cuticle and a relatively small surface area. In addition, gymnosperms depend on wind for pollination. Examples of gymnosperms are pines, firs, redwoods, junipers, and sequoia.

ANGIOSPERMS: THE FLOWERING PLANTS

Angiosperms are seed plants whose reproductive structures are flowers and fruits. Today, these are the most diverse plant species, including about 90 percent of all plants. The color and scent of a flower attracts animals that will carry pollen from one plant to another over great distances. After pollination and fertilization, the **ovary** becomes the **fruit** and the **ovule** becomes the **seed**. Fruit protects dormant seeds and aids in their dispersal. Maple trees have seeds with wings that enable them to be dispersed great distances by the wind. Some plants have burrs on their fruits that cling to an animal's fur or a person's clothing. Many plants have fruit that are brightly colored and very sweet. An animal eats and digests the fleshy part of the fruit while the tough seed passes through the animal's digestive tract and is deposited with its feces as a package of fertilizer. There are two groups of angiosperms: monocots and dicots.

The Principal Differences Between Monocots and Dicots		
Characteristic	Monocots	Dicots
Cotyledons (seed leaves)	One	Two
Vascular bundles in stem	Scattered	In a ring
Leaf venation	Parallel	Netlike
Floral parts	Usually in 3s	Usually in 4s or 5s
Roots	Fibrous Roots	Taproots

STRATEGIES THAT ENABLED PLANTS TO MOVE TO LAND

Plants began life in the seas and moved to land as competition for resources increased. The biggest problems a plant on land faces are supporting the plant body and absorbing and conserving water. Here are some modifications that evolved that enable plants to live on land.

- **Cell walls** made of cellulose lend support to the plant whose cells, unsupported by a watery environment, must maintain their own shape.
- **Roots** and **root hairs** absorb water and nutrients from the soil.
- **Stomates** open to exchange photosynthetic gases and close to minimize excessive water loss.
- The waxy coating on the leaves, **cutin**, helps prevents excess water loss from the leaves.
- In some plants, gametes and zygotes form within a protective jacket of cells called **gametangia** that prevents drying out.
- **Sporopollenin**, a tough polymer, is resistant to almost all kinds of environmental damage and protects plants in a harsh terrestrial environment. It is found in the walls of spores and pollen.
- **Seeds** and **pollen** are a means of dispersing offspring. They also have a protective coat that prevents dessication.
- The gametophyte generation has been reduced.

- Xylem and phloem vessels enable plants to grow tall.
- Lignin embedded in xylem and other plant cells provides support.

PRIMARY AND SECONDARY GROWTH

Plants continue to grow as long as they live because they contain tissue called **meristem** that continually divides and generates new cells.

Primary growth is the *elongation of the plant down into the soil and up into the air*. **Apical meristem** at the tips of the roots and in the buds of shoots is the source of primary growth. See Figure 11.5.

Lateral meristem provides **secondary growth** which is increase in girth. In **herbaceous** (nonwoody) plants, there is only primary growth. In **woody plants**, secondary growth is responsible for the gradual thickening of the roots and shoots formed from earlier primary growth.

PLANT TISSUE

A plant consists of three types of tissue, each with different functions: **dermal tissue**, **vascular tissue**, and **ground tissue**.

Dermal Tissue

Dermal tissue covers and protects the plant. It includes **epidermis** and modified cells like guard cells, root hairs, and cells that produce a waxy cuticle.

Vascular Tissue

Vascular tissue consists of **xylem** and **phloem**. These transport water and nutrients around the plant; see Figure 11.2.

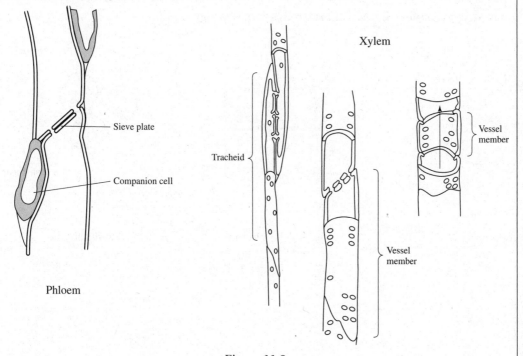

Figure 11.2

XYLEM

Xylem, the water- and mineral-conducting tissue, consists of two types of elongated cells: **tracheids** and **vessel elements**. (See Figure 11.2.) Both tracheids and vessels are dead at functional maturity. **Tracheids** are long, thin cells that overlap and are tapered at the ends. Water passes from one cell to another through **pits**, areas with no secondary cell wall. Because their secondary cell walls are hardened with **lignin**, tracheids function to support the plant as well as to transport nutrients and water. **Vessel elements** are generally wider, shorter, thinner walled, and less tapered than tracheids. Vessel elements are aligned end to end and differ from tracheids in that the ends are perforated to allow free flow through the vessel tubes. Seedless vascular plants and most gymnosperms have only tracheids; most angiosperms have both tracheids and vessel members. Xylem is what makes up wood.

PHLOEM

Phloem carries sugars from the photosynthetic leaves to the rest of the plant by active transport. The phloem vessels consist of chains of **sieve tube members** or **elements** whose end walls contain **sieve plates** that facilitate the flow of fluid from one cell to the next. In contrast to the xylem elements, these cells are alive at maturity, although they lack nuclei, ribosomes, and vacuoles. Connected to each sieve tube member is at least one **companion cell** that does contain a full complement of cell organelles and nurtures the sieve tube elements. (See Figure 11.2.)

Ground Tissue

By far the most common tissue type in a plant is the **ground tissue**, which functions mainly in support, storage, and photosynthesis. A major theme in biology is that form relates to function. For example, if a plant cell's function is support, you would expect that the cell walls would be thick and perhaps contain lignin. Think of this concept as you read about the different cell types of ground tissue in plants. Ground tissue consists of three cell types: **parenchyma**, **collenchyma**, and **sclerenchyma**; see Figure 11.3.

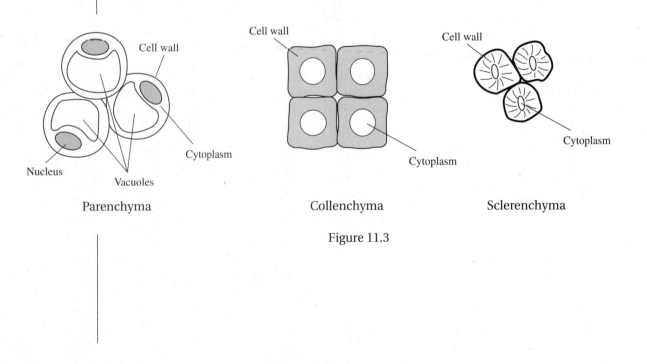

Parenchyma   Collenchyma   Sclerenchyma

Figure 11.3

PARENCHYMAL CELLS

Parenchymal cells look like classic plant cells. They have primary cell walls that are thin and flexible. They lack secondary cell walls. The protoplasm contains one large vacuole, and the cell carries out most metabolic functions. Some, like mesophyll cells in the leaf, contain chloroplasts and carry out photosynthesis. In contrast, parenchymal cells in roots contain plastids and store starch. When turgid (swollen) with water, they give support and shape to the plant. Most parenchymal cells retain the ability to divide and differentiate into other cell types after a plant has been injured in some way. In a laboratory, an entire plant can be regenerated or cloned from a single parenchymal cell.

COLLENCHYMAL CELLS

Collenchymal cells have unevenly thickened primary cell walls but lack secondary cell walls. Mature collenchymal cells are alive and their function is to support the growing stem. The "strings" of a stalk of celery, for example, consist of collenchymal cells.

SCLERENCHYMAL CELLS

Sclerenchymal cells have very thick primary and secondary cell walls fortified with **lignin**. Their function is to support the plant. There are two forms of these cells: fibers and sclereids. Fibers are long, thin, and fibrous like, and they usually occur in bundles. They are used commercially to make rope and flax fibers, which are used to make linen. Sclereids are short and irregular in shape. They make up tough seed coats and pits and they give the pear its gritty texture.

ROOTS
Function and Structure

The three **functions** of roots are to **absorb nutrients** from the soil, **anchor** the plant, and **store food**. The root consists of specialized tissues and structures organized to carry out these various functions of the roots. See the sketch of monocot and dicot roots in cross section in Figure 11.4.

STUDY TIP

The dicot root has a cross or star in the center.

- The **epidermis** covers the entire surface of the root and is modified for **absorption**. Slender cytoplasmic projections from the epidermal cells called **root hairs** extend out from each cell and greatly increase the absorptive surface area.
- The **cortex** consists of **parenchymal cells** that contain many **plastids** for the storage of starch and other organic substances.
- **Stele**: The **vascular cylinder** or **stele** of the root consists of **vascular tissues** (xylem and phloem) surrounded by one or more layers of tissue called the **pericycle**, from which **lateral roots** arise.
- **Endoderm**: The vascular cylinder is surrounded by a tightly packed layer of cells called the **endodermis**. Each endoderm cell is wrapped with the **Casparian strip**, a continuous band of **suberin**, a waxy material that is impervious to water and dissolved minerals. The function of the endoderm is to select what minerals enter the vascular cylinder and the body of the plant.

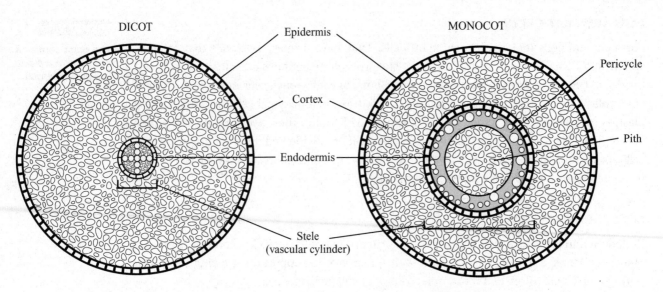

Figure 11.4 Root in Cross Section

Apical meristem, located at the tips of the roots provides **primary growth**, that is, *elongation of the plant down into the soil and up into the air*. Growth in length is concentrated near the root's tip. Three zones of cells at different stages of primary growth are located: the **zone of cell division** called **apical meristem**, the **zone of elongation**, and the **zone of differentiation**; see Figure 11.5. The root tip is protected by a **root cap**, which secretes a substance that helps digest the earth as the root tip grows through the soil.

- **Zone of cell division:** These are **meristem cells** that are actively dividing and are responsible for producing new cells that grow down into the soil. This is probably the region you observed under the microscope in lab when you were studying cells undergoing mitosis.
- **Zone of elongation:** Here the cells elongate and are responsible for pushing the root cap downward deeper into the soil.
- **Zone of differentiation:** Here cells undergo **specialization** into three primary meristems that give rise to three tissue systems in the plant. The **protoderm** becomes the epidermis, the **ground meristem** becomes the **cortex** (for storage), and the **procambium** becomes the primary xylem and phloem.

Types of Roots

A **taproot** is a single, large root that gives rise to lateral **branch roots**. In many **dicots**, the primary root is the **taproot**. Some taproots tap water deep in the soil. Others, like carrots, beets, and turnips, are modified for storage of food. A **fibrous root system**, common in monocots like grasses, holds the plant firmly in place. As a result, grasses make fine ground cover because they minimize soil erosion. **Adventitious roots** are roots that arise above ground. Trees that grow in swamps or salt marshes like mangroves have **aerial roots** that stick up out of the water and serve to aerate the root cells. English ivy has aerial roots that enable the ivy to cling to the sides of buildings. Some tall plants like corn have **prop roots** that grow above ground out from the base of the stem and help support the plant.

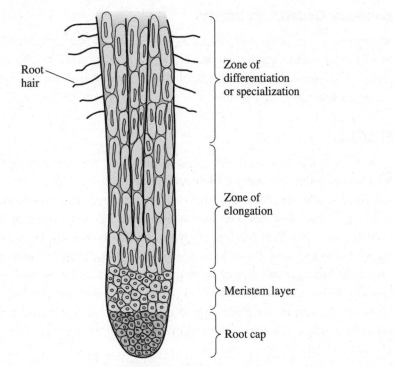

Root hair

Zone of differentiation or specialization

Zone of elongation

Meristem layer

Root cap

Figure 11.5 Longitudinal View of a Root Tip

STEMS

Primary Tissue of Stems

Vascular tissue runs the length of the stem in strands called **vascular bundles**. Each vascular bundle contains xylem on the inside, phloem on the outside, and meristem tissue in between the two. In monocots, the vascular bundles are *scattered throughout the stem*. In dicots, they are *arranged in a ring* around the edge of the stem. The ground tissue of the stem consists of **cortex** and **pith**, parenchymal tissues modified for storage. **Apical meristem**, located at the tips of the shoots and roots, supply cells for the plant to grow in length. See Figure 11.6.

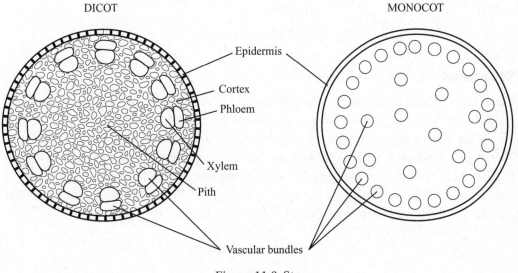

DICOT
MONOCOT

Epidermis

Cortex

Phloem

Xylem

Pith

Vascular bundles

Figure 11.6 Stem

Secondary Growth in Stems

Secondary growth in stems is produced by **lateral meristem**. Lateral meristem replaces the epidermis with a secondary dermal tissue, such as bark, which is thicker and tougher. A second lateral meristem adds layers of vascular tissue. **Wood** is secondary xylem that accumulates over the years. See Figure 11.6.

THE LEAF

The leaf is organized to maximize sugar production while minimizing water loss. The **epidermis** is covered by a **waxy cuticle** made of **cutin** to minimize water loss. **Guard cells** are modified epidermal cells that contain chloroplasts, are photosynthetic, and control the opening of the **stomates**. The inner part of the leaf consists of **palisade** and **spongy mesophyll** cells whose function is photosynthesis. The cells in the palisade layer are packed tightly, while the spongy cells are loosely packed to allow for diffusion of gases into and out of these cells. **Vascular bundles** or **veins** are located in the mesophyll and carry water and nutrients from the soil to the leaves and also carry sugar, the product of photosynthesis, from the leaves to the rest of the plant. Specialized mesophyll cells called **bundle sheath cells** surround the veins and separate them from the rest of the mesophyll. Figure 11.7 shows a sketch of a C-3 leaf.

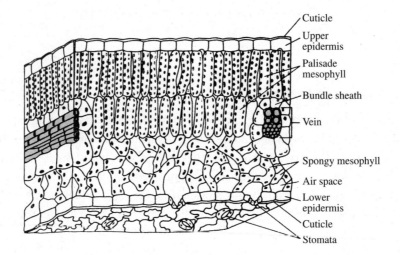

Figure 11.7 The C-3 Leaf

Stomates

Plants lose water by transpiration. About 90 percent of that water escapes through stomates, which account for only about 1 percent of the surface of the leaf. Guard cells are modified epithelium containing chloroplasts that control the opening and closing of the stomates by changing their shape. The cell walls of guard cells are not uniformly thick. Cellulose **microfibrils** are oriented in such a direction (radially) that when the guard cells absorb water by osmosis and become **turgid**, they curve like hot dogs, causing the stomate to open. When guard cells lose water and become **flaccid**, the stomate closes. See Figure 11.8.

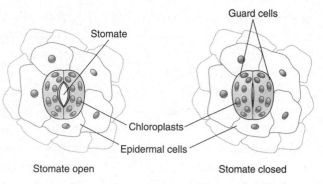

Figure 11.8 Stomates and Guard Cells

TRANSPORT IN PLANTS

Transport of Xylem

Xylem fluid rises in a plant against gravity but requires no energy. The fluid in the xylem can be *pushed up* by root pressure or *pulled up* by **transpirational pull**. **Root pressure** results from water flowing into the stele from the soil as a result of the high mineral content in the root cells. It can push xylem sap upward only a few yard (meters). Droplets of water that appear in the morning on the leaf tips of some herbaceous dicots, like strawberries, are due to root pressure. This is known as **guttation**.

Transpirational pull can carry fluid up the world's tallest trees. **Transpiration**, the **evaporation of water from leaves**, causes negative pressure (tension) to develop in the xylem tissue from the roots to the leaves. The **cohesion** of water due to strong attraction between water molecules makes it possible to pull a column of water from above within the xylem. The absorption of sunlight drives transpiration by causing water to evaporate from the leaf. **Transpirational pull–cohesion tension theory** states that *for each molecule of water that evaporates from a leaf by transpiration, another molecule of water is drawn in at the root to replace it.*

Several factors affect the rate of transpiration:

- High humidity slows down transpiration, while low humidity speeds it up.
- Wind can reduce humidity near the stomates and thereby increase transpiration.
- Increased light intensity will increase photosynthesis and thereby increase the amount of water vapor to be transpired and increase the rate of transpiration.
- Closing stomates stops transpiration.

PLANT REPRODUCTION

Asexual Reproduction

Plants can **clone** themselves or reproduce asexually by **vegetative propagation**. In this process, a piece of the **vegetative** part of a plant, the **root**, **stem**, or **leaf**, produces an entirely new plant genetically identical to the parent plant. Examples of asexual reproduction are grafting, cuttings, bulbs, and runners.

Sexual Reproduction in Flowering Plants

The flower is the sexual organ of a plant. Fertilization in a flower begins with **pollination** when one pollen grain containing three haploid nuclei (one **tube nucleus**, and two **sperm**

nuclei) lands on the sticky **stigma** of the flower. The pollen grain absorbs moisture and sprouts, producing a pollen tube that burrows down the style into the ovary. The two sperm nuclei travel down the pollen tube into the ovary. Once inside the ovary, the two remaining sperm nuclei enter the ovule through the **micropyle**. One sperm nucleus fertilizes the egg and becomes the **embryo** (2*n*). The other sperm nucleus fertilizes the **two polar bodies** and becomes the **triploid** (3*n*) **endosperm**, the food for the growing embryo. This process is known as **double fertilization** because two fertilizations occur. After fertilization, the ovule becomes the **seed** and the ripened ovary becomes the **fruit**. In monocots, food reserves remain in the endosperm. In dicots, the food reserves of the **endosperm** are transported to the **cotyledons** and consequently, the mature dicot seed lacks endosperm. In the monocot **coconut**, the endosperm is liquid. See Figure 11.9.

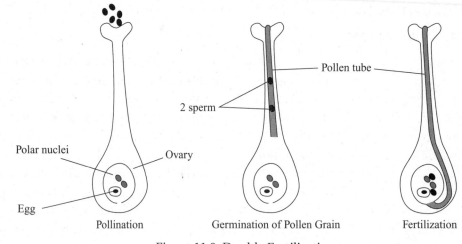

Figure 11.9 Double Fertilization

The Seed

The seed consists of a protective **seed coat**, an **embryo**, and the **cotyledon** or **endosperm**, food for the growing embryo. The embryo consists of the **hypocotyl**, **epicotyl**, and **radicle**. The hypocotyl becomes the lower part of the stem. The epicotyl becomes the upper part of the stem. The radicle, or **embryonic root**, is the first organ to emerge from the germinating seed. Figure 11.10 shows a dicot seed, like a peanut, split in half.

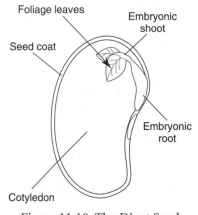

Figure 11.10 The Dicot Seed

DORMANCY AND GERMINATION

Seeds have the ability to remain dormant for hundreds of years because the seed coat keeps water and oxygen away from the embryo. In addition, the seed releases chemicals that directly inhibit germination. Changes in the environment cause a seed to break dormancy. If the seed coat is damaged as it passes through an animal's digestive system or if heavy rains wash away the chemical inhibitors, the seed will germinate.

The first step in germination is the absorption of water, a process called **imbibition**. The force exerted by imbibing seeds will greatly expand the seed and cause the seed coat to crack. As the seed takes up water, it undergoes metabolic changes as enzymes are activated. RNA also begins to direct the synthesis of proteins. As the seed begins to germinate, it uses carbohydrates, amino acids, and lipids stored in the cotyledon or endosperm. Germination is complete when the radicle (embryonic root) emerges from the seed coat. The plant is then a seedling.

ALTERNATION OF GENERATIONS

The sexual life cycle of plants is characterized by an **alternation of generations** in which **haploid (n)** and **diploid ($2n$)** generations alternate with each other. The **gametophyte (n)** produces gametes by mitosis. These gametes fuse during fertilization to yield $2n$ zygotes. Each zygote develops into a **sporophyte ($2n$)**, which produces haploid spores (n) by meiosis. Each haploid spore forms a new gametophyte, completing the life cycle; see Figure 11.11.

Dormancy and germination are examples of how the environment affects the expression of genes.

STUDY TIP

Gametophyte = n
Sporophyte = $2n$

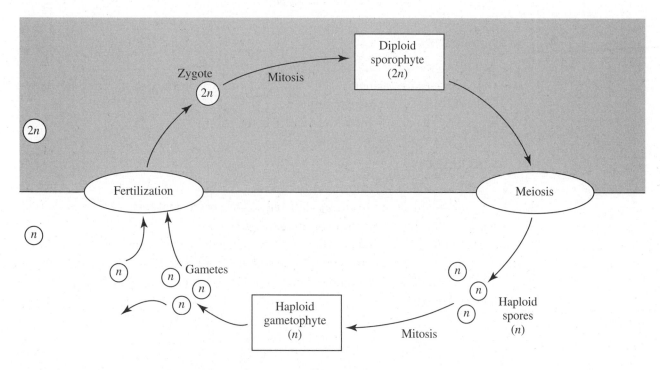

Figure 11.11 Alternation of Generations

PLANT RESPONSES TO STIMULI
Hormones

Plant hormones help coordinate growth, development, and responses to environmental stimuli. They are produced in very small quantities, but they have a profound effect on the plant because the hormone signal is amplified. **Signal transduction pathways** amplify the hormonal signal and connect it to specific cell responses. A plant's response to a hormone usually depends not so much on absolute quantities of hormones but on relative amounts. Hormones can have multiple effects on a plant, and they can work synergistically with other hormones or in opposition to them. Here is an overview of plant hormones and what they stimulate.

1. AUXINS

- Responsible for **phototropisms** due to *unequal distribution of auxin.*
- Enhance **apical dominance**, the preferential growth of a plant upward (toward the sun), rather than laterally. (The terminal bud actually suppresses lateral growth by suppressing development of axial buds.)
- Stimulate stem elongation and growth by softening the cell wall.
- **Indoleacetic acid** (**IAA**) is a naturally occurring auxin.
- A human-made auxin, 2,4-D, is used as a weed killer.
- When used as rooting powder, it causes roots to develop quickly in a plant cutting.
- A synthetic auxin sprayed onto tomato plants will induce fruit production with no pollination. This results in seedless tomatoes.

2. CYTOKININS

- *Stimulate cytokinesis and cell division.*
- Delay **senescence** (aging) by inhibiting protein breakdown. Florists spray cut flowers with cytokinins to keep them fresh.
- Are produced in roots and travel upward in the plant.

3. GIBBERELLINS

- *Promote stem and leaf elongation.*
- 100 different naturally occurring gibberellins have been identified.
- Work in concert with auxins to promote cell growth.
- Induce **bolting**, the rapid growth of a floral stalk. When a plant such as broccoli, which normally grows close to the ground, enters the reproductive stage, it sends up a very tall shoot on which the flower and fruit develop. This is a mechanism to ensure pollination and seed dispersal.

4. ABSCISIC ACID (ABA)

- *Inhibits growth.*
- Enables plants to withstand drought.
- Closes stomates during times of water stress.
- Counteracts the breaking of dormancy during a winter thaw.
- *Promotes seed dormancy.* This prevents seeds that have fallen on the ground in the fall from sprouting until the spring when environmental conditions are better. In some desert plants, seeds break dormancy only after heavy rains have washed all ABA out of the cells. (The ground is moist and can support plant growth.)

5. ETHYLENE

- This plant hormone is a gas.
- *Promotes fruit ripening.* Ethylene gas promotes ripening, which triggers increased production of ethylene gas. "One bad apple spoils the whole barrel."
- Commercial fruit sellers pick perishable fruit before they are ripe, while still hard. When they arrive at their destination, they are sprayed with ethylene gas to hasten the ripening. In contrast, apples are kept in an environment of CO_2 to eliminate exposure to ethylene gas and, thus, keep the apples from ripening or rotting. In this way, apples can be stored for long periods of time.
- Is produced in large quantities in times of stress such as drought, flooding, mechanical pressure, injury, and infection.
- *Facilitates **apoptosis**:* programmed cell death. Prior to death, cells break down many of their chemical components for the plant to salvage and reuse.
- *Promotes **leaf abscission**—*the leaf dies and falls from the plant. Subsequently, a scar forms at the abscission layer to prevent pathogens from entering the plant.

REMEMBER

Ethylene promoting fruit ripening is an example of *positive feedback*.

Tropisms

A tropism is the growth of a plant toward or away from a stimulus. Examples are **thigmotropisms** (touch), **geotropisms** or **gravitropisms** (gravity), and **phototropisms** (light). A growth of a plant toward a stimulus is known as a **positive tropism**, while a growth away from a stimulus is a **negative tropism**.

Phototropisms result from an *unequal distribution* of **auxins**, which accumulate on the side of the plant away from the light. Since auxins cause growth, the cells on the shady side of the plant enlarge and the stem bends toward the light.

Geotropisms result from an interaction of auxins and **statoliths**, specialized plastids containing dense starch grains.

Photoperiodism

The environmental stimulus a plant uses to detect the time of year is the **photoperiod**, the relative lengths of day and night. Plants have a biological clock set to a 24-hour day, known as a **circadian rhythm**. The physiological response to the photoperiod, such as flowering, is known as **photoperiodism**. Some plants will flower only when the light period is longer than a certain number of hours. These plants are called **long-day plants**. Some plants are **short-day plants**, and some are **day-neutral** and will flower regardless of the length of day. Plants actually respond to the length of darkness, not the length of light. So despite its name a *long-day plant* is actually a *short-night plant*.

The photoreceptor responsible for keeping track of the length of day and night is the pigment **phytochrome**. There are two forms of phytochrome, **Pr (red-light absorbing)** and **Pfr (infrared light absorbing)**. Phytochrome is synthesized in the Pr form. When the plant is exposed to light, Pr converts to Pfr. In the dark, Pfr reverts back to Pr. The conversion from one to the other enables the plant to keep track of time. The plant is able to sense the concentrations of the two phytochromes and respond accordingly.

(daylight)
Red light

Pr $\rightleftarrows$ Pfr (triggers germination)

IR light
(Slow conversion at night)

1. Which is true of mosses?

    (A) They have true xylem but lack phloem.
    (B) They are more advanced than the ferns.
    (C) Their cell walls consist of chitin instead of cellulose.
    (D) They lack vascular tissue.

2. The ancestors of land plants were most likely similar to modern

    (A) conifers
    (B) ferns
    (C) green algae
    (D) flowering plants

3. Which is true of seeds?

    (A) They contain the cotyledon.
    (B) They are characteristic of all plants.
    (C) They are a mechanism for dispersal of pollen.
    (D) They are not characteristic of the conifers.

4. Vascular plant tissue includes all of the following EXCEPT

    (A) meristem
    (B) sieve tube cells
    (C) vessels
    (D) tracheids

5. Which is CORRECT about roots?

    (A) Adventitious roots grow below water in certain saltwater plants.
    (B) Only the roots of monocots have a stele.
    (C) The function of the endoderm is to control what water and nutrients enter the body of the plant.
    (D) The Casparian strip is a modification of root hairs in monocots.

6. Which tissue makes up most of the wood of a tree?

    (A) primary phloem
    (B) secondary phloem
    (C) primary xylem
    (D) secondary xylem

7. Which is CORRECT about monocots?

    (A) Vascular bundles in the stem are in a ring.
    (B) Their floral parts are usually in 3 s.
    (C) They usually have taproots.
    (D) The veins in the leaves are netlike.

8. Cortex, mesophyll, epidermal cells, and pith all consist of these cells.

   (A) parenchyma
   (B) collenchyma
   (C) sclerenchyma
   (D) All of the above

9. All of the following evolved in plants in response to water shortage EXCEPT

   (A) large surface area of leaves to absorb sunlight
   (B) C-4 plants
   (C) Kranz anatomy
   (D) cutin

10. You cut your initials in a tree 5 feet above the ground this year. The tree grows 2 feet per year and you return in 8 years. How high will your initials be at that time?

   (A) 5 feet
   (B) 10 feet
   (C) 16 feet
   (D) 21 feet

11. To observe the process of mitosis in plant roots, a student should examine the plant's

   (A) root cap
   (B) zone of maturation
   (C) meristem tissue
   (D) pericycle

12. Guttation in a plant results directly from

   (A) transpirational pull
   (B) injury to the plant
   (C) condensation of water vapor onto the leaf
   (D) root pressure

13. The primary function of the cortex in a plant stem is

   (A) storage
   (B) transport
   (C) photosynthesis
   (D) absorption

14. Phototropisms are controlled by

   (A) excess levels of ethylene gas
   (B) an unequal distribution of auxins
   (C) circadian rhythm
   (D) degradation of phytochromes

15. Which is correct about plants?

    (A) The epicotyl becomes the upper stem.
    (B) The hypocotyl becomes the endosperm.
    (C) Monocots usually have food reserves in the cotyledon.
    (D) Pollen contains the sporophyte.

16. Which of the following occurs after fertilization?

    (A) The ovule becomes the seed, the ovary becomes the fruit.
    (B) The ovary becomes the seed, the ovule becomes the fruit.
    (C) The micropyle becomes the seed, the sepals become the fruit.
    (D) The stigma becomes the seed, the ovule becomes the fruit.

17. When you pinch off the terminal buds from a young plant to make it grow bushy, which of the following hormones is responsible?

    (A) cytokinins
    (B) auxins
    (C) gibberellins
    (D) abscisic acid

18. "One rotten apple spoils the whole bunch" is the work of which of the following?

    (A) cytokinin
    (B) auxin
    (C) gibberellins
    (D) ethylene

Answers to Multiple-Choice Questions

1. **(D)** Mosses are primitive, nonvascular plants, classified as bryophytes. They show an alternation of generations and produce gametes by meiosis and mitosis. The cell walls of fungi consist of chitin.

2. **(C)** The ancestor of the modern multicellular plant is the green algae, Chlorophyta.

3. **(A)** Seeds contain the embryo and the food for the embryo, the endosperm (in monocots) or the cotyledon (in dicots). All plants do not produce seeds; for example, bryophytes and the ferns do not. Conifers do produce seeds. They are located on cones.

4. **(A)** Meristem tissue is actively dividing tissue that gives rise to other tissue, such as xylem and phloem. Xylem consists of vessel and tracheids. Phloem consists of companion cells and sieve tube elements.

5. **(C)** The function of the endoderm is to control what enters the body of the plant. Endoderm cells are wrapped with a waterproof Casparian strip. Adventitious roots are roots that grow above ground. The roots of both monocots and dicots have a stele, a vascular cylinder. The Casparian strip surrounds the endodermis cells and helps control what enters the stele.

6. **(D)** Wood consists of secondary xylem; the primary xylem was formed first and is located in a small area at the center of a tree.

7. **(B)** Monocots have floral parts oriented in 3s and have parallel veins in the leaves. Dicots usually have taproots, and the veins in their leaves are netlike. Most trees are dicots. Palm trees are monocots.

8. **(A)** Parenchyma is the most common ground tissue in a plant. Parenchymal cells have primary cell walls that are thin and flexible and look like a traditional plant cell.

9. **(A)** All choices describe responses to water shortage except choice A. A large surface area of leaf to maximize the amount of sunlight absorbed would also increase water loss. Leaves in a tropical rain forest, where the environment is very humid, are typically broad. The needles on a conifer evolved to minimize water loss. Conifers are found in cold, dry northern regions.

10. **(A)** A tree grows upward from the apical meristem at the top of the tree, not from the bottom of the tree. Therefore, the initials do not move from their original spot.

11. **(C)** Meristem tissue is the only tissue of the choices given that is actively dividing tissue. Meristem tissue is growth tissue; it gives rise to other cell types. For example, meristem gives rise to xylem and phloem cells.

12. **(D)** Droplets of water that appear in the morning on the leaf tips of some herbaceous leaves are due to root pressure. This phenomenon is known as guttation.

13. **(A)** The cortex consists of parenchymal cells and stores starch, among other things.

14. **(B)** A phototropism is a plant growth toward or away from a stimulus and is the result of unequal distribution of auxins.

15. **(A)** The epicotyl is part of the embryo and becomes the upper part of the stem and the leaves; the hypocotyl becomes the lower part of the stem and the roots. Monocots store food reserves in the endosperm, not the cotyledon. Pollen consists of three haploid sperm nuclei; it is the gametophyte, not the sporophyte. The tube nucleus is one of the sperm nuclei in pollen. It grows the pollen tube.

16. **(A)** After fertilization, the ovule becomes the seed and the ovary becomes the fruit. The other choices make no sense.

17. **(B)** The growing tip of a young plant produces auxins and enhances apical dominance. Removal of the growing tip, known as pinching back the plant, removes the auxins, and the plant grows laterally, that is, bushier.

18. **(D)** Ethylene gas is given off by plants while they are ripening. To ripen hard fruit quickly, put them into a paper bag with a ripe banana, which gives off large amounts of ethylene gas.

Directions: Answer all questions. You must answer the question in essay—not outline—form. You may use labeled diagrams to supplement your essay, but diagrams alone are *not* sufficient. Before you start to write, read each question carefully so that you understand what the question is asking.

1. Discuss the movement of water from the roots to the leaves, naming all the cells and structures water passes through along the way.

2. Discuss the structural and functional strategies that enabled plants to move to land.

Typical Free-Response Answers

Note: The key words are in boldface. Remember, you get credit only when you state a correct fact and use the correct scientific term.

1. Water diffuses into a plant through **root hairs**, which are cytoplasmic extensions of epidermal cells. From the root hairs, water travels into the **parenchymal cells** of the **cortex** in two ways, along the **symplast** and the **apoplast**. The symplast is a continuous system of cytoplasm of cells interconnected by **plasmodesmata**. The apoplast is the network of cell walls and intercellular spaces within the plant body that permits extensive extracellular movement of water.

From the cortex, water moves to the **endodermis**, a tightly packed layer of cells that surrounds the **vascular cylinder** or **stele**. Each endoderm cell is encircled by a band of wax, the **Casparian strip**, which is not permeable to water. It controls water entering the plant by the apoplast route, which must diffuse into the endoderm cells before continuing. Once water has entered the endodermis, it freely passes into the vascular cylinder and into the xylem, which consists of **tracheids** and **vessel elements**. Once in the xylem, water moves upward toward the leaves by a combination of **transpirational pull** and **cohesion tension**. Water diffuses from the roots, where the **water potential** is highest, to the air spaces in the leaves, where the water potential is the lowest. From the veins in the leaf, water diffuses into **air spaces** within the **spongy mesophyll** and then into the **palisade** and **spongy mesophyll cells** as needed for photosynthesis. In the light reactions of photosynthesis, water molecules are broken down during photolysis, providing electrons for the light reactions and protons for the dark reactions. The **oxygen** from the water molecules is given off into the atmosphere as a waste product.

> *Note: This question demonstrates a common type of essay question on the AP Exam because it checks for broader understanding by integrating several topics. Key words are boldface to keep you focused on the need for scientific terms. If you need help, review sections in this book on plants, cells, and photosynthesis.*

2. Plants began life in the seas and moved to land as **competition** for resources increased. Some of the problems a plant living on land faces are **supporting the plant body**, **absorbing** and **conserving water**, and **reproducing outside of a watery environment**.

Strong stems (trunks) and branches hold leaves up toward the sunlight. Three specialized cells help support plants: parenchyma, collenchyma, and sclerenchyma. Although they lack a secondary cell wall, **parenchymal cells** lend support to a plant when the cell is **turgid**. **Collenchymal cells** have thick primary cells walls and provide flexible support without restraining growth. **Sclerenchymal cells** are very specialized for support. They have rigid and thick walls fortified with lignin. Even after these cells die, their rigid walls provide a skeleton for the plant.

Absorbing and conserving water are two other problems for plants living on land. **Roots**, in addition to anchoring the plant in the soil, absorb nutrients. **Root hairs** are slender cytoplasmic extensions from epidermal cells of the root that greatly increase the absorptive surface area. **Mycorrhizae** are symbiotic fungi that live on mature older regions of roots that lack root hairs. This **symbiont** increases the quantity of water and nutrients plants can absorb. The opening and closing of **stomates** in a leaf limit the loss of water from a plant by transpiration and are controlled by guard cells. A **waxy cuticle** made of **cutin** covers leaves, further minimizing water loss. Some plants have stomates nestled in **stomatal crypts** that minimize exposure of the stomate to air and further minimize transpiration.

Modifications in anatomy and physiology enable plants to be successful on dry land. **C-4 plants** have evolved **Kranz anatomy**, wherein bundle sheath cells sequester CO_2 deep within the leaf and away from stomates. The **Hatch-Slack pathway**, a biochemical pathway, also removes CO_2 from the air spaces and away from the stomates. Both strategies evolved in plants in dry environments to keep their stomates closed as much as possible, thus minimizing excessive water loss through transpiration.

Reproduction on land also offers great challenges for land organisms. In some plants, gametes and zygotes form within a protective jacket of cells called **gametangia** that prevent drying out. **Sporopollenin** is a tough polymer, found in the walls of spores and pollen, that resists harsh terrestrial environments. **Seeds** also have a very tough protective coating to prevent drying out. Seeds have been known to sprout after being dormant for 1,000 years.

Human Physiology 12

> **Much of this chapter contains background information, but focus on:**
>
> **Chemical Signals**
>
> **Nervous System**
>
> **Human Immune System (Chapter 13)**

INTRODUCTION

For the most part, this chapter focuses on human physiology. However, some review of other animals is included to help you gain a broader understanding and insight into adaptations common to all animals. The human immune system and human reproduction and development are presented as separate chapters with their own practice questions.

DIGESTION IN DIFFERENT ANIMALS

Hydra

In the **hydra** (cnidarians), digestion occurs in the **gastrovascular cavity**, which has only one opening. Cells of the **gastrodermis** (lining of the gastrovascular cavity) secrete digestive enzymes into the cavity for **extracellular digestion**. Some specialized nutritive cells have flagella that move the food around the gastrovascular cavity, and some have pseudopods that engulf food particles.

Earthworm

The digestive tract of the **earthworm** is a long, straight tube. As the earthworm burrows in the ground, creating tunnels that aerate the soil, the mouth ingests decaying organic matter along with soil. From the mouth, food moves to the esophagus and then to the **crop** where it is stored. Posterior to the crop, the **gizzard**, which consists of thick, muscular walls, grinds up the food with the help of sand and soil that were ingested along with the organic matter. The rest of the digestive tract consists of the intestines where chemical digestion and absorption occur. Absorption is enhanced by the presence of a large fold in the upper surface of the intestine, called the **typhlosole**, which greatly increases the surface area.

Grasshopper

Like the earthworm, the **grasshopper** has a digestive tract that consists of a long tube consisting of a **crop** and **gizzard**. However, there are several differences. The grasshopper has specialized mouth parts for tasting, biting, and crushing food and has a gizzard that contains **plates** made of **chitin** that help in grinding the food. In addition, in the grasshopper, the digestive tract is also responsible for removing nitrogenous waste (**uric acid**) from the animal.

DIGESTION IN HUMANS

The human digestive system has two important functions: **digestion**—breaking down large food molecules into smaller usable molecules and **absorption**—the diffusion of these smaller molecules in the body's cells. **Fats** get broken down into glycerol and fatty acids, **starch** into monosaccharides, **nucleic acids** into nucleotides, and **proteins** into amino acids. **Vitamins** and **minerals** are small enough to be absorbed without being digested. The digestive tract is about 30 feet (9 m) long and made of **smooth (involuntary) muscle** that pushes the food along the digestive tract by a process called **peristalsis**.

Mouth

In the mouth, the **tongue** and differently shaped **teeth** work together to break down food mechanically. Form relates to function, and the type of teeth a mammal has reflects its dietary habits. Humans are **omnivores** and have three different types of teeth: **incisors** for cutting, **canines** for tearing, and **molars** for grinding. **Salivary amylase** released by **salivary glands** begins the chemical breakdown of **starch**.

Esophagus

After swallowing, food is directed into the esophagus, and not the windpipe, by the **epiglottis**, a flap of cartilage in the back of the **pharynx** (throat). No digestion occurs in the esophagus.

Stomach

The **stomach** churns food mechanically and secretes **gastric juice** that begins the digestion of **proteins**. The lining of the stomach contains **gastric pits** that are themselves lined with three types of cells. **Chief cells** secret pepsinogen, the inactive form of **pepsin** that becomes activated by acid. **Parietal cells** secrete the hydrochloric acid that keeps the pH of gastric juices at 2–3 and activates pepsinogen. HCl also kills ingested microorganisms and breaks down protein. A third type of cell secretes mucus that protects the stomach lining from these two cell-digesting substances.

The stomach of all mammals also contains **rennin** to aid in the digestion of the protein in milk. The **lower esophageal sphincter** at the top of the stomach keeps food in the stomach from backing up into the esophagus and burning it. The **pyloric sphincter** at the bottom of the stomach keeps the food in the stomach long enough to be digested.

Excessive acid can cause an **ulcer** to form in the esophagus, the stomach, or the **duodenum** (the first 12 inches [30 cm] of the small intestine). However, scientists now know that a common cause of ulcers is a particular bacterium, *Helicobacter pylori*, which can be effectively treated with antibiotics.

Small Intestine

Digestion is completed in the **duodenum**. Intestinal enzymes and pancreatic amylases hydrolyze starch and glycogen into maltose. **Bile**, which is produced in the **liver** and stored in the **gallbladder**, is released into the small intestine as needed and acts as an **emulsifier** to break down fats, creating greater surface area for digestive enzymes. **Peptidases**, such as trypsin and chymotrypsin, continue to break down proteins. Nucleic acids are hydrolyzed by **nucleases**, and **lipases** break down fats. Once digestion is complete, the lower part of the small intestine is the site of **absorption**. Millions of fingerlike projections called **villi** absorb all the nutrients that were previously released from digested food. Each villus contains capillaries, which absorb amino acids, vitamins and monosaccharides, and a **lacteal**, a small vessel of the **lymphatic system**, which absorbs fatty acids and glycerol. Each epithelial cell of the villus has many microscopic cytoplasmic appendages called **microvilli** that greatly increase the rate of nutrient absorption by the villi. Figure 12.1 shows a villus with a lacteal, capillaries, and microvilli.

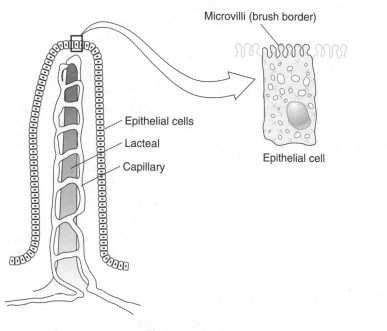

Figure 12.1

Large Intestine

The **large intestine** or **colon** serves three main functions: **egestion**, the removal of undigested waste; **vitamin production**, from bacteria symbionts living in the colon; and the **removal of excess water**. Together, the small intestine and colon reabsorb 90 percent of the water that entered the alimentary canal. If too much water is removed from the intestine, **constipation** results; if inadequate water is removed, **diarrhea** results. The last 7–8 inches (18–20 cm) of the gastrointestinal tract is called the **rectum**. It stores **feces** until their release. The opening at the end of the digestive tract is called the **anus**.

Hormones that Regulate the Digestive System

Hormones are released as needed as a person sees or smells food or as food moves along the gut. Table 12.1 summarizes the hormones involved in regulating digestion.

Table 12.1 Hormones That Regulate Digestion		
Hormone	**Site of Production**	**Effect**
Gastrin	Stomach wall	Stimulates sustained secretion of gastric juice
Secretin	Duodenum wall	Stimulates pancreas to release bicarbonate to neutralize acid in duodenum
Cholecystokinin (CCK)	Duodenum wall	Stimulates pancreas to release pancreatic enzymes and gall bladder to release bile into small intestine

GAS EXCHANGE IN DIFFERENT ANIMALS

Respiration is the exchange of respiratory gases, oxygen and carbon dioxide, between the external environment and the cell or body. It occurs passively by **diffusion**. Therefore, respiratory surfaces must be **thin**, be **moist**, and have **large surface areas**. Although all organisms must exchange respiratory gases, they have evolved different strategies to accomplish it.

- In simple animals, like **sponges** and **hydra**, gas exchange occurs over the entire surface of the organism wherever cells are in direct contact with the environment.
- **Earthworms** and **flatworms** have an **external respiratory surface** because diffusion of O_2 and CO_2 occurs at the skin. Oxygen is carried by **hemoglobin** dissolved in blood.
- The **grasshopper** and other **arthropods** and **crustaceans** have an **internal respiratory surface**. Air enters the body through **spiracles** and travels through a system of **tracheal tubes** into the body, where diffusion occurs in sinuses or hemocoels. In **arthropods** and in some **mollusks**, oxygen is carried by **hemocyanin**, a molecule similar to hemoglobin but with **copper**, instead of iron, as its core atom.
- Aquatic animals like **fish** have **gills** that take advantage of **countercurrent exchange** to maximize the diffusion of respiratory gases.

GAS EXCHANGE IN HUMANS

In **humans**, air enters the nasal cavity and is **moistened**, **warmed**, and **filtered**. From there, air passes through the **larynx** and down the **trachea** and **bronchi** into the tiniest **bronchioles**, which end in microscopic air sacs called **alveoli** where diffusion of respiratory gases occurs; see Figure 12.2. Humans have an **internal respiratory surface**. As the rib cage expands and the **diaphragm** contracts and lowers, the chest cavity expands, making the internal pressure lower than atmospheric pressure. Thus, air is drawn into the lungs by **negative pressure**.

The **medulla** in the brain, which contains the breathing control center, sets the rhythm of breathing and monitors CO_2 **levels** in the blood by sensing changes in pH of the blood. CO_2, the by-product of cell respiration, dissolves in blood to form **carbonic acid**. Therefore, the higher the CO_2 concentration in the blood, the lower the pH. Blood pH lower than 7.4 causes the medulla to increase the rate of breathing to rid the body of more CO_2.

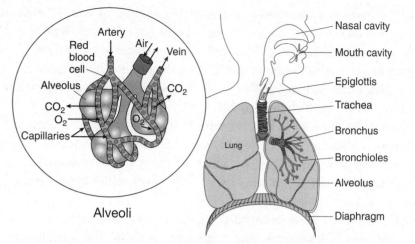

Figure 12.2 Adaptations for Human Respiration

The body is also sensitive to low O_2 levels but to a much lesser degree. O_2 sensors—chemoreceptors—are located in the nodes of neural tissue in the aorta and carotid arteries leaving the heart. If O_2 levels drop drastically, these chemoreceptors become activated and send nerve impulses to the medulla.

Hemoglobin

Oxygen is carried in the human blood by the respiratory pigment **hemoglobin**, which can combine loosely with four oxygen molecules, forming the molecule **oxyhemoglobin**. To function in the transport of oxygen, hemoglobin must be able to bind with oxygen in the lungs and unload it at the body cells. The more tightly the hemoglobin binds to oxygen in the lungs, the more difficult it is to unload the cells. Hemoglobin is an **allosteric** molecule and exhibits **cooperativity**. This means that once it binds to one oxygen molecule, hemoglobin undergoes a **shape change** and binds more easily to the remaining three oxygen molecules. In addition, hemoglobin's conformation is sensitive to pH. A drop in pH lowers the affinity of hemoglobin for oxygen (known as the Bohr shift). Because CO_2 dissolves in water to form carbonic acid, actively respiring tissue which releases large quantities of CO_2, will lower the pH of its surroundings and induce hemoglobin to release its oxygen at the cells where needed.

Figure 12.3 contains four graphs showing saturation-dissociation curves for hemoglobin (Hb). *The further to the right the curve is, the less affinity the hemoglobin has for oxygen.*

- Graph A: Here is a dissociation curve of adult hemoglobin at normal and low blood pH showing the Bohr shift. At a lower pH, the hemoglobin has less affinity for oxygen.

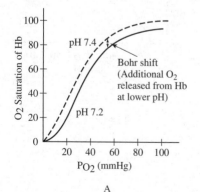

A

- Graph B: Here is a dissociation curve for hemoglobin of two different mammals, a mouse and an elephant. The mouse has a much higher metabolism, and its body cells have a correspondingly higher oxygen requirement. To accommodate the animal's oxygen needs, the mouse hemoglobin has a dissociation curve located to the right of the elephant hemoglobin. In other words, mouse hemoglobin drops off O_2 at cells more easily than does elephant hemoglobin.
- Graph C: Here is a dissociation curve for fetal and maternal hemoglobin. Fetal hemoglobin has a higher affinity for oxygen than adult hemoglobin so it can take oxygen from the maternal hemoglobin. Also note that the curve of fetal hemoglobin does not have the S-shape common to the other curves. It bonds to each oxygen atom with the same ease; there is no cooperativity.
- Graph D: Here is a dissociation curve for mammals evolved at sea level and mammals evolved at very high altitudes. Since less oxygen is available at high altitudes, mammals that evolved there must have hemoglobin with a greater affinity for oxygen.

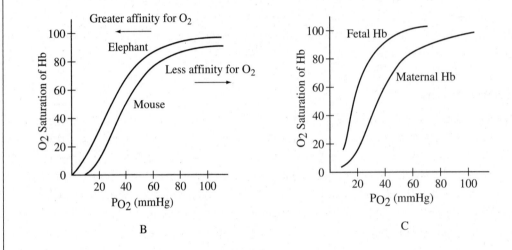

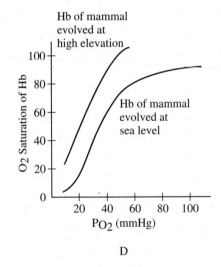

Figure 12.3

Transport of Carbon Dioxide

CO$_2$ is carried in the plasma. O$_2$ is carried by red blood cells

Very little carbon dioxide is transported by hemoglobin. Most carbon dioxide is carried in the **plasma** as part of the reversible blood buffering **carbonic acid–bicarbonate ion system**, which maintains the blood at a constant pH 7.4. The bicarbonate ion is produced in a two-stage reaction. First, carbon dioxide combines with water to form carbonic acid. This reaction is catalyzed by carbonic acid anhydrase found in red blood cells. Then carbonic acid dissociates into a bicarbonate ion and a proton. The protons can be given up into the plasma, which lowers the blood pH, or taken up by the **bicarbonate ion**, which raises the blood pH. Here is the equation:

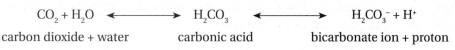

$$CO_2 + H_2O \longleftrightarrow H_2CO_3 \longleftrightarrow H_2CO_3^- + H^+$$

carbon dioxide + water   carbonic acid   bicarbonate ion + proton

CIRCULATION IN DIFFERENT ANIMALS

Primitive animals like the sponge and the hydra have no circulatory systems. All their cells are in direct contact with the environment, and such a system is unnecessary. The earthworm has a closed circulatory system where blood is pumped by the heart through arteries, veins, and capillaries. Oxygen is carried by hemoglobin that is dissolved in the blood. The grasshopper, as a representative animal of the arthropods, has an open **circulatory system**. After blood is pumped by the heart into an artery, it leaves the vessel and seeps through spaces called sinuses or hemocoels as it feeds body cells. The blood then moves back into a vein and circulates back to the heart. This system lacks capillaries. Arthropod blood is colorless and does not carry oxygen.

HUMAN CIRCULATION

Human circulation consists of a **closed circulatory system** with arteries, veins, and capillaries. Table 12.2 shows the components of blood.

Table 12.2

Components of Blood		
Component	**Scientific Name**	**Properties**
Plasma	— —	Liquid portion of the blood
		Contains clotting factors, hormones, antibodies, dissolved gases, nutrients, and wastes
		Maintains proper osmotic potential of blood, 300 mosm/L
Red Blood Cells	Erythrocytes	Carry hemoglobin and oxygen
		Do not have a nucleus and live only about 120 days
		Formed in the bone marrow and recycled in the liver
White Blood Cells	Leukocytes	Fight infection and are formed in the bone marrow
		Die fighting infection and are one component of pus
		One type of leukocyte—the B lymphocyte—produces antibodies
Platelets	Thrombocytes	These are not cells but cell fragments that are formed in the bone marrow from megakaryocytes
		Clot blood

Red blood cells, white blood cells, and platelets all develop in bone marrow from multipotent **stem cells**. These stem cells keep dividing and constantly replenish the population of blood cells throughout a person's life.

The Mechanism of Blood Clotting

Blood clotting is a complex mechanism that begins with the release of **clotting factors** from platelets and damaged tissue. It involves a complex set of reactions, including the activation of inactive **plasma proteins**. **Anticlotting factors** normally circulate in the plasma to prevent the formation of a clot or **thrombus**, which can cause serious damage in the absence of injury.

Here is the pathway of normal clot formation.

Damaged Tissue and Platelets
↓
Thromboplastin + Ca^{2+}
↓
Prothrombin $\longrightarrow$ Thrombin
(inactive)   (active)
↓
Fibrinogen $\longrightarrow$ Fibrin (Clot)
(inactive)   (active)

Structure and Function of Blood Vessels

Table 12.3 describes the various blood vessels.

Table 12.3

Blood Vessels		
Vessel	**Function**	**Structure**
Artery and Arteriole	Carry blood away from the heart under enormous pressures	Walls made of thick, elastic, smooth muscle
Vein and Venule	Carry blood back to the heart under very little pressure	Thin walls have valves to help prevent back flow; veins are located within skeletal muscle, which propels blood upward and back to heart as the body moves
Capillary	Allows for diffusion of nutrients and wastes between cells and blood	Walls are one-cell thick and so small that blood cells travel in single file

The Heart

The heart is located beneath the sternum and is about the size of a clenched fist. It beats about **70 beats per minute** and pumps about 5 quarts (5 L) of blood per minute, or the total volume of blood in the body each minute. Two **atria** receive blood from the body cells, and two **ventricles** pump blood out of the heart. Individual **cardiac muscle cells** have the ability to contract even when removed from the heart. The heart itself has its own innate **pacemaker**, the **sinoatrial (SA) node**, which sets the timing of the contractions of the heart.

Located in the wall of the right atrium, it generates and sends electrical signals to the atrio-ventricular (AV) node. The action potential of pacemaker cells is generated by voltage-gated Ca^{2+} channels. From the pacemaker, impulses are sent to the bundle of His and Purkinje fibers, which trigger the ventricles to contract. Electrical impulses travel through the cardiac and body tissues to the skin, where they can be detected by an **electrocardiogram** (**EKG**). The heart's pacemaker is influenced by a variety of factors: two sets of nerves that cause it to speed up or slow down, hormones such as adrenalin, and body temperature. **Blood pressure** is lowest in the veins and highest in the arteries when the ventricles contract. Blood pressure for all normal, resting adults is **120/80**. The **systolic number** (**120**) is a measurement of the pressure when the ventricles contract, while the **diastolic number** (**80**) is a measure of the pressure when the heart relaxes; see Figure 12.4.

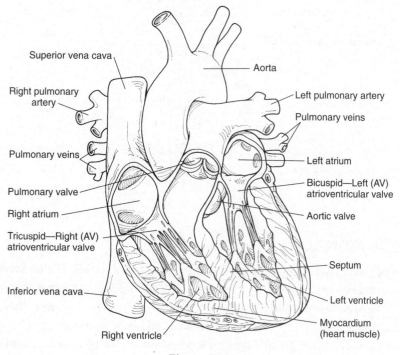

Figure 12.4

Pathway of Blood

Blood enters the heart through the vena cava. From there it continues to the

Right atrium
Right atrioventricular (AV) valve—tricuspid valve
Right ventricle
Pulmonary semilunar valve
Pulmonary artery
Lungs
Pulmonary vein
Left atrium
Bicuspid (left AV) valve
Left ventricle
Aortic semilunar valve
Aorta
To all the cells in the body

STUDY TIP

Learn the pathway of the blood in the body.

Blood circulates through the **coronary circulation** (heart), **renal circulation** (kidneys), and **hepatic circulation** (liver). The **pulmonary circulation** includes the pulmonary artery, lungs, and pulmonary vein. Figure 12.5 is a drawing of the human circulatory system.

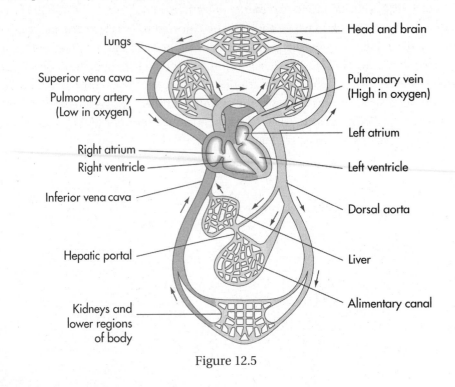

Figure 12.5

CHEMICAL SIGNALS

Animals have two major regulatory systems that release chemicals: the **endocrine system** and the **nervous system**. The endocrine system secretes **hormones**, while the nervous system secretes **neurotransmitters**. Even though the two systems are separate, there is overlap between them, and together they work to regulate the body. **Epinephrine** (adrenaline), for example, functions as the **fight-or-flight** hormone secreted by the adrenal gland as well as a neurotransmitter that sends a message from one neuron to another.

Hormones

Hormones are produced in **ductless (endocrine) glands** and move through the blood to a specific **target** cell, tissue, or organ that can be far from the original endocrine gland. They can produce an immediate short-lived response, the way **adrenaline** (epinephrine) speeds up the heart rate, and increases blood sugar. They can dramatically alter the development of an entire organism, the way **ecdysone** controls **metamorphosis** in insects. **Tropic hormones** have a far-reaching effect because they stimulate other glands to release hormones. For example, the anterior pituitary releases TSH (thyroid-stimulating hormone), which stimulates the thyroid to release thyroxin. Other types of chemical messengers reach their target by special means. **Pheromones** in the urine of a dog carry a message between different individuals of the same species. In vertebrates, **nitric oxide** (NO), a gas, is produced by one cell and diffuses to and affects only neighboring cells before it is broken down. Table 12.4 is an overview of the hormones of the endocrine system.

Table 12.4

Overview of Hormones		
Gland	**Hormone**	**Effect**
Anterior pituitary	Growth hormone (GH)	Stimulates growth of bones
	Luteinizing hormone (LH)	Stimulates ovaries and testes
	Thyroid-stimulating hormone (TSH)	Stimulates thyroid gland
	Adrenocorticotropic (ACTH) hormone	Stimulates adrenal cortex to secrete glucocorticoids
	Follicle-stimulating hormone (FSH)	Stimulates gonads to produce sperm and ova
	Prolactin	Stimulates mammary glands to produce milk
Posterior pituitary	Oxytocin	Stimulates contractions of uterus and milk production by mammary glands
	Antidiuretic hormone (ADH)	Promotes retention of water by kidneys
Thyroid	Thyroxin (T_3 and T_4)	Controls metabolic rate
	Calcitonin	Lowers blood calcium levels
Parathyroid	Parathormone	Raises blood calcium levels
Adrenal cortex	Glucocorticoids	Raises blood sugar levels
Adrenal medulla	Epinephrine (adrenaline) Norepinephrine (noradrenaline)	Raises blood sugar level by increasing rate of glycogen breakdown by liver
Pancreas—islets of Langerhans	Insulin—secreted by β cells	Lowers blood glucose levels
	Glucagon—secreted by α cells	Raises blood glucose levels by causing breakdown of glycogen into glucose
Thymus	Thymosin	Stimulates T lymphocytes
Pineal	Melatonin	Involved in biorhythms
Ovaries	Estrogen	Stimulates uterine lining, promotes development and maintenance of primary and secondary characteristics of female
	Progesterone	Promotes uterine lining growth
Testes	Androgens	Support sperm production and promote secondary sex characteristics

The Hypothalamus

The **hypothalamus** plays a special role in the body; it is the *bridge between the endocrine and nervous systems*. The hypothalamus acts as part of the nervous system when, in times of stress, it sends electrical signals to the adrenal gland to release adrenaline. It acts like a nerve when it secretes **gonadotropic-releasing hormone (GnRH)** from neurosecretory cells that stimulate the anterior pituitary to secrete **FSH** and **LH**. It acts as an endocrine gland when it produces **oxytocin** and **antidiuretic hormone** that it stores in the posterior pituitary. The hypothalamus also contains the body's thermostat and centers for regulating hunger and thirst.

The hypothalamus receives information from nerves throughout the body, including the brain. In response, it initiates *endocrine* signaling appropriate to the environmental conditions. For example in many vertebrates, nerve signals from the brain pass sensory information to the hypothalamus about seasonal changes. The hypothalamus, in turn, regulates the release of reproductive hormones necessary during the breeding season. Signals from the hypothalamus travel to the posterior pituitary and anterior pituitary.

Posterior Pituitary

Two hormones are synthesized in the hypothalamus and reach the posterior pituitary by hypothalamic axons. These hormones are released from the posterior pituitary upon receiving nerve impulses from the hypothalamus. One hormone is **oxytocin**, which regulates milk secretion by mammary glands and uterine contractions during labor. In addition, oxytocin targets the brain, influencing behaviors related to trust, maternal care, pair bonding, and sexual activity. The other hormone, **antidiuretic hormone** (**ADH**), also known as **vasopressin**, targets the nephron and increases water retention by the kidneys, thus decreasing urine volume. This regulates blood osmolarity.

Anterior Pituitary

All hormones released by the anterior pituitary are controlled by at least one hormone from the hypothalamus, either a releasing or an inhibiting hormone. For example, the hypothalamus releases **prolactin-releasing hormone** into tiny capillaries so it can immediately reach the anterior pituitary. This hormone stimulates the release of **prolactin** from the anterior pituitary, which stimulates milk production.

ADH also plays a role in the behavior in prairie voles whose brains have large numbers of vasopressin receptors. The male prairie voles hover over their young pups, care for them and act aggressively toward intruders. If, however, the males are injected with a drug that blocks the vasopressin receptor, the males fail to form pair-bonds after mating.

REMEMBER

Hormones can regulate behavior.

Feedback Mechanisms

A feedback mechanism is a self-regulating mechanism that increases or decreases the level of a particular substance. **Positive feedback** *enhances an already existing response*. During labor, for example, the pressure of the baby's head against sensors near the opening of the uterus stimulates more uterine contractions, which causes increased pressure against the uterine opening, which causes yet more contractions. This positive-feedback loop brings labor *to an end* and the birth of a baby. This is very different from negative feedback. **Negative feedback** is a common mechanism in the endocrine system (and elsewhere) that *maintains homeostasis*. A good example is how the body maintains proper levels of thyroxin. When the level of thyroxin in the blood is too low, the hypothalamus releases **thyrotropin-releasing hormone (TRH)**, which stimulates the anterior pituitary to release a hormone, thyroid-stimulating hormone (TSH), which stimulates the thyroid to release more thyroxin. Thyroxin exerts negative feedback on the hypothalamus and the anterior pituitary. When the level of thyroxin is adequate, the hypothalamus stops stimulating the pituitary. (See "Positive and Negative Feedback of Menstrual Cycle" on page 285.)

How Hormones Trigger a Response in Target Cells

There are two types of hormones, and they stimulate target cells in different ways. These are illustrated in Figure 12.6.

1. **Lipid** or **steroid hormones** diffuse directly through the plasma membrane and bind to a receptor inside the nucleus that triggers the cell's response.

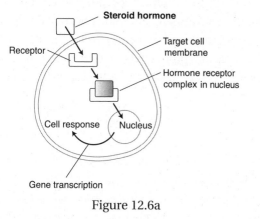

Figure 12.6a

2. **Protein** or **peptide hormones** (nonsteroidal) cannot dissolve in the plasma membrane, so they bind to a receptor on the surface of the cell. Once the hormone (the first messenger) binds to a receptor on the surface of the cell, it triggers a secondary messenger, such as c-AMP inside the cell, which converts the extracellular chemical signal to a specific response.

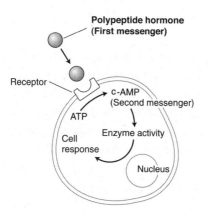

Figure 12.6b

You may wish to refer back to the chapter "Cell Communication" for the basics of how hormones influence cells.

Details of How Two Hormones Function

TESTOSTERONE

Testosterone is a good example of how a steroid hormone functions. Testosterone is secreted by cells in the testes and travels through the blood, entering cells all over the body. It readily passes through the cell membrane and binds to a receptor in the cytoplasm, activating it. *Only cells with testosterone receptors can respond to testosterone.* With the hormone attached, the activated receptor enters the nucleus and acts as a **transcription factor**, turning on specific genes in the nucleus that control male characteristics. However, different genes are turned on in different cells. In muscle cells, the genes that increase muscle mass are turned on. In hair cells, genes for the protein keratin (what hair is made of) are turned on.

To stress the importance of receptors, let's look at an example. There is a condition in humans called *androgen insensitivity syndrome* (**AIS**). People with this condition are born with the XY genotype but lack testosterone receptors in all their cells. (They do however, possess receptors for another hormone that inhibits normal development of internal female reproductive structures.) So although testosterone is produced in normal amounts, individuals with AIS do not develop male reproductive ducts, male genitals, or male secondary sexual characteristics. The result is a genetically XY individual who appears female but who has internal testes.

EPINEPHRINE

Epinephrine "tells" the liver to release glucose. The hormone epinephrine acts as the first messenger when it binds to and activates *G protein membrane receptors* on the surface of liver cells. The activated G protein then activates *adenylyl cyclase* in the membrane to produce cAMP (the second messenger). This begins a *kinase cascade* that does two things. First, it activates glycogen phosphorylase to convert glycogen into glucose. Second, it inhibits the conversion of glucose into glycogen. The cascade amplifies the original signal. For every molecule of epinephrine bound to a receptor on the cell surface, 20 molecules of cAMP are produced, and 10,000 molecules of glucose are released by the liver. The scientist who received the Nobel Prize in 1971 for discovering the role of cAMP in this pathway is Earl Sutherland.

BPA AND YOU

Bisphenol A (**BPA**) is an industrial compound used in the manufacture of plastics, such as plastic food containers. The danger of this compound is that it mimics the feminizing hormone estrogen and can set up an estrogen *signal transduction pathway* even though no estrogen is present. So a person exposed to BPA experiences the same effect as if he or she received a dose of estrogen. Imagine the consequence if a baby or young child were exposed to this substance over the long term! As of 2012, the FDA has banned BPA from baby bottles and children's drinking cups. However, the new prohibition does not apply more broadly to the use of BPA in other cups and food containers.

TEMPERATURE REGULATION

Most life exists within a fairly narrow range, from 0°C (the temperature at which water freezes) to about 50°C. Temperatures on land fluctuate enormously. Therefore, temperature regulation, like water conservation, became a problem for animals when they moved to the land. To stay alive, animals must generate their own body heat, seek out a more suitable climate, and/or change behavior. Here are some examples of behavioral changes that alter body temperature:

- A snake can warm itself in the sun or cool off by hiding under a rock.
- Animals on a cold winter prairie huddle together to decrease heat loss.
- Bees swarming in a hive raise the temperature inside the hive.
- Dogs pant and sweat through their tongues to cool themselves off.
- Elephants lack sweat glands, but they cool themselves off by squirting water onto their skins and flapping their ears like fans.
- Humans shiver, jump around, and develop goose bumps to keep warm.
- Birds migrate to a warmer climate and a better source of food to feed themselves and their young.

Ectotherms

Ectotherms are animals that gain most of their body heat from their environment. They have such a low metabolic rate that the amount of heat they can generate is too small to have any effect on body temperature. The body temperature of an ectotherm equilibrates to the environment. They must maintain adequate body temperature through behavioral means, like those previously described. Ectotherms include fish, amphibians, and reptiles. The term ectotherm is closest in meaning to the common term cold-blooded, which has no real scientific meaning.

Endotherms

Endotherms are animals that use metabolic processes (oxidizing sugar) to produce body heat. The body temperature of endotherms remains constant. In a cold environment, an endotherm can keep its body warmer than the environment. The temperature of the human body fluctuates during the course of a day, but the average temperature is 37°C (98.6°F). All mammals and birds are endotherms. Some scientists believe that certain dinosaurs may have been endotherms. The term endotherm is closest in meaning to the common term warm-blooded, which has no real scientific meaning.

In terms of energy consumption, endothermy is very costly. Humans can use 60% of what we eat to maintain our body heat. The metabolic rate of a mammal is much higher than that of a reptile of similar size. As a consequence, mammals must take in many more calories than a reptile of similar size. When considering the metabolic requirements of mammals, in general, the smaller the mammal the faster the metabolism is. Flying birds have an even higher metabolic requirement than do mammals.

Despite the fact that the requirement for being an endotherm is so great, it may have given birds and mammals a critical advantage during the time when ancient Earth was dominated by reptiles. Perhaps being able to maintain a high body temperature made it possible for mammals and birds to invade and colonize colder environments that reptiles found uninhabitable.

Poikilotherms and Homeotherms

You may see the terms **poikilotherm** (having a body temperature that varies with the environment) and **homeotherm** (having a constant body temperature despite fluctuations in environmental temperature). However, these terms are not always meaningful. When discussing thermoregulation, the real issue is not whether an animal has variable or constant body temperature but, rather, the *source of heat* used to maintain its body temperature. For example, some scientists classify a fish as a poikilotherm. However, many fish inhabit water that has such a constant temperature that the fish also have a constant body temperature, just like a homeotherm. Although mammals are classified as homeotherms, their temperature sometimes varies greatly. During **torpor**, **estivation**, summer torpor, or **hibernation**, mammals save energy by drastically decreasing their metabolic rate and body temperature.

Problems of Living on Land

Through natural selection, animals have evolved various anatomical and physiological adaptations for life in different environments. The size of the ears in a jackrabbit can be correlated to the climate in which it lives. Jackrabbits living in cold northern regions have small ears to minimize heat loss. In contrast, rabbits in warm, southern regions have long ears to dissipate heat from the many capillaries that make their ears appear pink. This anatomical difference across a geographic range is called a **north-south cline**.

Countercurrent heat exchange is a mechanism that has evolved in a variety of organisms. It helps to warm or cool extremities. See Figure 12.7. This can be seen in arctic animals such as the polar bear that must reach into icy waters to catch fish to eat. While its arm is in the frigid water, warm core blood is flowing out to the paw in the arteries, warming the chilled blood returning to the heart in veins, which lie directly next to the arteries.

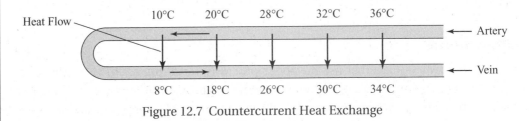

Figure 12.7 Countercurrent Heat Exchange

OSMOREGULATION

Osmoregulation is the management of the body's water and solute concentration. Organisms in different environments face different problems maintaining the proper concentration of body fluids.

For **marine vertebrates**, like bony fish, the ocean is a strongly dehydrating environment because it is very hypertonic to the organisms living in it. Fish constantly lose water through their gills and skin to the surrounding environment. To counteract the problem, they produce very little urine and drink large amounts of seawater. The extra salt that is taken in with the seawater is actively transported out through the gills.

The osmoregulation problems faced by **freshwater organisms** are opposite those living in salt water. The environment is hypotonic to the organisms, and they are constantly gaining water and losing salt. Freshwater fish excrete copious amounts of dilute urine.

Terrestrial organisms face entirely different problems. They have had to evolve mechanisms and structures that enable them to rid themselves of metabolic wastes while retaining as much water as possible.

EXCRETION

Different excretory mechanisms have evolved in various organisms for the purpose of osmoregulation and removal of metabolic wastes. Here are some organisms matched with their excretory structures.

Organism	Structures
Protista	Contractile vacuole
Platyhelminthes (planaria)	Flame cells
Earthworm	Nephridia (metanephridia)
Insects	Malpighian tubules
Humans	Nephrons

Excretion is the removal of metabolic wastes, which include **carbon dioxide** and **water** from cell respiration and **nitrogenous wastes** from protein metabolism. The organs of excretion in humans are the **skin, lungs, kidney,** and the **liver** (where urea is produced). There are three nitrogenous wastes: **ammonia, urea,** and **uric acid.** Which waste an organism excretes is the result of the environment it evolved and lives in. Here are some characteristics of each nitrogenous waste.

Nitrogenous Wastes

1. **AMMONIA**

   - Very soluble in water and highly toxic
   - Excreted generally by organisms that live in water, including the hydra and fish

2. **UREA**

   - Not as toxic as ammonia
   - Excreted by earthworms and humans
   - In mammals, it is formed in the liver from ammonia

3. **URIC ACID**

   - Pastelike substance that is not soluble in water and therefore not very toxic
   - Excreted by insects, many reptiles, and birds, with a minimum of water loss

The Human Kidney

The kidney functions as both an **osmoregulator** (regulates blood volume and concentration) and an organ of **excretion**. Humans have two kidneys supplied by blood from the **renal artery** and **renal vein**. The kidneys filter about 1,000–2,000 liters of blood per day and produce, on average, about 1.5 liters of urine (700–2,000 mL). As terrestrial animals, humans need to conserve as much water as possible. However, people must balance the need to conserve water against the need to rid the body of poisons. The kidney must adjust both the volume and the concentration of urine depending on the animal's intake of water and salt and the production of urea. If fluid intake is high and salt intake is low, the kidney will produce large volumes of **dilute (hyposmotic) urine**. In periods of high salt intake where water is unavailable, the kidney can produce **concentrated (hyperosmotic) urine**.

The Nephron

The functional unit of the kidney is the **nephron**; see Figure 12.8. The nephron consists of a cluster of capillaries, the **glomerulus**, which sits inside a cuplike structure called **Bowman's capsule** and connects to a long narrow tube, the **renal tubule**. Each human kidney contains about 1 million nephrons. The nephron carries out its job in four steps: **filtration**, **secretion**, **reabsorption**, and **excretion**.

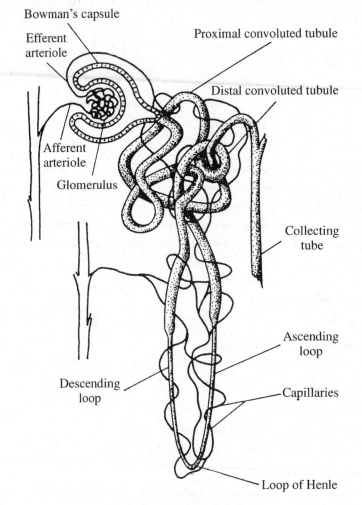

Bowman's capsule
Efferent arteriole
Afferent arteriole
Glomerulus
Proximal convoluted tubule
Distal convoluted tubule
Collecting tube
Ascending loop
Capillaries
Descending loop
Loop of Henle

Figure 12.8 The Nephron

Filtration occurs as blood pressure (about twice that of capillaries outside the kidney) forces fluid from the blood in the **glomerulus** into Bowman's capsule. Specialized cells of **Bowman's capsule** are modified into **podocytes**, which, along with **slit pores**, increase the rate of filtration. Filtration occurs by *diffusion* and is *passive* and *nonselective*. The filtrate contains everything small enough to diffuse out of the glomerulus and into Bowman's capsule, including glucose, salts, vitamins, waste such as urea, and other small molecules. From Bowman's capsule, the filtrate travels into the **proximal tubule**.

Secretion occurs in the **proximal** and **distal tubules**. It is the *active, selective* uptake of certain drugs and toxic molecules that did not get filtered into Bowman's capsule. The proximal tubule also secretes ammonia to neutralize the acidic filtrate.

Reabsorption is the process by which most of the water and solutes (glucose, amino acids, and vitamins) that initially entered the tubule during filtration are transported back into the peritubular capillaries and, thus, back to the body. This process begins in the proximal convoluted tubule and continues in the loop of Henle and collecting tubule. The main function of the loop of Henle is to move salts from the filtrate and accumulate them in the medulla surrounding the loop of Henle and the collecting tubule. In this way, the loop of Henle acts as a **countercurrent exchange mechanism** (see the chapter "The Cell"), maintaining a steep salt gradient surrounding the loop. This gradient ensures that water molecules will continue to flow out of the collecting tubule of the nephron, thus creating hypertonic urine and conserving water. *The longer the loop of Henle, the greater is the reabsorption of water.*

Excretion is the removal of metabolic wastes, for example, nitrogenous wastes. Everything that passes into the collecting tubule is **excreted** from the body. From the collecting tubule or duct, urine passes through the **ureter** to the **urinary bladder**. Urine is temporarily stored in the urinary bladder until it passes out of the body via the **urethra**. See Figure 12.9.

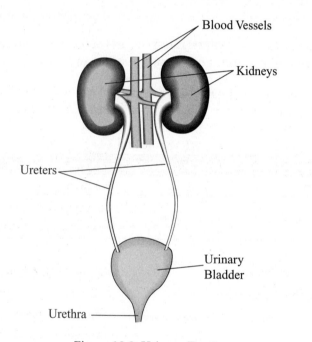

Figure 12.9 Urinary Tract

Hormone Control of the Kidneys

If blood pressure falls, the ability of the nephron to filter blood at Bowman's capsule is impaired, placing the body in danger. Therefore, the kidney must respond quickly by increasing blood pressure. It can do this because the kidney is under the control of three hormones: aldosterone, antidiuretic hormone (ADH), and renin.

Aldosterone is a hormone released by the adrenal glands in *response to a decrease in blood pressure or volume.* It acts on the distal tubules of the nephron to reabsorb more sodium ions and water, thus increasing blood volume and pressure. **ADH**, also known as **vasopressin**, is produced by the hypothalamus and both stored in and released from the posterior pituitary. It is released in response to dehydration due to excessive sweating or inadequate water intake, which causes the blood to become too concentrated (osmolarity to increase). ADH increases the permeability of the collecting tubules to water by opening *renal aquaporins* (transmembrane proteins that function as water channels). This opening allows more water to be reabsorbed and urine volume to be reduced. **Renin**, a hormone released from the kidneys, converts an inactive protein into active **angiotensin**, which stimulates the adrenal cortex to release aldosterone.

Physicians prescribe several effective drugs to treat excessively high blood pressure. The most common drug prescribed for high blood pressure is a *diuretic*, commonly called a water pill, which reduces blood volume by increasing the amount of water released in urine. The person simply urinates more. Another commonly prescribed drug is an *angiotensin-converting enzyme (ACE) inhibitor*, which prevents the formation of angiotensin II. ACE inhibitors inhibit the formation of aldosterone and thereby decrease blood volume. Both these drugs work in opposition to what the renin-angiotensin-aldosterone system normally does—raise blood pressure.

NERVOUS SYSTEM

The vertebrate nervous system consists of central and peripheral components.

1. **THE CENTRAL NERVOUS SYSTEM (CNS)** consists of the brain and spinal cord.
2. **THE PERIPHERAL NERVOUS SYSTEM (PNS)** consists of all nerves outside the CNS.

The peripheral nervous system is then further divided and subdivided into various systems. Table 13.5 gives an overview of the peripheral nervous system.

Table 13.5

Components of the Peripheral Nervous System	
System	**Function**
SENSORY	Conveys information from sensory receptors or nerve endings
MOTOR	
■ Somatic System	Controls the voluntary muscles
■ Autonomic System	Controls involuntary muscles
Sympathetic	• **Fight or Flight** response • Increase heart and breathing rate • Liver converts glycogen to glucose • Bronchi of lungs dilate and increase gas exchange • Adrenalin raises blood glucose levels
Parasympathetic	• **Opposes the sympathetic system** • Calms the body • Decreases heart/breathing rate • Enhances digestion

The Neuron

The neuron consists of a cell body, which contains the **nucleus** and other organelles, and two types of cytoplasmic extensions called **dendrites** and **axons**. **Dendrites** are **sensory**; they receive incoming messages from other cells and carry the electrical signal to the cell body. A neuron can have hundreds of dendrites. A neuron has only one axon, which can be several feet long in large mammals. **Axons** transmit an impulse from the cell body outward to another cell. Many axons are wrapped in a fatty **myelin sheath** that is formed by **Schwann cells**. Figure 12.10 shows a sketch of a neuron.

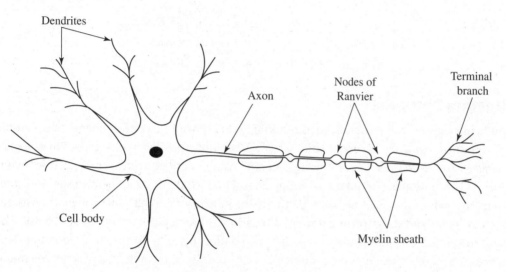

Figure 12.10 The Neuron

There are three types of neurons.

- **Sensory neurons** receive an initial stimulus from a sense organ, such as the eyes and ears, or from another neuron.
- A **motor neuron** stimulates **effectors** (**muscles** or **glands**). A motor neuron, for example, can stimulate a digestive gland to release a digestive enzyme or to stimulate a muscle to contract.
- The **interneuron** or **association neuron** resides within the spinal cord and brain, receives sensory stimuli, and transfers the information directly to a motor neuron or to the brain for processing.

The Reflex Arc

The simplest nerve response is a **reflex arc**. It is *inborn, automatic, and protective*. An example is the **knee-jerk reflex**, which consists of only two types of neurons: sensory and motor. A stimulus, a tap from a hammer, is felt in the sensory neuron of the kneecap, which sends an impulse to the motor neuron, which directs the thigh muscle to contract. A more complex reflex arc consists of three neurons, sensory, motor, and interneuron or association neurons; see Figure 12.11. A sensory neuron transmits an impulse to the interneuron in the **spinal cord**, which sends one impulse to the brain for processing and also one to the motor neuron to effect change immediately (at the muscle). This is the type of response that quickly jerks your hand away from a hot iron before your brain has figured out what occurred.

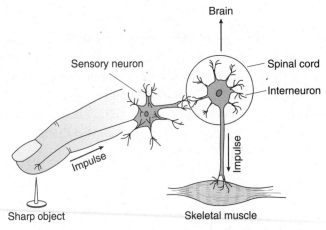

Figure 12.11 The Three-Neuron Reflex Arc

Resting Potential

All living cells exhibit a **membrane potential**, a difference in electrical charge between the cytoplasm (negatively charged) and extracellular fluid (positively charged). Physiologists measure this difference in membrane potential using microelectrodes connected to a voltmeter. This potential should be between −50 mV and −100 mV. The negative sign indicates that the inside of the cell is negative relative to the outside of the cell. A neuron in its unstimulated or **polarized state** (**resting potential**) has a membrane potential of about **−70 mV**. The **sodium-potassium pump** maintains the polarization by actively pumping ions that leak across the membrane. Potassium ions leak passively down the gradient across the cell membrane through "leak channels" in resting neurons. In order for the nerve to fire, a stimulus must be strong enough to overcome the **resting threshold** or **resting potential**. *The larger the membrane potential, the stronger the stimulus must be to cause the nerve to fire.*

Gated Channels

Neurons have **gated-ion channels** that open or close in response to a stimulus and play an essential role in transmission of electrical impulses. These channels allow only one kind of ion, such as sodium or potassium, to flow through them. If a stimulus triggers a **sodium ion-gated channel** to open, sodium flows into the cytoplasm, resulting in a decrease in polarization to about −60 mV. The membrane becomes somewhat **depolarized**, so it is easier for the nerve to fire. In contrast, if a **potassium ion-gated channel** is stimulated, the membrane potential increases and the membrane becomes **hyperpolarized**, to about −75 mV, so that it is harder for the neuron to fire.

Action Potential

An **action potential**, or **impulse**, can be generated only in the **axon** of a neuron. When an axon is stimulated sufficiently to overcome the threshold, the permeability of a region of the membrane suddenly changes and the impulse can pass. **Sodium channels** open and sodium ions flood into the cell, down the concentration gradient. In response, **potassium channels open** and potassium ions flood out of the cell. This rapid movement of ions or **wave of depolarization** reverses the polarity of the membrane and is called an **action potential**. The action potential is localized and lasts a very short time. The **sodium-potassium pump** restores the membrane to its original polarized condition by pumping sodium and potassium ions back to their original position. This period of repolarization, which lasts a few milliseconds, is called the **refractory period**, during which the neuron *cannot respond to another stimulus*. The refractory period ensures that an impulse moves along an axon in one direction only since the impulse can move only to a region where the membrane is polarized. Figure 12.12 shows the axon of a neuron as an impulse action potential passes from left to right, depolarizing the membrane in front of it.

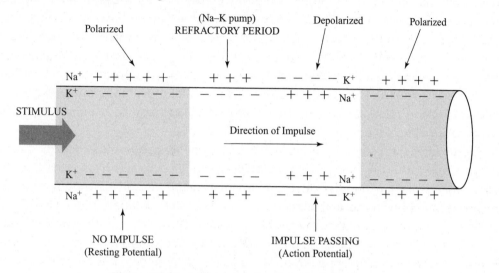

Figure 12.12 An Impulse Passing Along an Axon

The action potential is like a row of dominoes falling in order after the first one is knocked over. The first action potential generates a second action potential, which generates a third, and so on. The impulse moves along the axon propagating itself *without losing any strength*. If the axon is large or myelinated, the impulse travels faster. This occurs for two reasons. First, the myelin reduces ion leakage. Second, because voltage-gated channels are located on the nodes, the impulse leaps from node to node (Ranvier) in saltatory (jumping) fashion.

The action potential is an *all-or-none* event; either the stimulus is strong enough to cause an action potential or it is not. The body distinguishes between a strong stimulus and a weak one by the **frequency** of action potentials. A strong stimulus sets up more action potentials than a weak one does.

The graph in Figure 12.13 traces the events of a membrane experiencing sufficient stimulation to undergo an action potential.

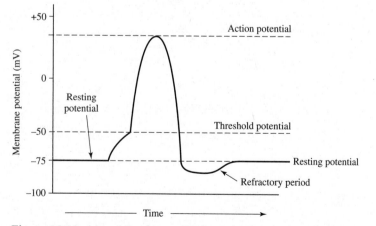

Figure 12.13 Axon Membrane Undergoing an Action Potential

The Synapse

Although an impulse travels along an axon electrically, it crosses a synapse **chemically**. The cytoplasm at the **terminal branch** of the **presynaptic neuron** contains many **vesicles**, each containing thousands of molecules of **neurotransmitter**. Depolarization of the presynaptic membrane causes **Ca++ ions** to rush into the terminal branch through **calcium-gated channels**. This sudden rise in Ca++ levels stimulates the vesicles to fuse with the presynaptic membrane and release the neurotransmitter by **exocytosis** into the synaptic cleft. Once in the synapse, the neurotransmitter bonds with receptors on the **postsynaptic side**, altering the membrane potential of the postsynaptic cell and resulting in a response. Depending on the type of receptors and the ion channels they control, the postsynaptic cell will be either *inhibited* or *excited*. In vertebrates, synapses between motor neurons and muscle cells are always excitatory. However, synapses between neurons can also be inhibitory by causing hyperpolarization of the postsynaptic cell.

Shortly after the neurotransmitter is released into the synapse, it is destroyed by an enzyme called **esterase** and recycled by the presynaptic neuron. The neurotransmitter at all neuromuscular junctions is **acetylcholine**. Others neurotransmitters are **serotonin**, **epinephrine**, **norepinephrine**, and **dopamine**. In addition, the neurotransmitter acetylcholine stimulates some cells to release the gas **nitric oxide** (**NO**), which, in turn, stimulates other cells. A single postsynaptic neuron may receive 1,000 impulses from presynaptic neurons, releasing a variety of different neurotransmitters. All of these impulses are integrated and summed up into either hyperpolarization (inhibited) or hypopolarization (stimulated).

Figure 12.14 shows the terminal branch of the neuronal axon and the synapse.

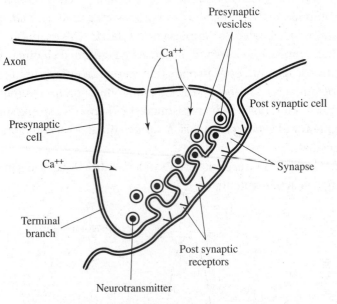

Figure 12.14 Terminal Branch of Neuron at Synapse

Organization of the Human Brain

Table 12.6 reviews the main parts of the brain. Clearly, each region receives information from other regions, integrates that information, and responds.

Table 12.6

The Brain	
Cerebrum	Learning, emotion, memory, perception
	Divided into right and left hemispheres
	The left side of the brain receives information from and controls the right side of the body and vice versa
Cerebellum	Coordinates movement and balance; helps in learning and remembering motor skills
	Receives sensory information about the position of joints and muscles
	Monitors motor commands from the cerebrum and integrates the information as it carries out coordination
Brainstem—includes medulla oblongata	Controls several automatic homeostatic functions, including breathing, heart and blood vessel activity, swallowing, and digestion
	Receives and integrates several types of sensory information and sends it to specific regions of the forebrain
	Transfers information between the PNS and other parts of the brain

For the AP Exam, focus on the basics. You need to know how the nervous system detects external and internal signals, that it transmits and integrates the signals, and that a response follows. A good model to use for this is the human eye. See Figure 12.15.

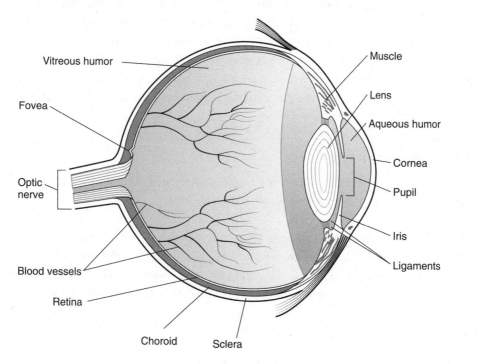

Figure 12.15 The Human Eye

VOCABULARY OF THE EYE

- **CONES:** Modified neurons; photoreceptors in the retina that distinguish different colors; embedded with visual pigments
- **CORNEA:** Tough, clear covering that protects the eye and allows light to pass through
- **HUMOR:** Fluid that maintain the shape of the eyeball
- **IRIS:** Colored part of the eye that controls the size of the pupil and how much light enters the eye
- **LENS:** Focuses light on the retina
- **OPSIN:** A membrane protein bound to a light-absorbing pigment molecule
- **PUPIL:** Small opening in the middle of the iris where light enters the eye
- **RETINA:** On the back of the eyeball; converts light to nerve impulses that are carried to the brain; consists of 5 layers of neurons
- **RETINAL:** Light-absorbing pigment in rods and cones
- **RHODOPSIN:** A visual pigment consisting of *retinal* and *opsin*; upon absorbing light, retinal changes shape and separates from opsin
- **RODS:** Modified neurons; photoreceptors in the retina that are extremely sensitive but distinguish only black and white; embedded with the visual pigment retinal

SIGNAL TRANSDUCTION PATHWAY AND AMPLIFICATION CASCADE

The sensory phase of vision begins when **photons** of light pass through the lens, which focuses the light onto the retina. The retina consists of several layers of neurons. When light strikes the retina it is absorbed by photoreceptors in some of those neurons—rods (black and white vision) and cones (color vision). These rods and cones are embedded with the light-absorbing pigment *retinal*. While in a photoreceptor, each photon causes a change in shape of retinal from *cis* to *trans*, which results in the excitement of the visual pigment **rhodopsin**. This triggers a familiar mechanism, a **signal transduction pathway** and **amplification cascade.** The stimulation of retinal activates a **G protein-signaling mechanism** that ultimately alters the membrane potential and closes Na^+ channels. (In the absence of light, the Na^+ channels are open.) Each single molecule of photoexcited rhodopsin activates several hundred enzyme molecules, each of which activates several hundred molecules of cGMP, closing hundreds of Na^+ channels, and causing an impulse to be sent to the optic nerve. From the optic nerve, impulses travel to the cortex of the brain where the messages are interpreted and seeing actually occurs.

COLOR VISION AND YOU

Mammals have poorer color vision than other vertebrates. This may result from our long evolutionary history of nocturnal living in which we have lost much of the color vision found in other vertebrates.

In humans, another opsin molecule has evolved by means of gene duplication. The gene for it is located on the X chromosome and has generally enhanced our color vision. However, in about 10% of men of European descent (and much higher in some other groups), the duplicated gene is nonfunctional. This results in red-green color blindness in those men.

MUSCLE

There are three types of muscle: smooth, cardiac, and skeletal. **Smooth** or **involuntary** muscle makes up the walls of the **blood vessels** and the **digestive tract**. Because of the arrangement of its actin and myosin filaments, it does not have a striated appearance. It is under the control of the autonomic nervous system. **Cardiac muscle** makes up the **heart**. It generates its own action potential; individual heart cells will beat on their own in a saline solution. **Skeletal** or **voluntary muscles** are very large and multinucleate. They work in pairs, one muscle contracts while the other relaxes. The biceps and triceps are one example. This discussion of muscle focuses on skeletal muscles; see Figure 12.16.

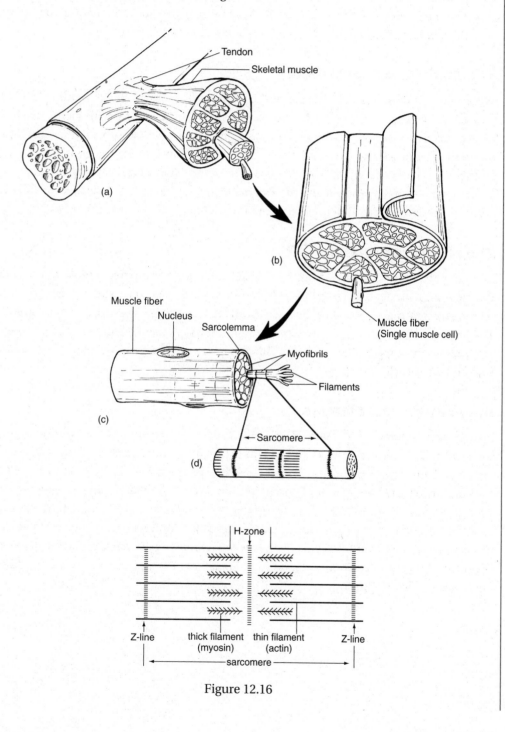

Figure 12.16

Every muscle consists of bundles of thousands of **muscle fibers** that are individual cylindrical **muscle cells**. Each cell is exceptionally **large** and **multinucleate**. Skeletal muscle consists of modified structures that enable the cell to contract.

- The **sarcolemma** is modified **plasma membrane** that surrounds each muscle fiber and can propagate an **action potential**.
- The **sarcoplasmic reticulum (SR)** is modified **endoplasmic reticulum** that contains sacs of **Ca⁺⁺** necessary for normal muscle contraction.
- The **T system** is a system of tubules that runs perpendicular to the SR and that connects the SR to the extracellular fluid.
- The **sarcomere** is the functional unit of the muscle fiber (cell). Its boundaries are the **Z lines**, which give skeletal muscle its characteristic striated appearance.

The Sliding Filament Theory

Within the cytoplasm of each muscle cell are thousands of fibers called **myofibrils** that run parallel to the length of the cell. Myofibrils consist of **thick** and **thin filaments**. Each **thin filament** consists of two strands of **actin proteins** wound around one another. Each **thick filament** is composed of two long chains of **myosin** molecules each with a globular head at one end. The heads serve two functions. The contraction of the sarcomere depends on two other molecules: **troponin** and **tropomyosin**, in addition to **Ca⁺⁺ ions** to form and break cross-bridges. Muscle contracts as thick and thin filaments slide over each other.

The Neuromuscular Junction

The axon of a motor neuron synapses on a skeletal muscle at the **neuromuscular junction**. The neurotransmitter **acetylcholine**, released by vesicles from the axon, binds to **receptors** on the **sarcolemma**, depolarizes the muscle cell membrane, and sets up an **action potential**. The impulse moves along the **sarcolemma**, into the **T system**, and stimulates the **sarcoplasmic reticulum** to release Ca⁺⁺. The **Ca⁺⁺ ions** then alter the troponin-tropomyosin relationship and the muscle contracts.

Summation and Tetanus

A single action potential in a muscle will cause the muscle to contract locally and minutely for a few milliseconds and then to relax. This brief contraction is called a **twitch**. If a second action potential arrives before the first response is over, there will be a **summation effect** and the contraction will be larger. If a muscle receives a series of overlapping action potentials, even further summation will occur. If the rate of stimulation is fast enough, the twitches will blur into one smooth, sustained contraction called **tetanus** (not related to the disease of the same name). Tetanus is what occurs when a large muscle such as the biceps muscle contracts. If the muscle continues to be stimulated without respite, it will eventually **fatigue** and relax; see Figure 12.17.

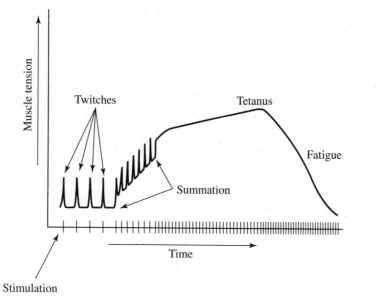

Figure 12.17

MULTIPLE-CHOICE QUESTIONS

DIGESTION

1. Gastric enzyme works best at a pH of

    (A) 2
    (B) 6
    (C) 8
    (D) 11

Questions 2–6

Questions 2–6 refer to the parts of the human digestive system listed below.

    (A) Large intestine
    (B) Esophagus
    (C) Stomach
    (D) Small intestine

2. Where does the digestion of fats occur?

3. Where does the reabsorption of water used during digestion occur?

4. In which structure is there no digestion?

5. In which structure is digestion completed?

6. Which structure contains the microvilli?

7. Which is TRUE of the stomach?

(A) The pyloric sphincter is at the top of the stomach.

(B) The stomach lining releases lipases to begin fat digestion.

(C) Hydrochloric acid activates the enzyme pepsinogen.

(D) The pH of the stomach varies from acid to basic depending on what must be digested.

8. The hormone gastrin is released by the _____ and has its effect on the _____.

(A) duodenum; stomach

(B) duodenum; pancreas

(C) stomach; gastric lining

(D) stomach; small intestine

9. The lacteal is found in the _____, and is involved with _____.

(A) stomach; the release of hormones

(B) duodenum; the hydrolysis of lipids

(C) small intestine; the absorption of fatty acids

(D) colon; the reabsorption of water

10. Absorption of nutrients occurs in the

(A) duodenum of the colon

(B) duodenum of the small intestine

(C) latter part of the small intestine

(D) latter part of the large intestine

GAS EXCHANGE

11. Which is CORRECT about gas exchange in humans?

(A) As humans inhale, the pressure in the chest cavity decreases and air is drawn into the lungs.

(B) Air is forced down the windpipe when a person inhales.

(C) The breathing rate is controlled by the hypothalamus in the brain.

(D) Hemoglobin carries carbon dioxide and oxygen in fairly equal amounts.

12. Tracheal tubes are found in

(A) earthworms

(B) hydra

(C) fish

(D) insects

13. Breathing in humans is usually regulated by

(A) the number of red blood cells

(B) the amount of hemoglobin in the blood

(C) inherent genetic control

(D) CO_2 levels and pH sensors

14. All of the following statements about the normal direction of the flow of blood are correct EXCEPT

    (A) lungs → pulmonary artery

    (B) right ventricle → tricuspid valve

    (C) aorta → aortic semilunar valve

    (D) vena cava → right atrium

15. The pacemaker of the heart is

    (A) sinoatrial node

    (B) the atrioventricular node

    (C) the diastolic node

    (D) the semilunar node

16. The Bohr effect on the oxyhemoglobin dissociation curve is produced directly by changes in

    (A) temperature

    (B) pH

    (C) CO_2 levels

    (D) oxygen concentration

17. Which is TRUE of the human circulatory system?

    (A) The right ventricle of the heart has the thickest wall.

    (B) Veins have thick walls consisting of smooth muscle cells to assist in returning blood to the heart.

    (C) Blood flow is slowest in capillaries to maximize the diffusion of nutrients and wastes.

    (D) The left and right ventricles contract alternately, which is responsible for the pulse sound.

18. In humans, the largest amount of the carbon dioxide produced by body cells is carried to the lungs as

    (A) CO_2 attached to hemoglobin in the red blood cells

    (B) the bicarbonate ion dissolved in the plasma

    (C) the bicarbonate ion attached to hemoglobin

    (D) CO_2 gas in solution in the plasma

19. During ventricular systole, the _____ valve(s) _____.

    (A) semilunar; close

    (B) semilunar; open

    (C) AV; open

    (D) AV and semilunar; close

20. All of the following are true about blood EXCEPT

    (A) red blood cells live for about 120 days

    (B) white blood cells are formed in the bone marrow

    (C) platelets are not cells but are cell fragments

    (D) platelets derive from specialized cells known as neutrophils

CHEMICAL SIGNALS

21. Which hormone acts opposite parathormone?

    (A) calcitonin
    (B) glucagon
    (C) insulin
    (D) adrenaline

22. The main target of antidiuretic hormone is the

    (A) heart
    (B) kidney
    (C) liver
    (D) spleen

23. The hormone secreted by the adrenal cortex

    (A) reduces inflammatory response
    (B) is an amino acid
    (C) stimulates the adrenal medulla
    (D) lowers blood sugar

The following questions refer to the list of chemical messengers below.

Questions 24–29

Questions 24–29 refer to the list of hormones below. You may use each one once, more than once, or not at all.

    (A) Glucagon
    (B) Adrenocorticotropic hormone
    (C) Oxytocin
    (D) Thyroxin

24. Induces labor

25. Released by the posterior pituitary

26. Stimulates the adrenal gland

27. Controls metabolic rate

28. Produced in the pancreas

29. Causes blood sugar levels to increase

30. Tropic hormones

    (A) are those that are secreted by the liver
    (B) are those that are secreted by the thyroid
    (C) stimulate other glands
    (D) are released only by the hypothalamus

EXCRETION

31. Which of the following processes of the kidney is the LEAST selective?

    (A) secretion
    (B) filtration
    (C) reabsorption
    (D) the target of ADH

32. In humans, urea is produced in the

    (A) kidneys
    (B) urinary bladder
    (C) urethra
    (D) liver

33. The main nitrogenous waste excreted by birds is

    (A) ammonia
    (B) urea
    (C) uric acid
    (D) nitrite

34. Which of the following is INCORRECTLY paired?

    (A) nephridia-earthworm
    (B) flame cell-bird
    (C) Malpighian tubules-insect
    (D) nephron-ape

35. Following a company picnic on a hot summer day where you ate traditional fare of hot dogs and hamburgers, ice cold beer, and played softball for an hour, you become very thirsty. Which would probably contribute LEAST to your thirst?

    (A) eating salty food
    (B) drinking an ice cold beer
    (C) drinking lots of ice water
    (D) sweating

36. Which nitrogenous waste requires the least water for its excretion?

    (A) ammonia
    (B) urea
    (C) nitrites
    (D) uric acid

Questions 37–40

Questions 37–40 refer to the parts of the kidney listed below.

    (A) Proximal and distal tubules
    (B) Ascending loop of Henle
    (C) Bowman's capsule
    (D) Collecting tubule

37. Where does filtration occur?

    (A) A
    (B) B
    (C) C
    (D) D

38. Identify where secretion occurs.

    (A) A
    (B) B
    (C) C
    (D) D

39. Identify the area that is impermeable to the diffusion of water.

    (A) A
    (B) B
    (C) C
    (D) D

40. Identify the target structure of ADH.

    (A) A
    (B) B
    (C) C
    (D) D

NERVOUS SYSTEM

41. Which of the following sequences describes the passage of a nerve impulse through a simple reflex arc in humans?

    (A) receptor → effector → interneuron → motor neuron → sensory neuron
    (B) receptor → sensory neuron → interneuron → effector → motor neuron
    (C) sensory neuron → effector → motor neuron → interneuron → receptor
    (D) receptor → sensory neuron → interneuron → motor neuron → effector

42. The function of a Schwann cell is to

    (A) produce esterases to break down neurotransmitters
    (B) form the myelin sheath around the axon of a neuron
    (C) act as an interneuron in the spinal cord
    (D) receive impulses and send them to the neuron

43. What is the function of cholinesterase in the transmission of an impulse?

   (A) It binds with postsynaptic receptors and blocks transmission of the impulse.
   (B) It prevents the release of any more neurotransmitter from presynaptic vesicles.
   (C) It enhances neurotransmission across the synapse.
   (D) It breaks down acetylcholine, preventing continuous neurotransmission.

44. Which would be associated with the parasympathetic system?

   (A) increase in blood sugar
   (B) increase in adrenaline
   (C) increase in breathing rate
   (D) increase in digestion

45. The threshold potential of a particular membrane measures –70 mV at time zero. After 10 minutes, it measures –90 mV. What is the best explanation for what has happened to the membrane?

   (A) It became depolarized.
   (B) The concentrations of Na^+ and K^+ became balanced.
   (C) The membrane became hyperpolarized.
   (D) The membrane became hypopolarized.

46. Which is an example of an effector?

   (A) hormone
   (B) gland
   (C) interneuron
   (D) brain

47. If the threshold potential were measured at –75 mV, what would be the value if the polarization across the membrane were increased?

   (A) –70 mV
   (B) –80 mV
   (C) zero
   (D) There would be no change because the membrane potential of a particular neuron can never change.

48. How do action potentials relay different intensities of information?

   (A) by changing the amplitude of the action potential
   (B) by changing the speed with which the impulse passes
   (C) by changing the frequency of the action potential
   (D) by changing the duration of the action potential

MUSCLE

49. Tendons connect

   (A) muscle to muscle
   (B) muscle to bone
   (C) ligaments to bones
   (D) ligaments to ligaments

50. The walls of arteries consist of

    (A) striated muscle and are under voluntary control

    (B) striated muscle and are not under voluntary control

    (C) smooth muscle and are controlled by the somatic nervous system

    (D) smooth muscle and are controlled by the autonomic nervous system

51. What is the basic unit of function of a skeletal muscle fiber?

    (A) myosin filaments

    (B) actin filaments

    (C) the sarcomere

    (D) Z line

52. What neurotransmitter at the synapse of a neuromuscular junction causes a muscle to contract?

    (A) GABA

    (B) norepinephrine

    (C) dopamine

    (D) acetylcholine

53. Which of the following is the name of a single muscle cell?

    (A) sarcomere

    (B) myofibril

    (C) muscle fiber

    (D) sarcolemma

54. All of the following are true about the contracting of skeletal muscle cells EXCEPT

    (A) thick filaments are composed of actin

    (B) Ca^{++} ions are necessary for normal muscle contraction

    (C) the sarcolemma can propagate an action potential

    (D) the T-system connects the sarcolemma to the sarcoplasmic reticulum

Answers to Multiple-Choice Questions

DIGESTION

1. **(A)** Gastric enzymes are strongly acidic and activate the gastric enzyme pepsinogen. Other enzymes, such as intestinal enzymes, are activated in an alkaline pH.

2. **(D)** Fats are digested by lipases in the small intestine.

3. **(A)** The large intestine is the site of egestion of undigested wastes, vitamin production, and water reabsorption.

4. **(B)** There is no digestion in the esophagus.

5. **(D)** All digestion and absorption of nutrients is completed in the small intestine.

6. **(D)** The microvilli are cytoplasmic extensions of the villi. Millions of villi line the endothelium of latter sections of the small intestine.

7. **(C)** The lower esophageal sphincter is at the top of the stomach; the pyloric sphincter is at the bottom of the stomach. The stomach releases inactive pepsinogen, which is activated by the acid environment, and digests proteins. The pH of the stomach is very acidic. The stomach is stimulated by the hormone gastrin.

8. **(C)** This is a statement of fact. See Table 13.1, "Hormones that Regulate Digestion."

9. **(C)** The lacteal is inside the villi that line the inside of the small intestine. The duodenum is the first 12 inches (30 cm) of the small intestine and is the site of digestion, not absorption. "Colon" is another name for large intestine.

10. **(C)** Digestion occurs in the duodenum—the first part of the small intestine, and absorption occurs in the latter part.

GAS EXCHANGE

11. **(A)** Humans breathe by negative pressure. When we inhale, the chest cavity expands as the diaphragm lowers. This increase in volume causes a decrease in internal pressure, and air is drawn into the lungs because the internal pressure is less than the external pressure.

12. **(D)** Spiracles, openings in the exoskeleton of the insects, connect to tracheal tubes that lead to the hemocoels where diffusion of respiratory gases occur. The respiratory surface is internal.

13. **(D)** The breathing rate is controlled by the medulla in the brain, which is primarily sensitive to CO_2 levels in the blood. The other choices are false.

14. **(A)** From the lungs, blood flows into the left atrium via the pulmonary vein.

15. **(A)** The pacemaker of the heart is the sinoatrial node. From the sinoatrial node, the impulse passes to the atrioventricular node, then to the bundle of His.

16. **(B)** When the pH becomes more acidic, hemoglobin has less affinity for oxygen, so the hemoglobin will drop off some oxygen at the cells that have become slightly acidic due to an accumulation of CO_2 from cell respiration.

17. **(C)** Diffusion occurs in the thin-walled capillaries where the blood circulates slowly. All other choices are false.

18. **(B)** Although a small amount of CO_2 is carried by the red blood cells, most is carried as the bicarbonate ion.

19. **(B)** Systole is the contraction of the ventricles of the heart. When the ventricles contract, blood is pushed out of the arteries through the semilunar valves while the bicuspid and tricuspid valves remain closed.

20. **(D)** Platelets are fragments of cells known as megakaryocytes. All the other choices are correct.

CHEMICAL SIGNALS

21. **(A)** Parathormone is released by the parathyroid and raises calcium levels in the blood; calcitonin is released by the thyroid and lowers calcium levels.

22. **(B)** The target of antidiuretic hormone is the collecting duct of the nephron. If ADH is released, less urine is excreted.

23. **(A)** The hormone released by the adrenal cortex is cortisol, which reduces inflammation. The pharmaceutical version of cortisol is cortisone, which has the same function.

24. **(C)** Oxytocin induces labor and the production of milk by the mammary glands.

25. **(C)** Oxytocin is produced in the hypothalamus and stored and released from the posterior pituitary.

26. **(B)** ACTH is a tropic hormone. A tropic hormone is one that stimulates another endocrine gland to release a substance. ACTH stimulates the adrenal cortex to release a glucocorticoid-like cortisol.

27. **(D)** The thyroid gland releases thyroxin, which controls the rate of metabolism.

28. **(A)** Glucagon is produced in the α cells of the pancreas and raises blood sugar by breaking down glycogen that is stored in the liver.

29. **(A)** See # 28.

30. **(C)** Tropic hormones are released by one gland and stimulate another gland to release its hormone. For example, TSH is released by the anterior pituitary and stimulates the thyroid to release thyroxin.

EXCRETION

31. **(B)** During filtration, all substances small enough to diffuse out of the glomerulus will do so. It is the least-selective process occurring in the nephron.

32. **(D)** Urea is a nitrogenous waste that is produced in the liver; urine is produced in the kidney.

33. **(C)** Uric acid is excreted as a crystal to conserve water by many terrestrial animals. However, humans and earthworms release nitrogenous waste in the form of urea.

34. **(B)** The flame cell is a primitive excretory structure in Platyhelminthes like planaria. All freshwater Protista have contractile vacuoles that pump out water that continually leaks in. The other statements are all true.

35. **(C)** All of the other options would contribute to thirst. Alcohol, particularly, is dehydrating because it blocks production of antidiuretic hormone, thus causing the release of extra water in the urine.

36. **(D)** Uric acid is not soluble in water, therefore, it does not require water in order to be excreted from the body.

37. **(C)** Filtration occurs in Bowman's capsule.

38. **(A)** Secretion occurs in the proximal and distal tubules.

39. **(B)** The ascending loop of Henle is impermeable to water.

40. **(D)** The target structure of ADH is the collecting tubule within the kidney.

NERVOUS SYSTEM

41. **(D)** An effector is a muscle or a gland. The interneuron is located in the spinal cord.

42. **(B)** The Schwann cell forms the myelin sheath around the axon and nourishes the neuron. There are more Schwann cells in the nervous system than there are neurons.

43. **(D)** There are several classes of esterases. They are all specific in action for a particular neurotransmitter.

44. **(D)** The parasympathetic system is calming. It is active when you are relaxing or digesting food and it slows down when you are excited.

45. **(C)** The gradient is steeper at –90 mV than it was at –75 mV. The membrane is now hyperpolarized. The threshold is higher. Depolarizing the membrane and passing an impulse are more difficult.

46. **(B)** An effector is a gland or muscle.

47. **(B)** The membrane potential reads zero when there is no differential between one side of the membrane and the other. The greater the differential, the more negative the potential.

48. **(C)** The only way a neuron distinguishes the intensity of the stimulus is by changing the frequency of the action potentials. A nerve response is all or none.

MUSCLE

49. **(B)** This is a fact.

50. **(D)** The autonomic nervous system controls those things that are not consciously controlled. It controls smooth and cardiac muscles.

51. **(C)** The sarcomere is the basic unit of contraction of skeletal muscle.

52. **(D)** Acetylcholine is the neurotransmitter at a neuromuscular junction.

53. **(C)** A single muscle cell is called a muscle fiber. It is a very large, multinucleated cell.

54. **(A)** Thick filaments are composed of myosin, not actin. The other statements are correct about muscles.

Directions: Answer all questions. You must answer the question in essay—**not** outline—form. You may use labeled diagrams to supplement your essay, but diagrams alone are *not* sufficient. Before you start to write, read each question carefully so that you understand what the question is asking.

1. Beginning at an axon at rest, describe the conduction of an impulse through a neuromuscular junction to the contraction of a skeletal muscle.

2. Regulation and homeostasis are critical to living things.
   a. Explain what a feedback mechanism is and how it helps an organism to maintain homeostasis.
   b. Explain the difference between positive feedback and negative feedback and give examples.

Typical Free-Response Answers

Note: The answer to this first essay question has three parts: what happens at the axon, what happens at the synapse, and what happens at the muscle. That means each section is worth from two to four points. You cannot get full credit unless you cover all three parts to some degree. If you write a masterpiece on the axon and the synapse and do not have time to discuss the muscle, you will not get the full ten points. Key words are all in boldface. This essay question is broad, and there is so much to write about. However, you can get full credit without writing an essay as detailed as this one.

1. An axon at rest is polarized and has a **resting potential** of about –70 mV. The **sodium-potassium pump** maintains this level of **polarization** by actively pumping ions that leak across the membrane. In order for the nerve to fire and an impulse to pass, a stimulus must be strong enough to overcome the resting threshold. The larger the resting potential, the stronger the stimulus must be to cause the nerve to fire.

   When an axon is stimulated sufficiently to overcome the **threshold**, the permeability of a region of the membrane suddenly changes. Sodium channels open and sodium floods into the cell, down the concentration gradient. In response, potassium channels open and potassium floods out of the cell. This **wave of depolarization** reverses the polarity of the membrane and is called an **action potential**. The action potential is localized and lasts a very short time. The sodium-potassium pump restores the membrane to its original polarized condition by pumping sodium and potassium ions back to their original positions. This period of repolarization, which lasts a few milliseconds, is called the **refractory period**, during which time the neuron cannot respond to another stimulus. The refractory period ensures that an impulse moves along an axon in **one direction only** since the impulse can move only to a region where the membrane is polarized. The impulse moves along the axon propagating itself without losing any strength. The action potential is an **all-or-none event**; either the stimulus is strong enough to cause an action potential or it is not. The body distinguishes between a strong stimulus and a weak one by the **frequency of action potentials**. A strong stimulus sets up more action potentials than a weak one does.

The action potentials end at the terminal branch of the axon where the cytoplasm contains many **vesicles**, each containing thousands of molecules of **neurotransmitter**. Depolarization of the **presynaptic membrane** causes Ca^{++} ions to rush into the terminal branch through **calcium-gated channels**. This sudden rise in Ca^{++} levels stimulates the vesicles to fuse with the presynaptic membrane and release neurotransmitter by **exocytosis** into the synapse. Once in the synapse, the neurotransmitter bonds with **receptors** on the postsynaptic side, altering the membrane potential of the postsynaptic cell. Shortly after the neurotransmitter is released into the synapse, it is destroyed by an enzyme called an **esterase** and recycled by the presynaptic neuron. The neurotransmitter at a neuromuscular junction is **acetylcholine**.

Neurotransmitter binds to receptors on the **sarcolemma** of the skeletal muscle membrane, depolarizes it, and sets up an action potential. This impulse moves along the sarcolemma, into the **T system**, and stimulates the **sarcoplasmic reticulum**, the modified endoplasmic reticulum, to release Ca^{++}. The Ca^{++} ions cause the thick **myosin** filaments and the thin **actin** filaments of the **sarcomere**, the basic functional unit of muscle, to slide over one another, and the muscle contracts.

2a. The body maintains homeostasis by acting through complex feedback mechanisms. A real-life example of a feedback system is the thermostat and a home heating system. You set the thermostat for a desired temperature. As soon as the temperature deviates from that set point, either the furnace or air conditioner turns on. In this way, the temperature in the building is maintained at a steady point.

2b. **Positive feedback** enhances an already existing response to complete something. During childbirth, the pressure of the baby's head against sensors near the opening of the uterus stimulates more uterine contractions, which causes increased pressure against the uterine opening, which causes yet more contractions. This positive feedback loop brings labor *to an end* and the birth of a baby. Another example of positive feedback can be seen in the helper T cells (T_h) in the human immune system. When a specific T_h cell is selected and activated by the MHC-antigen complex, it proliferates and makes thousands of copies of itself called clones. These clones secrete cytokines, which further stimulate T_h cell activity. In this case, an army of helper T cells is produced to fight a specific antigen. When the invading antigen is destroyed, the helper T population is no longer needed. Only a small number of memory helper cells remain circulating in the blood. This is very different from negative feedback.

Negative feedback is a common mechanism in the endocrine system (and elsewhere) that maintains homeostasis. A good example is how the body maintains proper levels of thyroxin. When the level of thyroxin in the blood is too low, the hypothalamus stimulates the anterior pituitary to release a hormone, thyroid-stimulating hormone (TSH), which stimulates the thyroid to release more thyroxin. When the level of thyroxin in the bloodstream is adequate again, the hypothalamus stops stimulating the pituitary.

The Human Immune System

13

→ **NONSPECIFIC INNATE DEFENSE MECHANISMS**
→ **ADAPTIVE IMMUNITY**
→ **TYPES OF IMMUNITY**
→ **BLOOD GROUPS AND TRANSFUSION**
→ **AIDS**
→ **POSITIVE FEEDBACK IN THE IMMUNE SYSTEM**
→ **OTHER TOPICS IN IMMUNITY**

INTRODUCTION

We live in a sea of germs. They are in the air we breathe, the food we eat, and the water we drink and swim in. Our body has three cooperating lines of defense that are able to respond to new threats as quickly as they appear. This chapter covers this complex topic in detail.

NONSPECIFIC INNATE DEFENSE MECHANISMS

First Line of Defense

The first line of **nonspecific defense** is a **barrier** that helps prevent **pathogens** (things that cause disease) from entering the body. These include:

- Skin
- Mucous membranes, which release mucus that contains antimicrobial substances including **lysozyme**
- Cilia in the respiratory system to sweep out mucus with its trapped microbes
- Stomach acid

Second Line of Defense

Microbes that get into the body encounter the second line of **nonspecific defense**, which is meant to limit the spread of invaders in advance of specific immune responses. This second line includes the following.

- **Inflammatory response includes:**

  ✔ **Histamine** triggers **vasodilation** (enlargement of blood vessels), which increases blood supply to the area, bringing more **phagocytes**. It is secreted by **basophils** (a type of circulating white blood cell) and **mast cells**, found in the connective tissue. Histamine is also responsible for the symptoms of the common cold: sneezing, coughing, redness, itchy eyes, and runny nose. Each attempts to rid the body of pathogens.

  ✔ **Prostaglandins** further promote blood flow to the area.

  ✔ **Chemokines** secreted by blood vessel endothelium and monocytes also attract phagocytes to the area.

  ✔ **Pyrogens**, released by certain leukocytes, increase body temperature to speed up the immune system and make it more difficult for microbes to function.

- **Phagocytes** ingest invading fungal and bacterial microbes. Two types of phagocytes, called **neutrophils** and **macrophages**, migrate to an infected site in response to local chemical attractants. This response is called **chemotaxis**. **Neutrophils** engulf microbes and die within a few days. **Monocytes** transform into **macrophages** ("giant eaters"), extend **pseudopods**, and engulf huge numbers of microbes over a long period of time. They digest the microbes with a combination of **lysozyme** and two toxic forms of oxygen: superoxide anion and nitric oxide.

- **Complement**, a group of proteins, that leads to the lysis (bursting) of invading cells.

- **Interferons** block against cell-to-cell viral infections.

- **Natural killer (NK) cells** destroy virus-infected body cells (as well as cancerous cells). They attack the cell membrane, causing it to **lyse** (burst open) and die.

ADAPTIVE IMMUNITY
Third Line of Defense

The **adaptive** third line of defense relies on **B lymphocytes** and **T lymphocytes**, which arise from **stem cells** in bone marrow. Once mature, both cell types circulate in the blood, lymph, and *lymphatic tissue* (spleen, lymph nodes, tonsils, and adenoids). Both cell types recognize different specific **antigens.***

The adaptive immune response is a specific response and involves three phases:

1. **RECOGNITION:** *Antigen receptors* on B and T lymphocytes recognize specific antigens or **epitopes** by binding to them. *Antigens* are any substance that elicits an immune response from B cell or T cells. *Epitopes* are an accessible piece of an antigen that elicits an immune response from a B or T cell. Each B cell displays specificity for one particular epitope. In order to recognize an antigen, it must be *presented* to a B or T cell by an *antigen-presenting cell.*

2. **ACTIVATION PHASE:** The binding of an antigen receptor activates B and T cells, causing them to undergo rapid cell division. The cells form populations of *effector cells* and *memory cells.*

3. **EFFECTOR PHASE:** After being activated, B cells produce a humoral response; they produce antibodies. T cells engage in a cell-mediated response. See Figure 13.3.

*Other blood cell types—eosinophils and basophils—play a lesser role in adaptive immunity and are beyond the scope of the AP course.

STUDY TIP

Each B or T cell displays specificity for one particular epitope only.

T Lymphocytes—T Cells

There are several types of T cells, but they all form in the bone marrow and mature in the thymus gland (from which they get their name). They fight pathogens in the **cell-mediated immune response**. In general, the activation process begins when T cell **antigen receptors** recognize and bind to antigens that are displayed on the surface of **antigen-presenting cells (APCs)** by a molecule called **MHC**, major histocompatibility complex. Examples of APCs are *macrophages, dendritic cells,* and sometimes *B cells.* Each T cell displays specificity for one particular **epitope**. Once activated, a T cell proliferates and forms a population of activated T *clones.* Some of the clones become **effector cells**, while the remainder become circulating **memory cells** that can rapidly respond to any exposure to the *same* antigen many years later. These memory cells are responsible for **immunological memory** and are the reason that a person does not generally get measles or chicken pox a second time.

The two main types of T cells are **helper T cells (T_h)** and **cytotoxic T cells (T_c)**. Helper T cells are activated by an interaction with an APC. Cytotoxic T cells are activated by helper T cells. Once activated, T_h **cells** announce to the immune system that foreign antigens have entered the body. They trigger both humoral and cell-mediated immune responses. In addition to activating other helper T cells, T_h cells activate cytotoxic T cells (T_c) that kill infected cells as well as B cells that produce antibodies. T_h cells activate these other cells by releasing **cytokines, interleukin-1 (Il-1), and interleukin-2 (Il-2).** Because T_h cells have CD4 accessory proteins on their cell surface, they are also referred to as **CD4 cells**. See Figure 13.3. The CD4 cells are specifically attacked by the human immune-deficiency virus (HIV).

Cytotoxic T cells, T_c, attack and kill *body cells* infected with pathogens as well as cancer cells. They do it by a cell-mediated immune response. An activated cytotoxic T cell proliferates and differentiates into an **effector cell** and a **memory cell**. Activated T_c cells attack and kill infected cells by releasing **perforin** (a protein that forms pores in the target cell's membrane) and granzymes (enzymes that break down proteins) that cause the cell to lyse and die. The infecting microbes are released into the blood or tissue and are disposed of by circulating antibodies. Because T_c cells have CD8 accessory protein on their surface, they are also referred to as **CD8 cells**. See Figure 13.1.

B Lymphocytes

B cells mature in bone marrow, from which they get their name. A typical B lymphocyte or B cell has about 100,000 identical **antigen receptors** on its surface that it uses to recognize pathogens. Each person produces more than a million different B cells, each with a different antigen receptor. Each antigen receptor is a Y-shaped molecule consisting of 4 polypeptide chains: two identical heavy chains and two identical light chains. Each chain has constant and variable regions. Antibodies secreted by B cells are soluble forms of these antigen receptors. Figure 13.2 shows a typical B lymphocyte.

B cells provide adaptive immunity through the **humoral immune response**, meaning they produce **antibodies (immunoglobins)**. When they become activated, B cells secrete about 2,000 antibodies per second over the cell's 4–5 day life span. A B cell becomes activated in several steps. First, an APC (**macrophage or dendritic cell**) presents an antigen or epitope on its cell surface using a **class II MHC** molecule. Next, a helper T cell that recognizes this **epitope-MHC molecule complex** is activated with the aid of **cytokines** secreted from the APC. See Figure 13.1, right side. Once activated, the B cell undergoes multiple cell divisions, producing thousands of clones. Some of these clones become **effector cells**, also called

plasma cells, that secrete antibodies. The remainder of the clones develop into long-term **memory cells** that can rapidly respond to any exposure to the *same* antigen many years later. These memory cells are responsible for **immunological memory** and are the reason that a person does not, for example, generally get measles a second time.

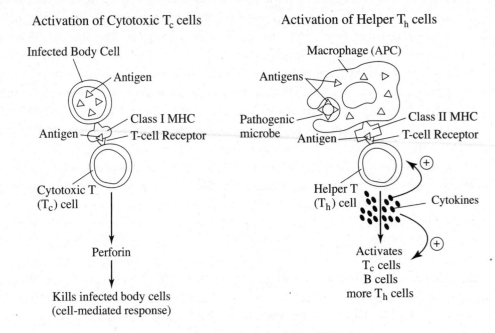

Figure 13.1 Activation of B and T cells

Regulator T Cells and Self-Tolerance

Normally, the immune system exhibits *self-tolerance*, meaning it does not attack body cells. However, sometimes immature lymphocytes develop that have antigen receptors specific for the body's own cells. If allowed to escape, these cells would attack the body's cells, a situation characteristic of *autoimmune disease*. To avoid this, lymphocytes are tested for *self-reactivity* as they mature in the bone marrow. B and T cells that are identified as self-reactive are destroyed by **apoptosis**, programmed cell death. Regulatory T cells, called T_{reg}, inhibit the activation of the immune system in response to *self-antigens*. T_{reg} secrete interleukin-10 (Il-10).

Major Histocompatibility Complex (MHC) Molecules

Major histocompatibility complex (MHC) molecules, also known as HLA (human leukocyte antigens), are a collection of cell surface markers that identify the cells as **self**. No two people, except identical twins, have the same MHC markers. There are two main classes of MHC markers, **class I** and **class II**.

- **Class I MHC molecules** are found on the surfaces of every nucleated body cell.
- **Class II MHC molecules** are found on specialized cells, including **macrophages**, **B cells**, dendritic cells, and **activated T cells**; see Figure 13.1.

APCs have both MHCI and MHCII on the cell surface.

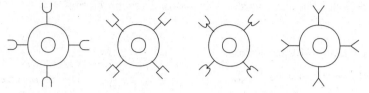

Antigen molecules

A variety of B lymphocytes

One particular B lymphocyte is "selected" — Clonal selection

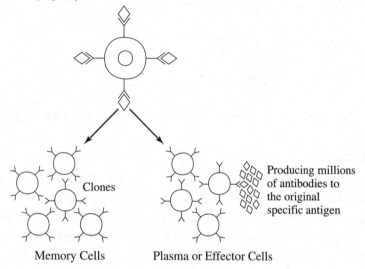

Clones

Producing millions of antibodies to the original specific antigen

Memory Cells   Plasma or Effector Cells

Figure 13.2 Clonal Selection

APCs: Antigen-Presenting Cells

APCs do just what their name says. They present an antigen or piece of an antigen, an *epitope*, to the immune system. In effect, they are saying, "Here is the enemy!" The process begins when an APC—a **macrophage**, **dendritic cell**, or **B cell**—takes in an antigen. Either the APC becomes infected with the antigen or engulfs it. Once inside the host, enzymes break apart the antigen into fragments and attach them to an MHC molecule in the cytoplasm. The MHC molecule, with the antigen fragment attached, then moves to the surface of the cell and displays it. Other cells of the immune system, such as T cells or other B cells, become activated if they can properly bind with the exposed antigen. See Figure 13.2, right side.

Clonal Selection

Clonal selection is a fundamental mechanism in the development of immunity. It is the means by which one particular lymphocyte that matches a specific antigen or epitope (also called **antigenic determinant**) is identified and activated. See Figure 13.2.

An antigen is presented to a steady stream of lymphocytes in lymph nodes by **APCs (antigen-presenting cells)** until a match is found. This triggers events that activate the lymphocyte. Once activated, the lymphocyte divide rapidly, forming a population of *clones* that develop into **effector** and **memory cells**. The effector cells are short-lived and begin battle immediately.

They neutralize or destroy all of those identified antigens and any pathogen that produces them. The remaining clones develop into long-term **memory cells** that remain in the body for the rest of one's life and can rapidly respond to any future exposure to the *same* antigen, even many years later. Memory cells are responsible for what is called **immunological memory** and the fact that you generally don't get measles, for example, a second time.

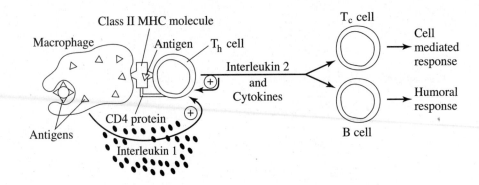

Figure 13.3 How a Macrophage Activates Immune System

Antibodies

Antibodies, also called **immunoglobins**, are a group of globular proteins. Each antibody molecule is a **Y-shaped molecule** consisting of four polypeptide chains, two identical heavy chains and two identical light chains joined by disulfide bridges. The molecule consists of four unchanging or **constant regions (C)** and four **variable regions (V)**. The tips of the Y have specific shapes and are the binding sites for different antigens.

It is now accepted that the blueprints for antibodies are made early in life, *prior* to any exposure to antigens. When you are exposed to an antigen, antibodies are chosen by **clonal selection** from a limitless variety as needed.

As a consequence, the variety in antibodies is *unlimited* and there is no viral disease for which humans cannot produce antibodies.

Figure 13.4 is a graph showing immunological memory.

Day 1: First exposure to antigen *A*.

Day 15: Maximum production of antibodies against *A*; the individual will fight off the infection.

Day 28: There is another exposure (a second challenge) to antigen *A*.

Day 28 to Day 35: There is a secondary immune response that is *more rapid* and *more intense* than the primary immune response. This is due to immunological memory.

Day 28: First exposure to antigen *B*. This is another primary immune response. The fact that there was an earlier exposure to antigen *A* has no bearing on antibodies against antigen *B* because adapted immune reactions are very specific.

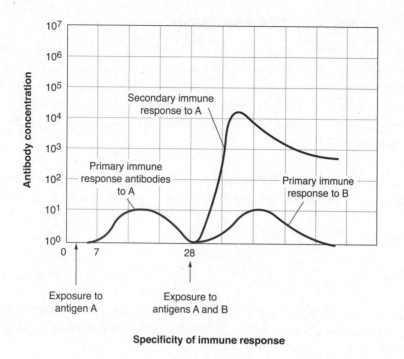

Specificity of immune response

Figure 13.4

TYPES OF IMMUNITY

Table 13.1 describes the two types of immunity.

Table 13.1

Types of Immunity	
Passive Immunity	**Active Immunity**
Temporary	**Permanent**
Antibodies are transferred to an individual from someone else. Examples are maternal antibodies that pass through the placenta to the developing fetus or through breast milk to the baby. Also, a person with a weak immune system often receives an injection of gamma globulin (IgG), which are antibodies culled from many people, to boost the weak immune system.	The individual makes his or her own antibodies after being ill and recovering or after being given an immunization or vaccine. A vaccine contains dead or live viruses or enough of the outer coat of a virus to stimulate a full immune response and to impart lifelong immunity.

BLOOD GROUPS AND TRANSFUSION

ABO antibodies circulate in the plasma of the blood and bind to ABO antigens on the surface of red blood cells. Giving the wrong blood type to someone can cause a transfusion reaction or even death. *The certain danger in a transfusion comes if the recipient has antibodies to the donor's antigens.* However, before someone receives a transfusion of blood, two samples of donor and recipient blood must be mixed to determine compatibility. This procedure is called a **cross-match**.

REMEMBER

The danger in transfusion is if the recipient has antibodies to the donor's antigens.

Blood Type	Antigens Present on the Surface of the Red Blood Cells	Antibodies Present Circulating in the Plasma
A	A	B
B	B	A
O	None	A and B
AB	A and B	None

Blood type O is known as the **universal donor** because it has no blood cell antigens to be clumped by the recipient's blood. Blood type AB is known as the **universal recipient** because there are no antibodies to clump the donor's blood.

Rh Factor

Rh factor is another antigen located on the surface of red blood cells. Most of the population—85 percent—has the antigen and are called Rh^+. Those without the antigen (15 percent) are Rh^-.

AIDS

AIDS stands for acquired immune deficiency disease. People with AIDS are highly susceptible to opportunistic diseases, infections, and cancers that take advantage of a collapsed immune system. The virus that causes AIDS, **HIV (human immunodeficiency virus)**, attacks cells that bear CD4 molecules on their surface, mainly helper T cells. HIV is a retrovirus. Once inside a cell, it reverse transcribes itself, using the enzyme **reverse transcriptase**, and integrates the newly formed DNA into the host cell genome. It remains in the nucleus as a provirus, directing the production of new viruses.

POSITIVE FEEDBACK IN THE IMMUNE SYSTEM

Positive feedback enhances an already existing process until some endpoint or maximum rate is reached. An example in the immune system can be seen in the activity of helper T cells. When a T_h cell becomes activated by a class II MHC molecule, it releases two cytokines, interleukin-I and interleukin-II. Il-2 stimulates B cells and other T cells into action. In this example, *interleukin-I enhances the activity of the already activated T_h cells, stimulating them more until they are activated to a maximum.* This is in contrast to negative feedback, which is a means to achieve stability.

OTHER TOPICS IN IMMUNITY

- Allergies are hypersensitive immune responses to certain substances called **allergens**. They involve the release of **histamine**, an anti-inflammatory agent, which causes blood vessels to dilate. A normal allergic reaction involves redness, a runny nose, and itchy eyes. Antihistamines can normally counteract these symptoms. However, sometimes an acute allergic response can result in a life-threatening response called **anaphylactic shock**, which can result in death within minutes.

- **Antibiotics** are medicines that kill bacteria or fungi. Where vaccines are given to prevent illness caused by viruses, antibiotics are administered after a person is sick.

- **Autoimmune diseases** such as **multiple sclerosis**, **lupus**, **arthritis**, and **juvenile diabetes** are caused by a terrible mistake of the immune system. The system cannot properly distinguish between **self** and **nonself**. It perceives certain structures in the body as nonself and attacks them. In the case of multiple sclerosis, the immune system attacks the **myelin sheath** surrounding certain neurons in the CNS.

- **Monoclonal antibodies** are antibodies produced by a single B cell that have been selected because they produce one specific antibody. They are important in research and in the treatment and diagnosis of certain diseases.

Overview of the Immune System

The immune system has four major characterisitics:

1. **SPECIFICITY:** The entire system depends on the matching of antigens to antigen receptors and the matching of antigens to antibodies.
2. **DIVERSITY:** A wide variety of different cell types defend our bodies from pathogens.
3. **MEMORY:** B and T memory cells circulate for a lifetime.
4. **CAPACITY TO DISTINGUISH SELF FROM NONSELF:** Fortunately, most of the time, the immune system does not attack healthy body cells. This phenomenon is known as self-tolerance.

MULTIPLE-CHOICE QUESTIONS

1. Which of the following is directly responsible for humoral immunity?

    (A) cytotoxic T cells
    (B) helper T cells
    (C) macrophages
    (D) B cells

2. Which of the following is correct about blood type?

    (A) Blood type O has O antigens on the surface of the red blood cells.
    (B) Blood type A has A antibodies circulating in the plasma.
    (C) The danger in a transfusion is if the donor has antibodies to the recipient.
    (D) A and B antigens are on the surface of the red blood cells.

3. All of the following are true of white blood cells EXCEPT

   (A) they are also called leukocytes
   (B) examples are neutrophils and thrombocytes
   (C) Lymphocytes are of two varieties: B type and T type
   (D) T lymphocytes engage in cell-mediated immunity

4. All of the following are true about major histocompatibility molecules EXCEPT

   (A) MHC molecules work in opposition to HLA markers
   (B) MHC molecules are found on every nucleated cell in the body
   (C) class II MHC molecules are found on macrophages
   (D) everybody except identical twins has different MHC molecules on their body cells

5. Which of the following is true about immunity?

   (A) T cells produce antibodies.
   (B) Another name for antibody is antibiotic.
   (C) Juvenile diabetes is an autoimmune disease.
   (D) Macrophages are one type of B lymphocyte.

6. Which of the following would be considered a safe transfusion?

   (A) type A donor, type B recipient
   (B) type AB donor, type O recipient
   (C) type O donor, type AB recipient
   (D) type B donor, type A recipient

Questions 7–11

For the questions 7–11 choose from the choices below. Use each one once and only once.

   (A) B cells
   (B) Macrophages
   (C) Helper T cells
   (D) Cytotoxic T cells

7. An antigen-presenting cell (APC) that is also part of innate immunity

8. Responsible for humoral immunity

9. Bonds to class I MHC molecules

10. A lymphocyte that secretes cytokines to stimulate other lymphocytes

11. Which is NOT part of the lymphatic system?

    (A) spleen
    (B) liver
    (C) tonsils
    (D) adenoids

12. Which is NOT part of the nonspecific immune defense?

    (A) macrophages
    (B) skin
    (C) immunoglobins
    (D) natural killer cells

13. Which is NOT part of the innate immune defense?

    (A) histamine
    (B) prostaglandins
    (C) pyrogens
    (D) cytokines

Question 14

Use the graph below to answer the following question.

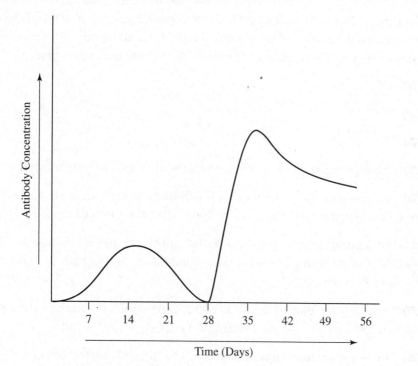

Time (Days)

14. What is happening at day 28?

    (A) The person begins to feel ill.
    (B) Antibody concentration begins to decline.
    (C) This is the first exposure to an antigen.
    (D) This is the second exposure to an antigen.

Answers to Multiple-Choice Questions

1. **(D)** Humoral immunity refers to the production of antibodies. Choice (D), B cells, is the only choice given that produces antibodies. Neutrophils are part of nonspecific immunity. Macrophages are antigen-presenting cells, and helper T cells activate B cells to produce antibodies.

2. **(D)** Blood type O has no antigens on the red blood cells. Blood type A has A antigens on the surface of the red blood cells and B antibodies circulating in the plasma. Type O is the universal donor because it has no antigens on the surface of the red blood cells to be clumped by the recipient's antibodies.

3. **(B)** is the only false statement because thrombocytes is another name for platelets.

4. **(A)** HLA, human leukocytic antigen, is another name for MHC. All other statements are correct.

5. **(C)** Antibodies are produced by B cells, not T cells. Another name for antibody is immunoglobin. Gamma globulin is the most common immunoglobin or antibody. Antibodies kill infected cells by bonding with the antibodies on the surface of the pathogens and clumping them. This makes them visible and easy for phagocytosing cells to gobble them up.

6. **(C)** The rule for safe transfusion is that the recipient must not have antibodies to the donor's blood. Type O blood is the universal donor because type O blood has no antigens on the surface of the red blood cells. Type AB blood is the universal recipient because type AB blood has no ABO antibodies circulating in the blood.

7. **(B)**

8. **(A)**

9. **(D)**

10. **(C)** Although macrophages secrete cytokines, they are not lymphocytes.

11. **(B)** The liver has many functions. It is part of the digestive system and is responsible for detoxifying the body. However, it does not function in the immune system.

12. **(C)** Immunoglobin is another name for antibody, and antibodies are key to the specificity of the immune system. Macrophages are part of both systems, nonspecific and specific immunity.

13. **(D)** Cytokines are part of the body's specific defenses because they are released by helper T cells, which activate T cells and B cells.

14. **(D)** This is the second exposure to the antigen that stimulates the secondary immune response. Antibody production is much more rapid than during the first immune response because antibodies are being produced from circulating memory B cells.

Directions: Answer all questions. You must answer the question in essay—not outline—form. You may use labeled diagrams to supplement your essay, but diagrams alone are *not* sufficient. Before you start to write, read each question carefully so that you understand what the question is asking.

Humans live in a sea of germs protected by three lines of defense.

a. Describe the three lines of defense and how each protects us.

b. Compare and contrast the role of T and B cells.

Typical Free-Response Answer

a. The immune system consists of three lines of defense. The first and second lines of defense are **nonspecific**, while the third line of defense is **specific**. The first line of defense is a barrier that prevents germs from entering the body. These include the **skin, mucous membranes** (which contain the hydrolytic enzymes, lysozyme), **cilia**, and **stomach acid**. The second line of defense, which is more complex, is engaged when microbes breach the first line of defense. The second line of defense is characterized by the **inflammatory response**, redness, swelling, sneezing, and coughing. This is brought about primarily by **histamine**, which triggers vasodilation and thus increases blood supply to the area. In addition, two types of phagocytes, **neutrophils** and **monocytes**, ingest invading microbes. Also, **natural killer cells** destroy virus-infected body cells, **interferons** block the advance of invading viruses, and complement helps destroy cells directly in a number of ways.

    The third line of defense is characterized by specificity and diversity. It is responsible for attacking a broad spectrum of microbes with a variety of specific responses. **B lymphocytes** produce **antibodies**, thousands per second, that fight germs in what is known as humoral response. **T lymphocytes** fight pathogens by what is called **cell-mediated response**, a kind of hand-to-hand combat. There are two types of T cells: cytotoxic T cells or killer T cells, and helper T cells. Both are activated by macrophages that present antigens on their surface for identification. T_c cells destroy body cells that have been infected with viruses by releasing perforin and lysing the cell. **Helper T** cells tell the immune system, with the help of macrophages and some B cells, that foreign antigens have entered the body. T_h cells stimulate T_c cells, B cells, and other T_h cells by releasing the **cytokines**, **interleukin-1 (Il-1)** and **interleukin-2 (Il-2)**.

b. Once a T or B lymphocyte is activated, it becomes metabolically active, proliferating and differentiating into plasma cells and memory cells. **Plasma cells** are short-lived and fight antigens immediately in what is called the primary immune response. This takes about 10–17 days. Memory cells are long-lived cells, bearing receptors specific to the same antigen as the plasma cells. **Memory cells** remain circulating in the blood in small numbers for a lifetime.

Animal Reproduction and Development

14

→ **ASEXUAL REPRODUCTION**

→ **SEXUAL REPRODUCTION**

→ **EMBRYONIC DEVELOPMENT**

→ **FACTORS THAT INFLUENCE EMBRYONIC DEVELOPMENT**

> **Much of this chapter contains background information, but focus on:**
>
> **Embyronic Development**
>
> **Factors that Influence Embryonic Development**

INTRODUCTION

From an evolutionary standpoint, reproduction and passing one's genes to the next generation is the goal of life. Nutrition, transport, excretion, and the other life functions are the processes that enable an organism to live long enough to reproduce. This chapter is primarily about how animals reproduce and develop, although initially there is a brief review of asexual reproduction. A small section covers spermatogenesis and oogenesis; the rest of meiosis is in the chapter entitled "Cell Division." Reflecting the AP Exam, the material on sexual reproduction is at a basic high school level. However, the material on embryonic development is at a college level.

ASEXUAL REPRODUCTION

Asexual reproduction produces offspring genetically identical to the parent. It has several advantages over sexual reproduction:

- It enables animals living in isolation to reproduce without a mate.
- It can create numerous offspring quickly.
- There is no expenditure of energy-maintaining reproductive systems or hormonal cycles.
- Because offspring are **clones** of the parent, it is advantageous when the environment is stable.

The following list shows the types of asexual reproduction in sample organisms.

- **Budding** involves the splitting off of new individuals from existing ones. (Hydra)
- **Fragmentation** and **regeneration** occur when a single parent breaks into parts that regenerate into new individuals. (Sponges, planaria, starfish)
- **Parthenogenesis** involves the development of an egg without fertilization. The resulting adult is haploid. (Honeybees, whiptail lizards)

SEXUAL REPRODUCTION

Sexual reproduction has one major advantage over asexual reproduction: **variation**. Each offspring is the product of both parents and may be better able to survive than either parent, especially in an environment that is changing.

The Human Male Reproductive System

- **Epididymis** is the tube in the testes where sperm gain motility.
- **Leydig cells** are clusters of cells located between seminiferous tubules that produce testosterone.
- The **prostate gland** is the large gland that secretes **semen** directly into the urethra.
- The **scrotum** is the sac outside the abdominal cavity that holds the testes. The cooler temperature there enables sperm to survive.
- **Seminal vesicles** secrete mucus, fructose sugar (which provides energy for the sperm), and the hormone **prostaglandin** (which stimulates uterine contractions) during sexual intercourse.
- **Seminiferous tubules** are the site of sperm formation in the testes.
- **Sertoli cells** provide nutrients for developing sperm.
- **Testes** (*testis*, singular) are the male gonads, where sperm are produced.
- The **urethra** is the tube that carries semen and urine.
- The **vas deferens** is the muscular duct that carries sperm during ejaculation from the epididymis to the urethra in the penis.
- See Figure 14.1.

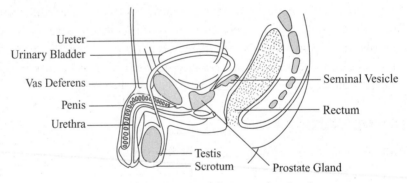

Figure 14.1 The Human Male Reproductive System

The Human Female Reproductive System

- The ovaries are where meiosis occurs.
- The **oviducts** or **Fallopian tubes** are where fertilization occurs. After ovulation, the egg moves through the oviduct to the uterus.
- The **uterus** is where the **blastocyst** will implant and where the embryo will develop during the nine-month **gestation** if fertilization occurs.
- The **endometrium** is the lining of the uterus that thickens monthly in preparation for implantation of the blastocyst.
- The **vagina** is the birth canal. During labor and delivery, the baby passes through the **cervix** (the mouth of the uterus) and into the vagina.
- See Figure 14.2.

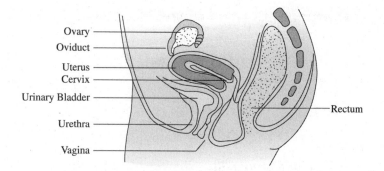

Figure 14.2 The Human Female Reproductive System

The Menstrual Cycle of the Human Female

Follicular phase   Several follicles in the ovaries grow and secrete increasing amounts of estrogen in response to **follicle-stimulating hormone** (**FSH**) from the **anterior pituitary**.

Ovulation   The **secondary oocyte** ruptures out of the ovaries in response to **luteinizing hormone**.

Luteal phase   The **corpus luteum** forms in response to luteinizing hormone. It is the follicle left behind after ovulation and secretes **estrogen** and **progesterone**, which thicken the **endometrium** of the **uterus**.

Menstruation   The monthly shedding of the lining of the uterus when implantation of an embryo does not occur.

Figure 14.3 illustrates the menstrual cycle.

The Menstrual Cycle

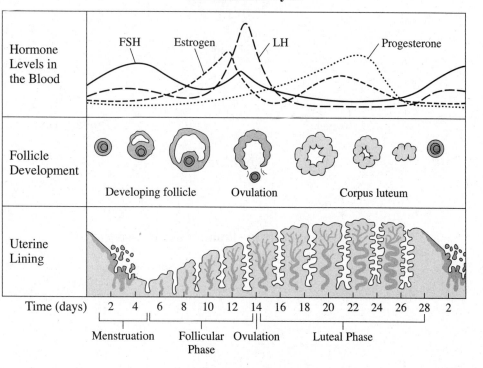

Figure 14.3

Hormonal Control of the Menstrual Cycle

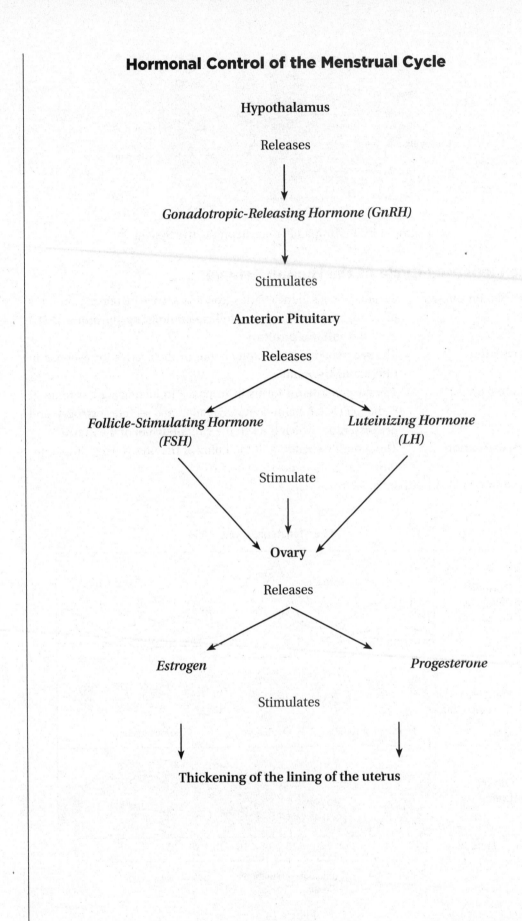

Hypothalamus

Releases

Gonadotropic-Releasing Hormone (GnRH)

Stimulates

Anterior Pituitary

Releases

Follicle-Stimulating Hormone (FSH)   *Luteinizing Hormone (LH)*

Stimulate

Ovary

Releases

Estrogen   *Progesterone*

Stimulates

Thickening of the lining of the uterus

Positive and Negative Feedback of Menstrual Cycle

STUDY TIP

Understanding positive and negative feedback may help you answer essay questions.

- **Positive feedback in the menstrual cycle**: *Positive feedback enhances a process until it is completed.* During the **folliclular phase**, estrogen released from the follicle stimulates the release of **LH** from the **anterior pituitary**. The increase in LH stimulates the follicle to release *even more* estrogen. The hormone levels continue to rise until the follicle matures and ovulation occurs.

- **Negative feedback in the menstrual cycle**: *Negative feedback stops a process once homeostasis is reached.* During the **luteal phase**, **LH** stimulates the **corpus luteum** to secrete **estrogen** and **progesterone**. Once the levels of estrogen and progesterone reach sufficiently high levels, they trigger the **hypothalamus** and **pituitary** to *shut off*, thereby **inhibiting** the secretion of LH and FSH.

Spermatogenesis

Spermatogenesis, the process of sperm production, is a continuous process that starts at puberty. It begins as **luteinizing hormone** (**LH**) induces the **interstitial cells** of the testes to produce **testosterone**. Together with **FSH**, **testosterone** induces maturation of the **seminiferous tubules** and stimulates the beginning of sperm production. In the seminiferous tubules, each **spermatogonium cell** ($2n$) divides by mitosis to produce two **primary spermatocytes** ($2n$). Each undergoes meiosis I to produce two **secondary spermatocytes** (n). Each secondary spermatocyte then undergoes meiosis II, which yields 4 **spermatids** (n). These spermatids **differentiate** and move to the **epididymis** where they become motile. See Figure 14.4.

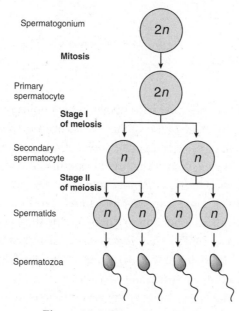

Figure 14.4 Spermatogenesis

Oogenesis

STUDY TIP

Know which cells are haploid and which are diploid in oogenesis and spermatogenesis.

Oogenesis, the production of ova, begins prior to birth. Within the embryo, an **oogonium cell** ($2n$) undergoes *mitosis* to produce **primary oocytes** ($2n$). These remain quiescent within small follicles in the ovaries until puberty, when they become reactivated by hormones. **FSH** periodically stimulates the follicles to complete meiosis I, producing **secondary oocytes** (n), which are released at ovulation. Meiosis II then stops again and does not continue until **fer-**

tilization, when a sperm penetrates the **secondary oocyte**. See Figure 14.5. Oogensis differs from sperm formation in three ways. First, it is a **stop-start process**. It begins prior to birth and is completed after fertilization. Second, cytokinesis divides the cytoplasm of the cell unequally, producing one large cell and *two small polar bodies* which will disintegrate. Third, one primary oogonium cell produces only *one active egg cell.*

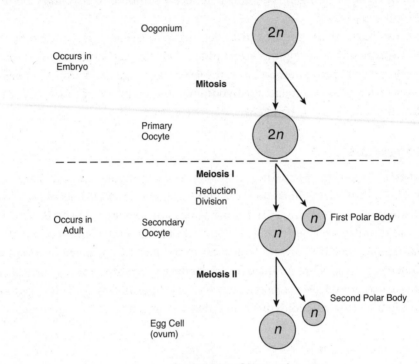

Figure 14.5 Oogenesis

Fertilization

Fertilization, the fusion of sperm and ovum nuclei, is a complex process. It begins with the **acrosome reaction**, when the head of the sperm, the **acrosome**, releases hydrolytic enzymes that penetrate the *jelly coat* of the egg. Specific molecules from the sperm bind with *receptor molecules* on the *vitelline membrane* before the sperm come in contact with the ovum's plasma membrane. This *specific recognition* ensures that the egg will be fertilized by only sperm from the same species. Once a sperm binds to receptors on the egg, the membrane is dramatically depolarized and no other sperm can penetrate the egg membrane. This change in the membrane is known as the *fast block to polyspermy*. This fast block lasts only a minute, just long enough to allow the *slow block to polyspermy*. The slow block converts the vitelline membrane into an impenetrable *fertilization envelope*.

Although the fusion of the sperm and egg triggers this activation, an unfertilized egg can be activated artificially by electrical stimulation or by injection with Ca^{++}. The development of an unfertilized egg is called **parthenogenesis**. The adult that results is haploid. **Drone honeybees** develop by natural parthenogenesis from unfertilized eggs and are haploid males.

EMBRYONIC DEVELOPMENT

Embryonic development consists of three stages: **cleavage**, **gastrulation**, and **organogenesis**. The stages below generally describe the development of all animal eggs but are typical of the **sea urchin** that has almost no yolk. In eggs with more yolk, such as those of the **frog**, cleavage is unequal, with very little cell division in the yolky region. In eggs with a great deal of yolk, such as a **bird egg**, cleavage is limited to a small, nonyolky disc at the top of the egg.

Cleavage *is the rapid mitotic cell division of the zygote that occurs immediately after fertilization.* In general, cleavage produces a fluid-filled ball of cells called a **blastula**. In mammals, the embryo at this stage is called a **blastocyst**. Clustered at one end of the **blastocoel cavity** is a group of cells called the **inner cell mass** that will develop into the **embryo**. The cells of the very early blastocoel stage are **pluripotent** and are the source of *embryonic stem cell lines*.) The cells that surround the inner cell mass are the **trophoblast**. They secrete fluid, creating the **blastocoel**, and also form structures that will attach the embryo to the mother's uterus. See Figure 14.6.

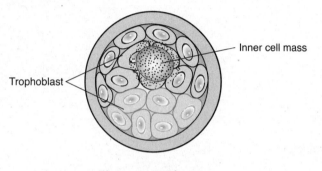

Figure 14.6 Blastocyst

During the next stage of development, the cells of the blastocyst or blastula communicate with each other and begin to differentiate. In many animals, the movement of the cells is so regular that it is possible to label a specific **blastomere** (individual cells of a blastocyst) and identify the tissue that results as embryonic development proceeds. Labeling these cells produces a fate map.

Gastrulation is a process that involves rearrangement of the blastula or blastocyst and begins with the formation of the **blastopore**, an opening into the blastula. In some animals, the blastopore becomes the mouth; in other animals (deuterostomes), the blastopore becomes the anus. Some of the cells on the surface of the embryo migrate into the blastopore to form a new cavity called the **archenteron** or primitive gut. As a result of this cell movement, gastrulation forms a **three-layered** embryo called a **gastrula**. These three differentiated layers, called **embryonic germ layers**, are the **ectoderm**, **endoderm**, and **mesoderm**. They will develop into all the parts of the adult animal.

- The **ectoderm** will become the **skin** and the **nervous system**.
- The **endoderm** will form the **viscera** including the lungs, liver, and digestive organs, and so on.
- The **mesoderm** will give rise to the **muscle, blood,** and **bones**. Some primitive animals (sponges and cnidarians) develop a noncellular layer, the **mesoglea,** instead of the mesoderm. See Figure 14.7.

Organogenesis is organ building. It is the process by which cells continue to **differentiate**, producing organs from the three embryonic germ layers. Three kinds of morphogenetic changes—folds, splits, and dense clustering called condensation—are the first evidence of organ building. Once all the organ systems have been developed, the embryo simply increases in size.

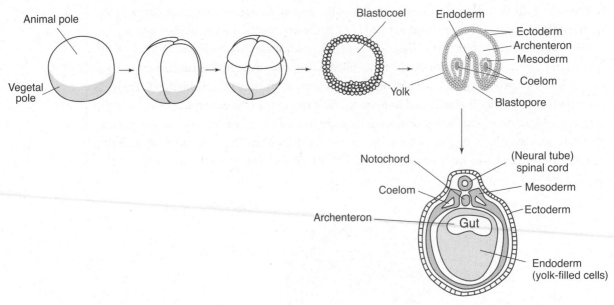

Figure 14.7

The Frog Embryo

Fertilization: One third of the frog egg is yolk, which is massed in the lower portion of the egg. This yolky portion of the egg is called the **vegetal pole**. The top half of the egg is called the **animal pole** and has a **pigmented cap**. The animal-vegetal asymetry dictates where the anterior-posterior axis forms in the embryo. Eggs are laid directly into water and fertilization is external. When the sperm penetrates the egg, the pigmented cap rotates toward the point of penetration and a **gray crescent** appears on the side opposite the point of entry of the sperm. The gray crescent is a marker of the future doral side and is critical to normal development of the growing embryo. (See "The Gray Crescent" on page 289.)

Cleavage and gastrulation: Because of the presence of yolk, cleavage is uneven. The **blastopore** forms at the border of the gray crescent and the *vegetal pole*. Cells at the **dorsal lip** above (dorsal to) the blastopore begin to stream over the dorsal lip and into the blastopore in a process called involution. As these ectoderm cells stream inward by what is called *epibolic movement*, the blastocoel disappears and is replaced by another cavity called the **archenteron**. The region of mesoderm lining the archenteron that formed opposite the blastopore is called the **dorsal mesoderm**.

Organogenesis: In chordates, the organs to form first are the **notochord**, the skeletal rod characteristic of all chordate embryos, and the **neural tube**, which will become the **central nervous system**. The neural tube forms from the **dorsal ectoderm** just above the notochord. Both form by **embryonic induction** (see "Embryonic Induction" on page 290). After the blueprints of the organs are laid down, the embryo develops into a larval stage, the tadpole. Later, **metamorphosis** will transform the tadpole into a frog.

The Bird Embryo

Cleavage and gastrulation: The bird's egg has so much yolk that development of the embryo occurs in a flat disc or **blastodisc** that sits on top of the yolk. A **primitive streak** forms instead of a gray crescent. Cells migrate over the primitive streak and flow inward to form the **archenteron**. As cleavage and gastrulation occur, the yolk gets smaller.

Extraembryonic membranes: Tissue outside the embryo forms four **extraembryonic membranes** necessary to support the growing embryo inside the shell. They are the **yolk sac, amnion, chorion,** and **allantois**. The **yolk sac** encloses the yolk, food for the growing embryo. The **amnion** encloses the embryo in protective **amniotic fluid**. The **chorion** lies under the shell and allows for the diffusion of respiratory gases between the outside and the growing embryo. The **allantois** is analogous to the placenta in mammals. It is a conduit for respiratory gases between the environment and the embryo. It is also the repository for **uric acid**, the **nitrogenous waste** from the embryo that accumulates until the chick hatches. See Figure 14.8.

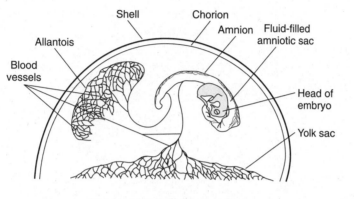

Figure 14.8 The bird embryo

FACTORS THAT INFLUENCE EMBRYONIC DEVELOPMENT
Cytoplasmic Determinants

STUDY TIP

This topic is important because it's about cell communication.

When an eight-ball sea urchin embryo is separated into two halves, the future development of the two halves depends on the plane in which they are cut. If the dissection is longitudinal, producing embryos containing cells from both animal and vegetal poles, subsequent development is normal. If, however, the plane of dissection is horizontal, the result is four abnormally developing embryos. *This demonstrates that the cytoplasm surrounding the nucleus has profound effects on embryonic development.* The importance of the cytoplasm in the development of the embryo is known as **cytoplasmic determinants**; see Figure 14.9.

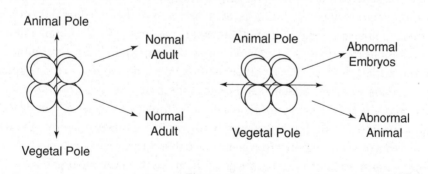

Figure 14.9

The Gray Crescent

Hans Spemann, in his now-famous experiment, demonstrated the importance of the **cytoplasm** associated with the **gray crescent** in the normal development of the animal. He dissected embryos in the two-ball stage in different ways. Only the cell containing the gray crescent developed normally. In addition, these experiments provide more proof that the cytoplasm plays a major role in determining the course of embryonic development; see Figure 14.10.

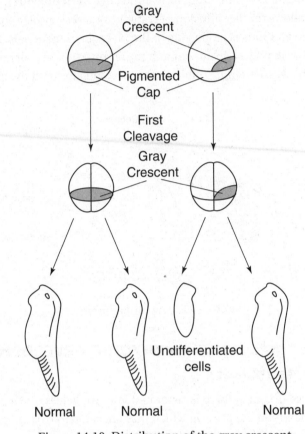

Figure 14.10 Distribution of the gray crescent affects the development of embryos

Embryonic Induction

Embryonic induction is the ability of one group of embryonic cells to influence the development of another group of embryonic cells. Spemann proved that the **dorsal lip** of the blastopore normally initiates a chain of inductions that results in the formation of a **neural tube**. In the now famous experiment, he grafted a piece of **dorsal lip** from one amphibian embryo onto the ventral side of a second amphibian embryo. What developed on the recipient was a complete secondary embryo attached at the site of the graft. *The dorsal lip induced the abdomen tissue above it to become neural tissue.* Because it plays a crucial role in development, Spemann named the dorsal lip the **primary embyronic organizer** or simply the organizer. After many years of study, scientists have identified that the protein β–catenin is a likely candidate for the transcription factor that triggers which cells become the organizers.

Homeotic, Homeobox, or Hox Genes

Homeotic, **homeobox**, or **Hox genes** are **master regulatory genes** that control the expression of genes that regulate the placement of specific anatomical structures. They play a critical role in normal embryonic development. A **homeotic gene** might give the instruction "place legs here" in the developing embryo.

Questions 1–4

Questions 1–4 refer to the list below of primary germ layers.

    (A) Ectoderm

    (B) Endoderm

    (C) Mesoderm

1. Gives rise to the lining of the digestive tract

2. Gives rise to the brain and eye

3. Gives rise to the blood

4. Gives rise to the bone

---

5. In human females, fertilization normally occurs in the _____ and implantation occurs in the _____.

    (A) ovary; uterus

    (B) Fallopian tube; uterus

    (C) ovary; oviduct

    (D) oviduct; vagina

Questions 6–10

Questions 6–10 refer to the list of terms below about hormones. Choose the best answer that fits each of the following descriptions.

    (A) Follicle-stimulating hormones (FSH)

    (B) Oxytocin

    (C) Gonadotropic-releasing hormone (GnRH)

    (D) Estrogen

6. Causes labor

7. Released by the hypothalamus and stimulates the anterior pituitary

8. Stimulates the ovary to mature a secondary oocyte

9. Responsible for thickening the endometrial lining of the uterus

10. Stimulates sperm production

Questions 11–13*

How many chromosomes are in each of the following human cells?

11. Primary spermatocyte?

    (A) 23
    (B) 46
    (C) 96

12. Spermatogonium cell?

    (A) 23
    (B) 46
    (C) 96

13. Spermatid?

    (A) 23
    (B) 46
    (C) 96

---

14. During the menstrual cycle, what is the main source of progesterone in human females?

    (A) anterior pituitary
    (B) posterior pituitary
    (C) hypothalamus
    (D) corpus luteum

15. Sperm gain motility in the

    (A) vas deferens
    (B) epididymis
    (C) seminiferous tubules
    (D) interstitial cells

16. In vertebrate animals, one primary oogonium develops into _____ active egg cell(s).

    (A) 1
    (B) 2
    (C) 4
    (D) 8

17. Which is FALSE about embryonic development?

    (A) Early embryonic division in deuterostomes is spiral.
    (B) The hollow ball stage is called the blastula.
    (C) The end of gastrulation is defined by the formation of primary germ layers.
    (D) The archenteron is the primary gut.

*These questions have only 3 answer choices.

For questions 18–21 match the descriptions with the extraembryonic membranes below.

    (A) yolk sac

    (B) allantois

    (C) amnion

    (D) chorion

18. It is analogous to the placenta in mammals. It is for the diffusion of nutrients and wastes.

19. It provides food for the growing embryo.

20. It lies beneath the shell and allows for the exchange of O_2 and CO_2 between the egg and the outside.

21. It protects the developing embryo from physical trauma.

---

22. What does gastrulation accomplish?

    (A) It changes a solid ball into a hollow ball.

    (B) It changes a blastula into a morula.

    (C) It creates the neural tube.

    (D) It creates a hollow embryo with three tissue layers.

23. Spemann bisected a two-ball stage of a frog embryo and found that only the cell containing the gray crescent developed normally. This demonstrated that

    (A) the gray crescent contains the DNA necessary for normal development

    (B) the gray crescent is really the nucleus

    (C) the gray crescent is necessary for proper development under only certain conditions

    (D) the cytoplasm plays a major role in determining the course of embryonic development

24. Spemann grafted a piece of dorsal lip from one amphibian embryo onto the ventral side of a second embryo. A second notochord and neural tube developed at the location of the graft. This experiment proved that

    (A) embryonic development does not follow any particular developmental pathway and can be easily altered

    (B) the dorsal lip can transform into the archenteron

    (C) the dorsal lip can transform into any organ or structure

    (D) the dorsal lip is an inducer that causes adjacent tissue to transform into some structure

25. Master genes that control the expression of other genes responsible for anatomical structures are called

    (A) mesoglea

    (B) acrosomes

    (C) Hox genes

    (D) cortical genes

26. Which of the following is TRUE about parthenogenesis?

   (A) It is the form of asexual reproduction carried out by Protista.
   (B) It is a primitive form of sexual reproduction.
   (C) It involves the development of the egg into an adult without fertilization.
   (D) The adult that results from parthenogenesis is diploid.

27. Embryonic induction is best illustrated by which of the following?

   (A) development of the chorion in a developing chick embryo
   (B) replacement of cartilage with bone in a developing human embryo
   (C) development of the ectoderm into skin
   (D) development of the neural tube after contact with the dorsal mesoderm

Answers to Multiple-Choice Questions

1. **(B)** The endoderm gives rise to the viscera, the internal organs.

2. **(A)** The ectoderm gives rise to the skin and the nervous system. The eye is part of the peripheral nervous system.

3. **(C)** The mesoderm gives rise to the blood, bones, and muscle.

4. **(C)** The mesoderm gives rise to the blood, bones, and muscle.

5. **(B)** Fact.

6. **(B)** Oxytocin is produced in the hypothalamus and released from the posterior pituitary.

7. **(C)** Gonadotropic-releasing hormone from the hypothalamus stimulates the anterior pituitary to release hormones, such as FSH and LH.

8. **(A)** FSH is released by the anterior pituitary and stimulates the follicle in the ovary to mature a secondary oocyte.

9. **(D)** Estrogen and progesterone are responsible for thickening the lining of the uterine wall in preparation for implantation of an embryo.

10. **(A)** FSH is active in males as well as females. In males, it stimulates sperm production; in females, it stimulates the maturation of a secondary oocyte in the ovary.

11. **(B)** See #13 below.

12. **(B)** See #13 below.

13. **(A)** The spermatogonium cell has 46 chromosomes. It undergoes mitosis and produces two primary spermatocytes, each containing 46 chromosomes. Each primary spermatocyte can undergo meiosis I, producing two secondary spermatocytes each containing 23 chromosomes. Each secondary spermatocyte then undergoes meiosis II, which yields 4 spermatids (n). Each spermatid undergoes differentiation and becomes an active sperm.

14. **(D)** The corpus luteum produces progesterone, which thickens the uterine wall.

15. **(B)** Sperm gain motility in the epididymis.

16. **(A)** In vertebrate animals, one primary oogonium develops into one active egg cell and two polar bodies. The polar bodies disintegrate.

17. **(A)** Early embryonic division in deuterostomes is radial. All the other choices are correct statements.

18. **(B)**

19. **(A)**

20. **(D)**

21. **(C)**

22. **(D)** The end of gastrulation is marked by the formation of the three embryonic layers, the ectoderm, endoderm, and mesoderm.

23. **(D)** During embryonic development, the cytoplasm of the egg plays a major role in influencing the development of the embryo.

24. **(D)** Embryonic induction is the ability of one group of cells to influence the development of another group of embryonic cells. The dorsal lip has the ability to induce tissue to which it is adjacent to become a neural tube.

25. **(C)** Hox genes are master genes that control the expression of other genes that regulate the placement of specific anatomical parts. Hox stands for homeotic genes.

26. **(C)** In honeybees, the queen bee fertilizes some eggs and does not fertilize others. The ones she fertilizes become female ($2n$) worker bees. The ones she does not fertilize become male (n) drones.

27. **(D)** Embryonic induction is the ability of one group of embryonic cells to influence the development of another group of embryonic cells.

FREE-RESPONSE QUESTIONS

> **Directions:** Answer all questions. You must answer the question in essay—**not** outline—form. You may use labeled diagrams to supplement your essay, but diagrams alone are *not* sufficient. Before you start to write, read each question carefully so that you understand what the question is asking.

1. Discuss the embryonic development of a frog egg.
   a. Describe the embryonic stages and explain what is accomplished at each stage.
   b. Give an example of embryonic induction.
   c. Describe one experiment in embryonic induction, and explain what it demonstrated.

2. Compare and contrast the formation of sperm and eggs.

Typical Free-Response Answers

1a. One-third of the frog egg is yolk, which is massed in the lower portion of the egg and is called the **vegetal pole**. The top half of the egg, the **animal pole**, has a **pigmented cap** and is where most of the cell division will occur. When the sperm penetrates the egg, the pigmented cap rotates toward the point of penetration and a **gray crescent** appears on the side opposite the point of entry of the sperm. The gray crescent exerts major influence on the normal development of the growing embryo. Because the frog is a **deuterostome**, the symmetry during early cleavage is **radial** and **indeterminate**, meaning the fate of the individual cells is not yet decided. Cleavage ends with the formation of the hollow-ball (blastula) stage. The cells of the blastula are called **blastomeres**, and the fluid-filled center of the ball is called the **blastocoel**. A blastopore (the first opening) forms at the border of the gray crescent and the vegetal pole. Because the frog is a deuterostome, the blastula will become the anus and the mouth will develop elsewhere.

   Gastrulation is the process that begins with the formation of the blastopore and ends with the formation of three germ layers: **ectoderm**, **endoderm**, and **mesoderm**. The ectoderm will become the skin and the nervous system. The endoderm will form the viscera, including the lungs, liver, digestive organs, and so on. The mesoderm will give rise to the muscles, blood, and bones.

1b. After the three germ layers are formed, organs begin to form in a process known as **organogenesis**. In chordates, like the frog, the organs to form first are the **notochord**, the skeletal rod characteristic of all chordates, and the neural tube, which will become the central nervous system. The neural tube is formed by **embryonic induction** by the endoderm layer lying under it called the dorsal endoderm. After the blueprints of the organs are laid down, the embryo develops into a larval stage, the tadpole. Later, meta-morphosis will transform the tadpole into a frog.

1c. In an experiment, **Spemann** grafted a piece of **dorsal lip** from one amphibian embryo onto the ventral side of a second embryo. A second notochord and neural tube developed at the location of the graft. This experiment proved the mechanics of embryonic induction. It showed that the dorsal lip is an **inducer** that causes adjacent tissue to transform into some structure that, in this case, it was not originally destined to be.

> *Note: Compare means discuss those things that are similar. Contrast means discuss what is different about the two processes.*

2. There are two stages in meiosis: **meiosis I** (**reduction division**), in which homologous chromosomes separate, and meiosis II, which is like **mitosis**. In meiosis I, each chromosome pairs up precisely with its homologue into a synaptonemal complex by a process called **synapsis** and forms a structure known as a tetrad or bivalent. Synapsis is important for two reasons. First, it ensures that each daughter cell will receive one homologue from each parent. Second, it makes possible the process of crossing-over by which homologous chromatids exchange genetic material. The two stages of meiosis are further divided into phases. At the beginning of meiosis, cells have the diploid chromosome number ($2n$). By the end of meiosis, cells contain the haploid or haploid chromosome number (n). Each meiotic cell division consists of the same four stages as mitosis: prophase, metaphase, anaphase, and telophase.

   In spermatogenesis, each spermatogonium cell undergoes meiosis to produce four active sperm cells. A **diploid spermatogonium** ($2n$) divides by mitosis to produce **primary spermatocytes** ($2n$) of equal size, which undergo meiosis I and yield two secondary spermatocytes (n), also of equal size. These spermatocytes undergo meiosis II yielding four **spermatids** (n), which differentiate in the epididymis to form four active motile sperm.

   During **oogenesis**, a **primary oogonium cell** ($2n$) undergoes mitosis to produce **primary oocytes** ($2n$). These remain quiescent within small follicles in the ovaries until puberty, when they become reactivated by hormones. **FSH** periodically stimulates the follicles to complete meiosis I, producing secondary oocytes (n), which are released at ovulation.

   Oogenesis differs from sperm formation in three ways. First, oogenesis is a stop-start process. The first part occurs prior to birth, the second part after fertilization. Second, cytokinesis divides the cell unequally. Almost all the cytoplasm remains in the egg, the other cells produced, polar bodies, have very little cytoplasm and disintegrate. Third, one primary oogonium cell produces only one active egg cell.

Ecology

15

INTRODUCTION

Ecology is the study of the interactions of organisms with their physical environment and with each other. Here is some introductory vocabulary for the topic.

1. A **population** is a group of individuals of one species living in one area who have the ability of interbreeding and interacting with each other.
2. A **community** consists of all the organisms living in one area.
3. An **ecosystem** includes all the organisms in a given area as well as the abiotic (nonliving) factors with which they interact.
4. **Abiotic factors** are nonliving and include temperature, water, sunlight, wind, rocks, and soil.
5. **Biosphere** is the global ecosystem.

PROPERTIES OF POPULATIONS

Here are 5 properties of populations you should know.

1. Size

Size is the total number of individuals in a population and is represented by N, on the next page.

2. Density

Density is the number of individuals per unit area or volume. Counting the number of organisms inhabiting a certain area is often very difficult, if not impossible. For example, imagine trying to count the number of ants in 1 acre (0.5 ha) of land. Instead, scientists use **sampling**

techniques to estimate the number of organisms living in one area. One sampling technique commonly used to estimate the size of a population is called **mark and recapture**. In this technique, organisms are captured, tagged, and then released. Some time later, the same process is repeated and the following formula is used for the collected data.

$$N = \frac{\text{(number marked in first catch)} \cdot \text{(total number in second catch)}}{\text{number of recaptures in second catch}}$$

Suppose 50 zebra mussels are captured, marked, and released. One week later, 100 zebra mussels are captured and 10 are found to have markings already. When using the formula, the total population would be about 500 zebra mussels.

3. Dispersion

Dispersion is the pattern of spacing of individuals within the area the population inhabits; ee Figure 15.1. The most common pattern of dispersion is **clumped**. Fish travel this way in schools because there is safety in numbers. Some populations are spread in a **uniform** pattern. For example, certain plants may secrete toxins that keep away other plants that would compete for limited resources. **Random** spacing occurs in the absence of any special attractions or repulsions. Trees can be spaced randomly in a forest.

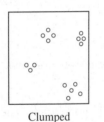
Clumped

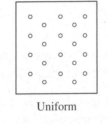
Uniform

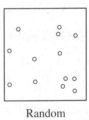
Random

Figure 15.1

4. Survivorship Curves

Survivorship or **mortality curves** show the size and composition of a population. There are three types of survivorship curves.

- **Type 1** curves show organisms with low death rates in young and middle age and high mortality in old age. There is a great deal of parenting, which accounts for the high survival rates of the young. This is characteristic of humans.
- **Type 2** curves describe a species with a death rate that is constant over the life span. This describes the hydra, reptiles, and rodents.
- **Type 3** curves show a very high death rate among the young but then shows that death rates decline for those few individuals that have survived to a certain age. This is characteristic of fish and invertebrates that release thousands of eggs, have external fertilization, and have no parenting; see Figure 15.2.

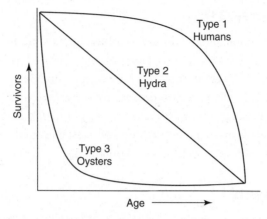

Figure 15.2 Survivorship Curves

5. Age Structure Diagrams

Another important parameter of populations is age structure. An age structure diagram shows the relative numbers of individuals at each age. Figure 16.3 shows two age structure diagrams. Country I shows the age structure of the human population of India; the pyramidal shape is characteristic of developing nations with half the population under the age of 20. Even after taking into account the disease, famine, natural disaster and emigration that will occur, the population in 20 years will be enormous. Country II shows an age structure for a developed nation like the United States with a stable population, **zero population growth**, where the number of people at each age group is about the same and the birth rates and the death rates are about equal.

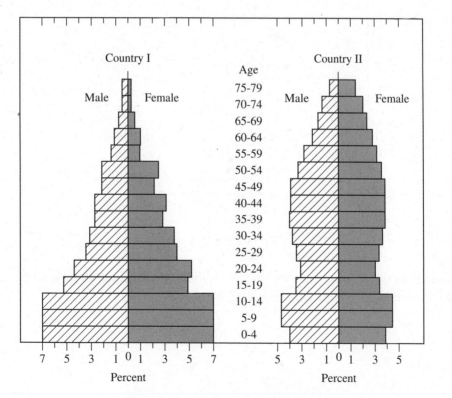

Figure 15.3 Age Structure Diagrams

POPULATION GROWTH

Every population has a characteristic **biotic potential**, the maximum rate at which a population could increase under ideal conditions. Different populations have different biotic potentials, which are influenced by several factors. These factors include *age at which reproduction begins, the life span during which the organisms are capable of reproducing, the number of reproductive periods in the lifetime, and the number of offspring the organism is capable of reproducing*. Regardless of whether a population has a large or small biotic potential, certain characteristics about growth are common to all organisms.

Exponential Growth

The simplest model for population growth is one with unrestrained or **exponential growth**. This population has no predation, parasitism, or competition. It has no immigration or emigration and is in an environment with unlimited resources. This is characteristic of a population that has been recently introduced into an area, such as a sample of bacteria newly inoculated onto a petri dish. Although exponential growth is usually short-lived, the human population has been in the exponential growth phase for over 300 years.

Carrying Capacity

Ultimately, there is a limit to the number of individuals that can occupy one area at a particular time. That limit is called the **carrying capacity (K)**. Each particular environment has its own carrying capacity around which the population size oscillates; see Figure 15.4.

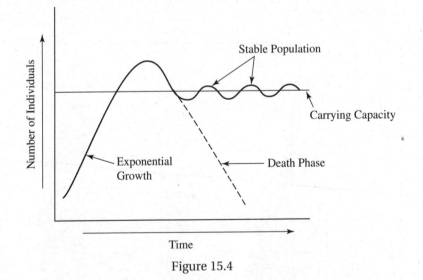

Figure 15.4

In addition, the carrying capacity changes as the environmental conditions change. Perhaps a fire destroyed several acres of forest habitat. See Figure 15.5.

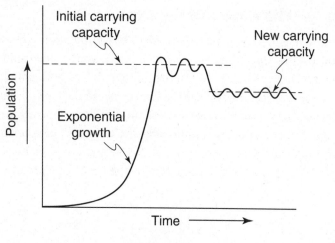

Figure 15.5

Limiting Factors

Limiting factors are those factors that limit population growth. They are divided into two categories: **density-dependent** and **density-independent** factors.

- **Density-dependent factors** are those factors that increase directly as the population density increases. They include competition for food, the buildup of wastes, predation, and disease.
- **Density-independent factors** are those factors whose occurrence is unrelated to the population density. These include earthquakes, storms, and naturally occurring fires and floods.

Growth Patterns

Some species are opportunistic; they reproduce rapidly when the environment is uncrowded and resources are vast. They are referred to as **r-strategists**. Other organisms, the **K-strategists**, live at a density near the carrying capacity (K). Table 15.1 is a chart comparing the two life strategies.

Table 15.1

Comparison of Two Life Strategies	
r-strategists	**K-strategists**
Many young	Few young
Little or no parenting	Intensive parenting
Rapid maturation	Slow maturation
Small young	Large young
Reproduce once	Reproduce many times
Example: insect	Example: mammals

A Case Study—The Hare and the Lynx

An excellent study in population growth involves the populations of **snowshoe hare** and **lynx** at the Hudson Bay Company, which kept records of the pelts sold by trappers from 1850–1930. The data reveal fluctuations in the populations of both animals. The hare feeds on the grass, and the lynx feeds on the hare. So the cycles in the lynx population are probably caused by cyclic fluctuations in the hare population. The hare population experiences cycles of **exponential growth** and **crashes**. Additionally, cycles in the hare population are probably due to a **limited food supply** for the hare due to a combination of malnutrition from cyclical overcrowding and overgrazing and of predation by the lynx; see Figure 15.6.

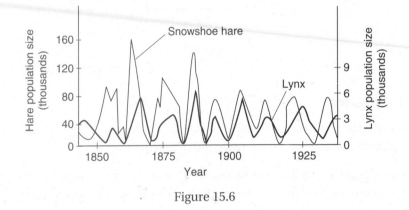

Figure 15.6

COMMUNITY STRUCTURE AND POPULATION INTERACTIONS

Communities are made of populations that interact with the environment and with each other. Communities are characterized by how diverse they are and how dense they are. Species diversity has two components. One is **species richness**, the number of different species in the community. The other is **relative abundance** of the different species. In general, diverse communities are more productive because they are more stable and survive for longer periods of time. They are also better able to withstand and recover from environmental stresses such as drought or an incursion by invasive species.

Interactions within a community are very complex but can be divided into five categories: **competition, predation, parasitism, mutualism,** and **commensalism.**

1. Competition

The Russian scientist **G. F. Gause** developed the **competitive exclusion principle** after studying the effects of interspecific competition in a laboratory setting. He worked with two very similar species, *Paramecium caudatum* and *Paramecium aurelia*. When he cultured them separately, each population grew rapidly and then leveled off at the carrying capacity. However, when he put the two cultures together, *P. aurelia* had the advantage and drove the other species to extinction. His principle states that *two species cannot coexist in a community if they share a* **niche**, *that is, if they use the same resources.*

In nature, there are two related outcomes, besides extinction, if two species *inhabit the same niche and therefore compete for resources.* One of the species will evolve through natural selection to exploit different resources. This is called **resource partitioning.** Another possible outcome is what occurred on the **Galapagos Islands**. Finches evolved different beak sizes through natural selection and were able to eat different kinds of seeds and avoid competition. This divergence in body structure is called **character displacement.**

2. Predation

Predation can refer to one animal eating another animal, or it can also refer to animals eating plants. For their protection, animals and plants have evolved defenses against predation.

- **Plants** have evolved **spines** and **thorns** and chemical **poisons** such as **strychnine**, **mescaline**, **morphine**, and **nicotine** to fend off attack by animals**.**
- **Animals** have evolved **active defenses** such as **hiding**, **fleeing,** or **defending** themselves. These, however, can be very costly in terms of energy. Animals have also evolved **passive defenses** such as **cryptic coloration** or **camouflage** that make the prey difficult to spot. Here are three examples.

  - ✔ **Aposematic coloration** is the very bright, often red or orange, coloration of poisonous animals as a warning that possible predators should avoid them.
  - ✔ **Batesian mimicry** is copycat coloration where one harmless animal mimics the coloration of one that is poisonous. One example is the **viceroy butterfly** which is harmless but looks very similar to the **monarch butterfly,** which stores poisons in its body from the milkweed plant.
  - ✔ In **Müllerian mimicry**, two or more poisonous species, such as the cuckoo bee and the yellow jacket, resemble each other and gain an advantage from their combined numbers. Predators learn more quickly to avoid any prey with that appearance.

3. Mutualism

Mutualism is a symbiotic relationship where both organisms benefit (+/+). An example is the bacteria that live in the human intestine and produce vitamins.

4. Commensalism

Commensalism is a symbiotic relationship where one organism benefits and one is unaware of the other organism (+/o). Barnacles that attach themselves to the underside of a whale benefit by gaining access to a variety of food sources as the whale swims into different areas. In addition, the whale is unaware of the barnacles.

5. Parasitism

Parasitism is a symbiotic relationship (+/−) where one organism, the parasite, benefits while the host is harmed. A tapeworm in the human intestine is an example.

ENERGY FLOW AND PRIMARY PRODUCTION

Every day, Earth is bombarded with enough sunlight to supply the needs of the entire human population for the next 25 years. Most solar radiation, though, is absorbed, scattered, or reflected by the atmosphere. Only a small fraction actually reaches green plants, and less than 1% is actually converted to chemical bond energy by photosynthesis. However, that energy is the basis for almost all of Earth's food chains and fuels all life on Earth. (An example of a food chain that does not rely on solar energy is one located around deep-ocean thermal vents.) Ecologists use two terms when they discuss energy flow on Earth: gross primary productivity and net primary productivity. **Gross primary productivity (GPP)** is the amount of light energy that is converted to chemical energy by photosynthesis per unit time. **Net primary productivity (NPP)** is equal to the GPP minus the energy used by producers for their own cellular respiration.

Different ecosystems vary in their NPP as well as what they contribute to the total or **global NPP** of Earth. Tropical rain forests are among the most productive terrestrial ecosystems and contribute a large portion of Earth's overall net primary production. (Unfortunately, that number is shrinking as we cut down rain forests to make way for farming.) Coral reefs, on the other hand, have a very high NPP but contribute relatively little to the global NPP because they occupy such a tiny part of the planet. The open oceans are just the opposite of coral reefs. Their NPP is very low per unit area. Because they occupy three-fourths of the globe, their global PNN is higher than that of any other biome.

ENERGY FLOW AND THE FOOD CHAIN

The **food chain** is the pathway along which food is transferred from one **trophic** or feeding **level** to another. Energy, in the form of food, moves from the **producers** to the **herbivores** to the **carnivores**. Only about **10 percent** of the energy stored in any trophic level is converted to organic matter at the next trophic level. This means that if you begin with 10,000 kJ of plant matter, the food chain can support 1,000 kJ of herbivores (primary consumers), 100 kJ of secondary consumers, and only 10 kJ of tertiary consumers. As a result of the loss of energy from one trophic level to the next, food chains are rather short. They never have more than four or five trophic levels. As you might expect, long food chains are less stable than short ones. This is because population fluctuations at lower trophic levels are magnified at higher levels, causing local extinction of top predators. A good model to demonstrate the interaction of organisms in the food chain and the loss of energy is the **food pyramid**; see Figure 15.7.

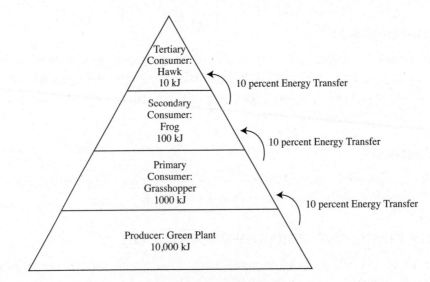

Figure 15.7 Food Pyramid

Food chains are not isolated; they are interwoven with other food chains into a **food web**. An animal can occupy one trophic level in one food chain and a different trophic level in another food chain. Humans, for example, can be primary consumers when eating vegetables but are tertiary consumers when eating a steak. Here are two sample food chains, each with four trophic levels.

Producers → Primary Consumers → Secondary Consumers → Tertiary Consumers

Terrestrial Food Chain
Green Plant → Grasshopper → Frog → Hawk

Marine Food Chain
Phytoplankton → Zooplankton → Small Fish → Shark

- **Producers**
  - ✔ **Autotrophs**
  - ✔ **Green plants**
  - ✔ Convert light energy to chemical bond energy
  - ✔ Have the greatest biomass of any trophic level
  - ✔ Examples: diatoms and phytoplankton

- **Primary consumers**
  - ✔ **Heterotrophs**
  - ✔ **Herbivores**
  - ✔ Eat the producers
  - ✔ Examples: grasshoppers, zooplankton

- **Secondary consumers**
  - ✔ **Heterotrophs**
  - ✔ **Carnivores**
  - ✔ Eat the primary consumers
  - ✔ Examples: frogs, small fish

- **Tertiary consumers**
  - ✔ **Heterotrophs**
  - ✔ **Carnivores**
  - ✔ Eat the secondary consumers
  - ✔ Top of the food chain
  - ✔ Have the **least biomass** of any other trophic level in the food chain.
  - ✔ Least stable trophic level and most sensitive to fluctuations in populations of the other trophic levels
  - ✔ Example: hawk

Dominant and Keystone Species

Dominant species in a community are the species that are the most abundant or that collectively have the highest biomass. They exert control over the abundance and distribution of other species. Sugar maples in North American forests are an example. They affect the abiotic factors, such as shade and soil nutrients (from rotting leaves), which in turn, provide special habitats for many other species.

Keystone species are not abundant in a community. However, they exert major control over other species in the community. Sea otters in the North Pacific are a perfect example.

They are high in the food chain and feed on sea urchins, which feed mainly on kelp. Where the sea otters are abundant, there are few sea urchins and kelp forests are abundant. In contrast, where orcas feed on sea otters, sea urchins are abundant and kelp is rare.

Biological Magnification

Organisms at higher trophic levels have greater concentrations of accumulated toxins stored in their bodies than those at lower trophic levels. This phenomenon is called **biological magnification**. The bald eagle almost became extinct because Americans sprayed heavily with the pesticide DDT in the 1950s, which entered the food chain and accumulated in the bald eagle at the top of the food chain. Because DDT interferes with the deposition of calcium in eggshells, the thin-shelled eggs were broken easily and few eaglets hatched. DDT is now outlawed, and the bald eagle was saved from extinction by human intervention.

Decomposers

Decomposers—**bacteria** and **fungi**—are usually not depicted in any diagram of a food chain. However, without decomposers to recycle nutrients back to the soil to nourish plants, there would be no food chain and no life.

ECOLOGICAL SUCCESSION

Most communities are dynamic, not stable. The size of a population increases and decreases around the carrying capacity. Migration of a new species into a habitat can alter the entire food chain. Major disturbances, whether natural or human-made, like volcanic eruptions, strip mining, clear-cutting a forest, and forest fires, can suddenly and drastically destroy a community or an entire ecosystem. What follows this destruction is the process of sequential rebuilding of the ecosystem called **ecological succession**. See Figure 15.8.

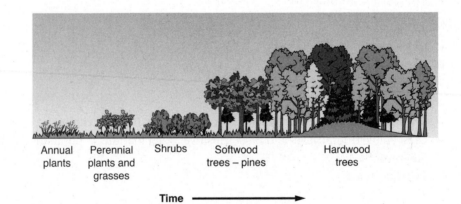

Figure 15.8. Ecological Succession

If the rebuilding begins in a lifeless area where even soil has been removed, the process is called **primary ecological succession**. *The essential and dominant characteristic of primary succession is soil building.* After an ecosystem is destroyed, the first organisms to inhabit a barren area are **pioneer organisms** like **lichens** (a symbiont consisting of algae and fungi) and **mosses**, which are introduced into the area as spores by the wind. Soil develops gradually as rocks weather and organic matter accumulates from the decomposed remains of the pioneer organisms. Once soil is present, pioneer organisms are overrun by other larger organisms: grasses, bushes, and then trees. The final stable community that remains is called the **climax community**. It remains until the ecosystem is once again destroyed by a **blowout**, a disaster that destroys the ecosystem once again.

One example of **primary succession** that was studied in detail is at the southern edge of Lake Michigan. As the lakeshore gradually receded northward after the last ice age (10,000 years ago), it left a series of new beaches and sand dunes exposed. Today, someone who begins at the water's edge and walks south for several miles will pass through a series of communities that were formed in the last 10,000 years. These communities represent the various stages beginning with bare, sandy beach and ending with a climax community of old, well-established forests. In some cases, the climax community is a beech-sugar maple forest, in other areas the forest is a mix of hickory and oak.

The process known as **secondary succession** occurs when an existing community has been cleared by some disturbance that leaves the soil intact. This is what happened in 1988 in Yellowstone National Park when fires destroyed all the old growth that was dominated by lodgepole pine but left the soil intact. Within one year, the burned areas in Yellowstone were covered with new vegetation.

BIOMES

Biomes are very large regions of the earth whose distribution depends on the amount of **precipitation** and **temperature** in an area. Each biome is characterized by **dominant vegetation** and **animal** life. There are many biomes, including freshwater, marine, terrestrial, and acquatic. In the northern hemisphere, from the equator to the most northerly climes, there is a trend in terrestrial biomes: from tropical rain forest, desert, grasslands, temperate deciduous forest, taiga, and finally, tundra in the north. Changes in altitude produce effects similar to changes in latitudes. On the slopes of the Appalachian Mountains in the east and the Rockies and coastal ranges in the west, there is a similar trend in biomes. As elevation increases and temperatures and humidity decrease, one passes through temperate deciduous forest to taiga to tundra. Here is an overview of the major biomes of the world.

STUDY TIP

Know the characteristics of each biome.

Tropical Rain Forest

- Found near the equator with abundant rainfall, stable temperatures, and high humidity.
- Although these forests cover only 4 percent of the earth's land surface, they account for more than 20 percent of the earth's net carbon fixation (food production).
- The most diversity of species of any biome on earth. May have as many as 50 times the number of species of trees as a temperate forest.
- Dominant trees are very tall with interlacing tops that form a dense canopy, keeping the floor of the forest dimly lit even at midday. The canopy also prevents rain from falling directly onto the forest floor, but leaves drip rain constantly.
- Many trees are covered with **epiphytes**, photosynthetic plants that grow on other trees rather than supporting themselves. They are not parasites but may kill the trees inadvertently by blocking the light.
- The most diverse animal species of any biome, including birds, reptiles, mammals, and amphibians.
- Some are biodiversity **hotspots**, meaning that many species are endangered.

Desert

- Less than 10 inches (25 cm) of rainfall per year; not even grasses can survive.
- Experiences the most extreme temperature fluctuations of any biome. Daytime *surface* temperatures can be as high as 158°F (70°C). With no moderating influence of vegetation, heat is lost rapidly at night. Shortly after sundown, temperatures drop drastically.
- Characteristic plants are the drought-resistant cactus with shallow roots to capture as much rain as possible during hard and short rains, which are characteristic of the desert.
- Other plants include sagebrush, creosote bush, and mesquite.
- There are many small annual plants that germinate only after a hard rain, send up shoots and flowers, produce seeds, and die, all within a few weeks.
- Most animals are active at night or during a brief early morning period or late afternoon, when heat is not so intense. During the day, they remain cool by burrowing underground or hiding in the shade.
- Cacti can expand to hold extra water and have modified leaves called spines, that protect against animals attacking the cactus for its water.
- As an example of how severe conditions in a desert can be, in the Sahara Desert are regions hundreds of miles across that are completely barren of any vegetation.
- Characteristic animals include rodents, kangaroo rats, snakes, lizards, arachnids, insects, and a few birds.

Temperate Grasslands

- Covers huge areas in both the temperate and tropical regions of the world.
- Characterized by low total annual rainfall or uneven seasonal occurrence of rainfall, making conditions inhospitable for forests.
- Principal grazing mammals include bison and pronghorn antelope in the United States and wildebeest and gazelle in Africa. Also, burrowing mammals, such as prairie dogs and other rodents, are common.

Temperate Deciduous Forest

- Found in the northeast of North America, south of the taiga, and characterized by trees that drop their leaves in winter.
- Includes many more plant species than does the taiga.
- Shows **vertical stratification** of plants and animals; that is, there are species that live on the ground, the low branches, and the treetops.
- Soil is rich due to decomposition of leaf litter.
- Principal mammals include squirrels, deer, foxes, and bears, which are dormant or hibernate through the cold winter.

Conifer Forest—Taiga

- Located in northern Canada and much of the world's northern regions.
- Dominated by conifer (evergreen) forests, like spruce and fir.
- Landscape is dotted with lakes, ponds, and bogs.
- Very cold winters.
- This is the largest terrestrial biome.

- Characterized by heavy snowfall, trees are shaped with branches directed downward to prevent heavy accumulations of snow from breaking their branches.
- Principal large mammals include moose, black bear, lynx, elk, wolverines, martens, and porcupines.
- Flying insects and birds are prevalent in summer.
- Has greater variety in species of animals than does the tundra.

Tundra

- Located in the far northern parts of North America, Europe, and Asia.
- Characterized by **permafrost**, permanently frozen subsoil found in the farthest point north including Alaska.
- Commonly referred to as the **frozen desert** because it gets very little rainfall and what rainfall occurs cannot penetrate the frozen ground.
- Has the appearance of gently rolling plains with many lakes, ponds, and bogs in depressions.
- Insects, particularly flies, are abundant. As a result, vast numbers of birds nest in the tundra in the summer and migrate south in the winter.
- Principal mammals include reindeer, caribou, Arctic wolves, Arctic foxes, Arctic hares, lemmings, and polar bears.
- Though the number of individual organisms in the tundra is large, the number of species is small.

Aquatic Biomes

Aquatic biomes cover about 75% of Earth. Unlike terrestrial biomes, they are not characterized by a single dominant group of organisms. The primary distinction among aquatic biomes is *salinity*. There are freshwater, estuary, and marine biomes. **Freshwater biomes** have a salinity of less than 0.1% and include rivers, streams, ponds, and wetlands. Some of our freshwater reserves are stored in groundwater. Freshwater makes up less than 4% of Earth's aquatic biomes. **Estuaries** are located at the mouths of rivers where saltwater and freshwater mix. Salt marshes and mangrove forests are estuaries that support enormous populations of animal life. However, the largest biome on Earth is the marine biome with a salinity of 3% on average. Here are the characteristics of marine biomes:

- The largest biome, covering three-fourths of the earth's surface
- The most stable biome with temperatures that vary little because water has a high heat capacity and there is such enormous volume of water.
- Provides most of the earth's food and oxygen
- The marine biome is itself divided into different regions classified by the amount of sunlight they receive, the distance from the shore, and the water depth and whether it is open water or ocean bottom.

CHEMICAL CYCLES

Although the earth receives a constant supply of energy from the sun, chemicals must be recycled. You must know several chemical cycles: **carbon**, **nitrogen**, **water cycles**.

The Water Cycle

Water evaporates from the earth, forms clouds, and rains over the oceans and land. Some rain percolates through the soil and makes its way back to the seas. Some evaporates directly from the land, but most evaporates from plants by **transpiration**.

The oceans contain 97% of the water in the biosphere. About 2% is locked in glaciers and polar ice caps, and the remaining 1% is in lakes, rivers, and ground water. A negligible amount is in the atmosphere.

The Carbon Cycle

The basis of this are the reciprocal processes of **photosynthesis** and **respiration**.

- Cell respiration by animals and bacterial decomposers adds CO_2 to the air and removes O_2.
- Burning of fossil fuels adds CO_2 to the air.
- Photosynthesis removes CO_2 from the air and adds O_2.

The major reservoir of carbon is fossil fuels, plant and animal biomass. Carbon is also found in the soil, in dissolved carbon compounds in the oceans, in sediments in aquatic ecosystems, and in the atmosphere as CO_2 (carbon dioxide) and CO (carbon monoxide).

The Nitrogen Cycle

Very little nitrogen enters ecosystems directly from the air. Most of it enters ecosystems by way of bacterial processes.

- **Nitrogen-fixing bacteria** live in the nodules in the roots of legumes and convert **free nitrogen** into the **ammonium ion** (NH_4^+).
- **Nitrifying bacteria** convert the ammonium ion into **nitrites** and then into **nitrates**.
- **Denitrifying bacteria** convert **nitrates** (NO_3) into **free** atmospheric **nitrogen**.
- **Bacteria of decay** decompose **organic matter** into **ammonia**.

The main reservoir of nitrogen is the atmosphere, which contains about 79% nitrogen gas (N_2). Nitrogen is also found bound in the soil and in lake, river, and ocean sediments. It is also fixed into animal and plant biomass. See Figure 15.9.

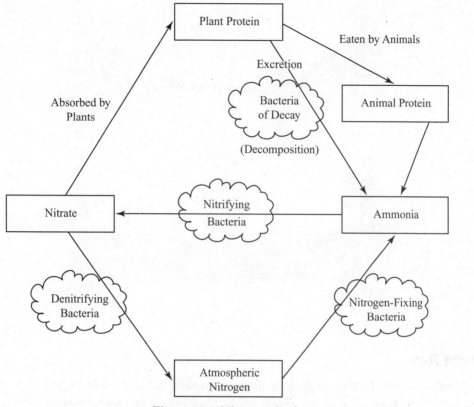

Figure 15.9 Nitrogen Cycle

HUMANS AND THE BIOSPHERE

Humans threaten to make Earth uninhabitable as the population increases exponentially and as people waste natural resources, destory animal habitats, and pollute the air and water. Here are several examples.

Eutrophication of the Lakes

Humans have disrupted freshwater ecosystems, causing a process called **eutrophication**. Runoff from sewage and manure from pastures increase nutrients in lakes and cause excessive growth of algae and other plants. Shallow areas become choked with weeds, and swimming and boating become impossible. As these large populations of photosynthetic organisms die, two things happen. First, organic material accumulates on the lake bottom and reduces the depth of the lake. Second, **detrivores** use up oxygen as they decompose the dead organic matter. Lower oxygen levels make it impossible for some fish to live. As fish die, decomposers expand their activity and oxygen levels continue to decrease. The process continues, more organisms die, the oxygen levels decrease, more decomposing matter accumulates on the lake bottom, and ultimately, the lake disappears. See Figure 15.10.

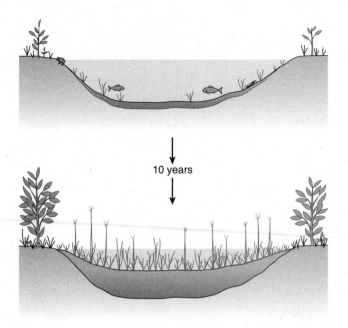

10 years

Figure 15.10

Acid Rain

Acid rain is caused by pollutants in the air from **combustion of fossil fuels**. Nitrogen and sulfur pollutants in the air turn into **nitric**, **nitrous**, **sulfurous**, and **sulfuric acids**, which cause the pH of the rain to be less than 5.6. This kills the organisms in lakes and damages ancient stone architecture.

Toxins

Toxins from industry have gotten into the **food chain**. Most cattle and chicken feed contain **antibiotics** and **hormones** to accelerate animal growth but may have serious ill affects on humans who eat the chicken and beef. Any **carcinogens** or **teratogens** (causing birth defects) that get into the food chain accumulate and remain in the human body's fatty tissues because we occupy the top of the food chain. This process is called biological magnification.

STUDY TIP

One of the four essays will most likely be a mix of ecology and evolution.

Global Warming

To understand global warming, we first must talk about the **greenhouse effect**. CO_2 and water vapor in the atmosphere absorb and retain much of the light and heat that comes to Earth from the sun. If the greenhouse effect did not occur, the average temperature on the surface of Earth would be much colder and life as we know it would not exist. However, atmospheric CO_2 levels have increased by more than 40% during the last 150 years due to the burning of fossil fuels and deforestation. Scientists link increased CO_2 levels to **global warming**. According to NASA, the top four warmest years since the 1890s were the last few years, and global temperatures continue to climb.

The region where global warming has already had a great impact is in the far north, which includes the Arctic tundra and northern coniferous forests. As temperatures rise, snow and ice melt, uncovering darker and more absorbent surfaces. As a result, more radiation is absorbed and Earth is warmed even more. In the summer of 2007, Arctic sea ice covered the smallest area on record. Melting ice and the resulting decrease in habitat endanger Arctic animals such as polar bears, seals, and seabirds. Higher temperatures also increase the likelihood of fires, which destroy even more animal habitats. Melting polar ice causes the seas to rise, resulting in more coastal and inland flooding, flash floods, and erosion.

The only solution to global warming is to reduce CO_2 emissions by industrialized nations and to reduce deforestation, particularly in the tropics. Forests absorb CO_2 from the atmosphere as they grow. They also store carbon in their wood, leaves, and soil. Forests release the CO_2 when they are cut down. Deforestation accounts for about 12% of greenhouse gas emissions. One possible solution to the problem of deforestation would be to pay countries *not* to cut down their forests. This would slow global warming as well as preserve biodiversity.

Global warming could have disastrous effects for the world's population. An increase of 1.0°C on average temperature worldwide would cause the polar ice caps to melt, raising the level of the seas. As a result, major coastal cities in the United States, including New York, Los Angeles, and Miami, would be under water.

Coral reefs like the Great Barrier Reef in Australia are under increased physiological stress due to an increase in global warming. This stress makes it more difficult for coral to build their skeletons. Oysters and sea urchins are also suffering because of increased acidification of the oceans due to an increase in CO_2 dissolved in the oceans.

Acidification of the Oceans

Carbon dioxide from the atmosphere normally dissolves in the oceans by combining with H_2O to form carbonic acid. With increased atmospheric carbon dioxide from burning fossil fuels, the oceans are rapidly becoming more acidic. The acidification of the oceans also directly results in a decrease in the concentration of carbonate ions (CO_3^{2-}) in the oceans. This compound is required by many marine organisms, including reef-building corals and animals that build shells. Since coral reefs provide shoreline protection and support a great diversity of commercial fish species, their destruction would be a great loss. In addition to affecting the shell and reef-building organisms directly, many food chains that include these shell-building animals are also negatively affected.

Depleting the Ozone Layer

The accumulations in the air of **chlorofluorocarbons**, chemicals used for refrigerants and aerosol cans, have caused the formation of a hole in the protective **ozone layer**. This allows more ultraviolet (UV) light to reach the earth, which is responsible for an increase in the incidence of **skin cancer** (melanoma) worldwide.

Introducing New Species

Humans have moved species from one area to another with serious consequences. Two examples are the "killer" honeybees and the zebra mussel.

- The **"killer" honeybee**: The African honeybee is a very aggressive subspecies of honeybee that was brought to Brazil in 1956 to breed a variety of bee that would produce more honey in the tropics than the Italian honeybee. The African honeybees escaped by accident and have been spreading throughout the Americas. By the year 2000, ten people were killed by these bees in the United States.

- The **zebra mussel**: In 1988, the zebra mussel, a fingernail-sized mollusk native to Asia, was discovered in a lake near Detroit. No one knows how the mussel got transplanted there, but scientists infer it was accidentally carried by a ship from a freshwater port in Europe to the Great Lakes. Without any local natural predator to limit its growth, the mussel population exploded. They were first discovered when they were found to have clogged the water intake pipes of those cities whose water is supplied by Lake Erie. To date, the zebra mussel has caused millions of dollars of damage. In addition, the influx of the zebra mussel threatens several native species with extinction by outcompeting indigenous species.

Pesticides vs. Biological Control

Scientists have developed a variety of pesticides, chemicals that kill organisms that we consider to be undesirable. These pesticides include insecticides, herbicides, fungicides, and mice and rat killers. On the one hand, these pesticides save lives by increasing food production and by killing animals that carry and cause diseases like bubonic plague (diseased rats) and malaria (anopheles mosquitoes). On the other hand, exposure to pesticides can cause cancer in humans. Moreover, spraying with pesticides ensures the development of resistant strains of pests through natural selection. The pests come back stronger than before. This problem requires that we spray more and more, which means more people will be exposed to these toxic chemicals.

An alternative to widescale spraying with pesticides is called biological control. The following are some biological methods of getting rid of pests without using dangerous chemicals.

1. Use crop rotation—change the crop planted in a field.
2. Introduce natural enemies of the pests—you must be careful, however, that you do not disrupt a delicate ecological balance by introducing an invasive species.
3. Use natural plant toxins instead of synthetic ones.
4. Use insect birth control—male insect pests can be sterilized by exposing them to radiation and then releasing them into the environment to mate unsuccessfully with females.

1. The high level of pesticides in birds of prey is an example of

   (A) the principle of exclusion
   (B) cycling of nutrients by decomposers
   (C) exponential growth
   (D) biological magnification

2. Which of the following best explains why there are usually no more than five trophic levels in a food chain?

   (A) There are not enough organisms to fill more than five levels.
   (B) There is too much competition among the organisms at the lower levels to support more animals at higher levels.
   (C) The statement is not true; there can be unlimited trophic levels.
   (D) Energy is lost at each trophic level.

3. Which of the following is NOT an important characteristic of the marine biome?

   (A) the largest biome
   (B) provides most of the earth's food
   (C) temperatures vary tremendously
   (D) the largest source of oxygen

4. The most important factors affecting the distribution of biomes are

   (A) temperature and rainfall
   (B) amount of sunlight and human population size
   (C) latitude and longitude
   (D) altitude and water supply

5. Which of the following is NOT an abiotic factor?

   (A) air
   (B) water
   (C) decomposers
   (D) temperature

6. Which of the following encompasses all the others?

   (A) ecosystem
   (B) community
   (C) population
   (D) individual

7. Which of the following lists the biomes as they appear as you move from the equator to the North Pole in the northern hemisphere?

   (A) tropical rain forest—desert—temperate deciduous forest—taiga—tundra
   (B) desert—tundra—taiga—temperate deciduous forest—tropical rain forest
   (C) taiga—temperate deciduous forest—tundra—desert—tropical rain forest
   (D) tundra—taiga—temperate deciduous forest—desert—tropical rain forest

Questions 8–11

Questions 8–11 refer to the survivorship curve shown below.

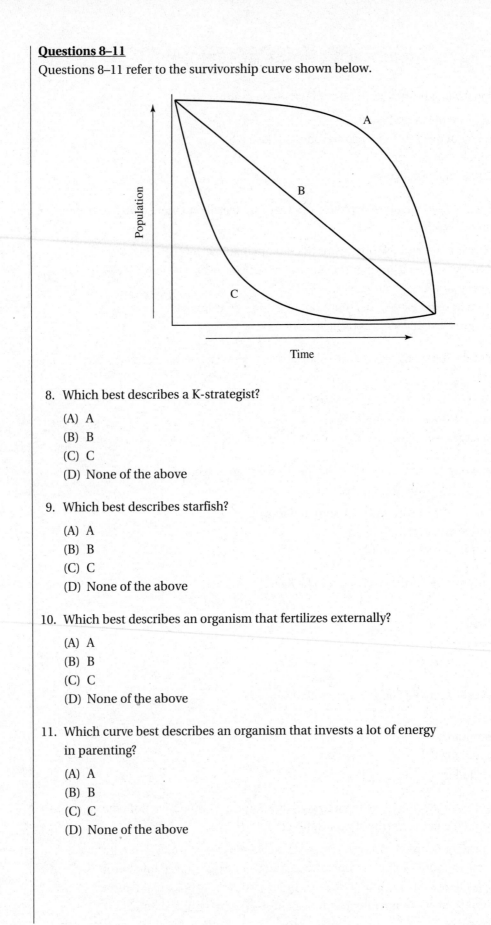

8. Which best describes a K-strategist?

   (A) A
   (B) B
   (C) C
   (D) None of the above

9. Which best describes starfish?

   (A) A
   (B) B
   (C) C
   (D) None of the above

10. Which best describes an organism that fertilizes externally?

    (A) A
    (B) B
    (C) C
    (D) None of the above

11. Which curve best describes an organism that invests a lot of energy in parenting?

    (A) A
    (B) B
    (C) C
    (D) None of the above

12. A species described as r-strategist would definitely NOT have which of the following characteristics?

    (A) clumped population pattern
    (B) much parenting
    (C) many offspring
    (D) random population pattern

13. Which of the following is a density-independent factor limiting human population growth?

    (A) famine
    (B) disease
    (C) competition for food
    (D) naturally occurring fires

14. What would most likely be the cause of bushes of one species growing in one area in a uniform spacing pattern?

    (A) random distribution of seeds
    (B) interactions among individuals in the population
    (C) chance
    (D) the varied nutrient supplies in that area

15. Animals from two different species utilize the same source of nutrition in one area. It is most accurate to say that the animals

    (A) will learn to get along
    (B) will compete for food
    (C) will die because there will not be enough food for both of them
    (D) will learn to eat different foods

16. Two animals live together in close association. One benefits, while the other is unaware of the first animal. The relationship is best described as

    (A) parasitism
    (B) mutualism
    (C) commensalism
    (D) predation

Questions 17–20 refer to the following depiction of a food web for a terrestrial ecosystem.

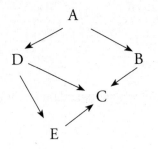

17. Which species is the producer?

   (A) A
   (B) B
   (C) C
   (D) D

18. A toxic pollutant would be found in highest concentrations in which species?

   (A) A
   (B) B
   (C) C
   (D) D

19. Which would have the greatest biomass?

   (A) A
   (B) B
   (C) C
   (D) D

20. Which would have the smallest biomass?

   (A) A
   (B) B
   (C) C
   (D) D

21. Eutrophication in lakes results from

   (A) an increase in ambient temperatures
   (B) a decrease in temperatures
   (C) an increase in carbon dioxide in the air
   (D) an increase in nutrients in the lake

Questions 22–23

Questions 22–23 refer to the graph below that shows changes in population over time.

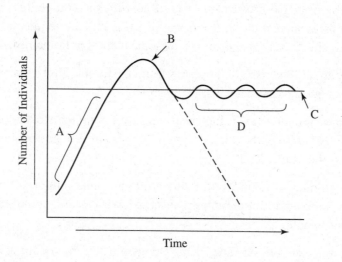

22. Which letter shows a mature, well-established population in favorable conditions?

    (A) A

    (B) B

    (C) C

    (D) D

23. Which letter shows the carrying capacity of the environment?

    (A) A

    (B) B

    (C) C

    (D) D

24. Many poisonous animals are brightly colored (red or orange) as a warning to predators. This special coloration is called

    (A) Müllerian mimicry

    (B) Batesian mimicry

    (C) aposematic coloration

    (D) mutualistic coloration

Answers to Multiple-Choice Questions

1. **(D)** Organisms at higher trophic levels have a greater concentration of accumulated toxins. The principle of exclusion has to do with the competition that arises when two organisms share a niche. Ecological succession is the sequential rebuilding of an ecosystem after it has been destroyed by some natural or human-made disaster.

2. **(D)** Only about 10 percent of the energy from one trophic level is transferred to the next level. The other choices do not make any sense.

3. **(C)** Because water has a very high heat capacity, it requires a lot of time to heat up and cool down. Therefore, the temperature of the oceans remains relatively constant and moderates the nearby land.

4. **(A)** The amount of rainfall and temperature are most important in distribution of biomes. Choice B must be eliminated because it is irrelevant. Altitude does not, in itself, determine climate.

5. **(C)** Decomposers are bacteria, living things. All the others are abiotic factors, not living.

6. **(A)** From the most specific to the most general: individual, population, species, community, and ecosystem.

7. **(A)** As you go toward the North Pole or up in elevation from the equator, the temperatures decrease.

8. **(A)** There is high survival rate of the young and death occurs in old age. The curve is flat in the beginning, and then drops at old age.

9. **(C)** Fertilization in the sea star (starfish) is external, so survival of the young is poor. That is why the curve dips steeply initially.

10. **(C)** Fertilization is external with no parenting, so survival of the young is poor. This is why the curve dips steeply initially.

11. **(A)** This is a K-strategist. K-strategists are characterized by intensive parenting of a few young whose maturation is slow.

12. **(B)** Population pattern is not tied to r-strategists or K-strategists. Think of *r* as standing for "risky." There is little or no parenting and high mortality of the young in the r-strategists.

13. **(D)** Fires are a natural occurrence and are independent of population.

14. **(B)** Plants may secrete toxins that keep other plants from growing nearby. This minimizes competition for limited resources.

15. **(B)** This is a restatement of Gause's principle of exclusion.

16. **(C)** An example of commensalism is the barnacle that attaches to the bottom of the whale. The whale is unaware of the barnacle, which gains a varied food source as the whale swims to different areas.

17. **(A)** Species A is the producer because both B and D feed on it.

18. **(C)** Species C would accumulate the most toxins because it is at the top of the food chain. The longest chain runs from A to D to E to C.

19. **(A)** The producer always has the greatest biomass.

20. **(C)** The top consumer always has the least biomass.

21. **(D)** Eutrophication means "true feeding" and results from runoff of sewage and manure from pastures; phosphates, nitrates, and sulfates are the major contaminants.

22. **(D)** The population at D fluctuates around the carrying capacity (C) for that environment.

23. **(C)** The carrying capacity is the maximum population size that can be supported by the available resources. It is symbolized as K.

24. **(C)** This is the definition of aposematic coloration.

FREE-RESPONSE QUESTIONS

Directions: Answer all questions. You must answer the question in essay—**not** outline—form. You may use labeled diagrams to supplement your essay, but diagrams alone are *not* sufficient. Before you start to write, read each question carefully so that you understand what the question is asking.

1. Describe the process of ecological succession.
2. Name five properties of populations and explain what they tell us about a population.
3. Explain the water, nitrogen, and carbon cycles.
4. Discuss four environmental issues that affect humans today.
5. Discuss the flow of energy through a food web. Include the recycling of energy.

Typical Free-Response Answers

Note: The essays in this section are simple and straightforward. Answering these essays is good practice to help you learn the vocabulary and the concept in the study of ecology. One essay below has the answer written out. Some other questions are listed below but with no answers, only key words to guide you as you write your essay. It would be good practice for you to answer those on your own after reviewing the material in this chapter. Once again, terms that you should include are written in bold.

1. **Ecological succession** is a sequential building or rebuilding of an entire ecosystem. The process is called **primary succession** if it begins in a virtually lifeless area where soil has yet to form. The first organisms to inhabit a barren area are **pioneer organisms** like **lichens** (a symbiont consisting of algae and fungi) and mosses, which are introduced into the area as spores by the wind. Soil develops gradually as rocks weather and organic matter accumulates from the decomposed remains of the pioneer organisms. Once soil is present, pioneer organisms are overrun by other larger organisms: grasses, bushes,

and then trees, the prevalent form of vegetation for that community. Primary succession can take hundreds or thousand of years to reach the **climax community**, the final stable community that develops. **Secondary succession** occurs when an existing community has been severely damaged by some disturbance that leaves the soil intact. Examples are the fire in Yellowstone Park in 1988 or the massive volcanic explosion of Mount Saint Helens in 1980 or the **clear-cutting** of forests that is ongoing in the Pacific Northwest, where all trees are removed. This process does not take as long as primary succession. Often, though, the same community does not return because the new environment is very different from the way it was when the climax community formed years before.

2. Key words/topics:
   Size
   Density
   Dispersion
   Survivorship curves
   Age structure diagrams

3. Key words/topics:
   Interdependence of organisms on earth
   Water—evaporation, condensation
   Photosynthesis—requires CO_2 and releases O_2
   Cellular respiration—requires O_2 and releases CO_2
   Bacteria of decay
   Nitrogen-fixing bacteria
   Nitrifying bacteria
   Denitrifying bacteria

4. Key words/topics:
   Interdependence of organisms on earth
   Eutrophication of lakes
   Acid rain
   Toxins in the food chain
   Global warming
   Depleting the ozone layer
   Introducing new species

5. Key words/topics:
   Interdependence of organisms in an ecosystem and on earth
   Sunlight/energy
   Producers
   Primary consumers
   Secondary consumers
   Tertiary consumers
   Food webs
   Food chains
   Biological magnification or amplification of toxins
   Decomposers and the cycling of nutrients

Animal Behavior

16

→ **FIXED ACTION PATTERN**
→ **MIGRATION**
→ **ANIMAL SIGNALS AND COMMUNICATION**
→ **LEARNING MODIFIES BEHAVIOR**
→ **SOCIAL BEHAVIOR**
→ **NATURAL SELECTION AND REPRODUCTIVE SUCCESS**
→ **EVOLUTION OF BEHAVIOR**

INTRODUCTION

Animal behavior is carried out in response to internal or external *stimuli*. It can be solitary or social, fixed or variable. Whatever the behavior, it enables organisms either to search for food or to find a mate. It evolved because of substantial evolutionary pressures. When we speak about behavior, we refer to either the proximate or the ultimate causes of behavior. The **proximate causes** are the immediate, genetic, physiological, neurological, and developmental mechanisms that determine how an individual behaves. The **ultimate causes** result from the evolutionary pressures that have fashioned an animal's behavior.

The study of behavior and its relationship to its evolutionary origins is called **ethology**. Foremost in the field of ethology are three scientists who shared the Nobel Prize in Physiology and Medicine in 1973: Karl von Frisch, Konrad Lorenz, and Niko Tinbergen. **Karl von Frisch** is known for his extensive studies of communication in honeybees and his famous description of their **waggle dance**. **Niko Tinbergen** is known for his elucidation of the **fixed action pattern**. **Konrad Lorenz** is famous for his work with **imprinting**.

As you study each topic, think of how a particular behavior benefits one animal and makes it more fit than other animals. Also think about how the behavior enhances the animal's survival and how the behavior might have evolved.

FIXED ACTION PATTERN

A **fixed action pattern (FAP)** is an innate, *highly stereotypic behavior,* that once begun is continued to completion, no matter how useless. FAPs are initiated by external stimuli called **sign stimuli**. When these stimuli are exchanged between members of the same species, they are known as **releasers**. An example of an FAP studied by **Tinbergen** involves the **stickleback fish**, which attacks other males that invade its territory. The **releaser** for the attack is the red belly of the intruder. The stickleback will not attack an invading male stickleback lacking a red underbelly, but it will readily attack a nonfishlike wooden model as long it has a splash of red visible.

MIGRATION

Animals **migrate** in response to environmental stimuli, like changes in day length, precipitation and temperature. The environment also provides cues to navigation.

- Some migrating animals track their position relative to the sun. Although the sun's position changes throughout the day, animals can monitor changes in the position of the sun against an internal **circadian clock** to keep track of where they are.
- Nocturnal animals keep track of their position using the North Star, which has a fixed position in the sky.
- Pigeons track their positions relative to Earth's magnetic field.
- Gray whales migrate seasonally between the Bering Sea and the coastal lagoons of Mexico. They do this by knowing and remembering elements in their environment. Orienting by landmarks is called **piloting**.

ANIMAL SIGNALS AND COMMUNICATION

Animals mark their territory with chemical signals called **pheromones.** These chemicals can also act as alarm signals. For example, if a catfish is injured, a substance is released from its skin that disperses in the water and induces a fright *response* in other fish. The frightened fish become hypervigilant and form into tightly packed schools for protection at the lake or river bottom.

Visual signals are effective in open places in daylight. They provide information about many factors including the sex, strength, and social status of an individual. An example of symbolic communication is the waggle dance in bees that gives details about the location of a food source.

LEARNING MODIFIES BEHAVIOR

Learning is a sophisticated process in which the responses of the organism are modified as a result of experience. The capacity to learn can be tied to length of life span and complexity of the brain. If the animal has a very short life span, like a fruit fly, it has no time to learn, even if it had the ability. It must therefore rely on fixed action patterns. In contrast, if the animal lives for a long time and has a complex brain, then a large part of its behavior depends on prior experience and learning.

Habituation

Habituation is one of the simplest forms of learning. An animal comes to ignore a persistent stimulus so it can go about its business. If you tap the dish containing a hydra, it will quickly shrink and become immobile. If you keep tapping, after a while the hydra will begin to ignore the tapping, elongate, and continue moving about. It has becomes **habituated** to the stimulus.

Associative Learning

Associative learning is one type of learning in which one stimulus becomes linked to another through experience. Examples of associative learning are classical conditioning and operant conditioning.

- **Classical conditioning**, a type of **associative learning**, is widely accepted because of the ingenious work of **Ivan Pavlov** in the 1920s. Normally, dogs salivate when exposed to food. Pavlov trained dogs to associate the sound of a bell with food. The result of this conditioning was that dogs would salivate upon merely hearing the sound of the bell, even though no food was present.
- **Operant conditioning**, also called **trial and error** learning, is another type of **associative learning**. An animal learns to associate one of its own behaviors with a reward or punishment and then repeats or avoids that behavior. The best-known studies involving operant conditioning were done by **B. F. Skinner** in the 1930s. In one study, a rat was placed into a cage containing a lever that released a pellet of food. At first, the rat would depress the lever only by accident and would receive food as a reward. The rat soon learned to associate the lever with the food and would depress the lever at will. Similarly, an animal can learn to carry out a behavior to avoid punishment. Such systems of rewards and punishment are the basis of most animal training.

STUDY TIP

The most commonly asked questions concern:

- **Imprinting**
- **Fixed action pattern**
- **Sign stimulus**
- **Releasers**

Imprinting

Imprinting is learning that occurs during a **sensitive** or **critical period** in the early life of an individual and is **irreversible** for the length of that period. When you see ducklings following closely behind their mother, you are seeing the result of successful imprinting. Mother-offspring bonding in animals that depend on parental care is critical to the safety and development of the offspring. If the pair does not bond, the parent will not care for the offspring and the offspring will die. At the end of the juvenile period, when the offspring can survive without the parent, the response disappears.

Classic imprinting experiments were carried out by **Konrad Lorenz** with geese. Geese hatchlings will follow the first thing they see that moves. Although the object is usually the mother goose, it can be a box tied to a string or, in the case of the classic experiment, it was Konrad Lorenz himself. Lorenz was the first thing the hatchlings saw, and they became **imprinted** on the scientist. Wherever he went, they followed.

Examples of Learning and Problem Solving

If a chimpanzee is placed into a room with several boxes on the floor and a banana hanging high out of reach, the chimp will stack the boxes on top of each other until he reaches the banana. This problem-solving behavior is highly developed in some mammals, especially in primates. It has also been observed in some birds, especially ravens. In one study, a raven was able to fly to a branch and step on a string in order to bring some hanging food within reach. Interestingly enough, some ravens were not able to solve the problem, which shows that *problem-solving ability varies with individual ability and experience.*

Learning Sometimes Happens in Stages

Some birds learn songs in stages from other members of their species. The white-crowned sparrow hears the species song within the first 50 days of life. Although it does not sing during this *sensitive period*, the young bird memorizes the song and chirps in response to hearing it. This sensitive period is followed by a second learning phase when the juvenile birds sings tentative notes of the song and compares it with what he hears around him. Once its own song matches what he has heard, the song is "crystallized" as the final white-crowned sparrow song.

Learning From Others

Many animals learn to solve problems by observing the behavior of other individuals. Young wild chimps learn how to crack open oil palm nuts with stones by copying experienced chimps. Wild vervet monkeys in Kenya learn to make alarm calls. At first, they make indiscriminate calls in response to danger and later fine tune the call as they learn from older monkeys in the group.

SOCIAL BEHAVIOR

Social behavior is any kind of interaction among two or more animals, usually of the same species. It is a relatively new field of study, developed in the 1960s. Types of social behaviors are **cooperation**, **agonistic**, **dominance hierarchies**, **territoriality**, and **altruism**.

Cooperation

Cooperation enables the individuals to carry out a behavior, such as hunting, that they can do as a group more successfully than they can do separately. Lions or wild dogs will hunt in a pack, enabling them to bring down a larger animal than an individual could ever bring down alone.

Agonistic Behavior

Agonistic behavior is aggressive behavior. It involves a variety of threats or actual combat to settle disputes among individuals. These disputes are commonly held over access to food, mating, or shelter. This behavior involves both real aggressive behavior as well as ritualistic or symbolic behavior. One combatant does not have to kill the other. The use of symbolic behavior often prevents serious harm. A dog shows aggression by baring its teeth and erecting its ears and hair. It stands upright to appear taller and looks directly at its opponent. If the aggressor succeeds in scaring the opponent, the loser engages in submissive behavior that says, "You win, I give up." Examples of submissive behaviors are looking down or away from the winner. Dogs or wolves put their tail between their legs and run off. Once two individuals have settled a dispute by agonistic behavior, future encounters between them usually do not involve combat or posturing.

Dominance Hierarchies

Dominance hierarchies are pecking order behaviors that dictate the social position an animal has in a culture. This is commonly seen in hens where the alpha animal (top-ranked) controls the behaviors of all the others. The next in line, the beta animal, controls all others except the alpha animal. Each animal threatens all animals beneath it in the hierarchy. The top-ranked animal is assured of first choice of any resource, including food after a kill, the best territory, or the most fit mate.

Territoriality

A **territory** is an area an organism defends and from which other members of the community are excluded. Territories are established and defended by *agonistic behaviors* and are used for capturing food, mating, and rearing young. The size of the territory varies with its function and the amount of resources available.

Altruism

Altruism is described as a behavior that reduces an individual's reproductive fitness (the animal might die) but increases the fitness of the colony or family. How does the trait of altruism remain in a population if the individual carrying the gene or genes that is responsible for the behavior dies?

An explanation can be seen in honeybees, where worker females share 75% of the same genes. When a worker honeybee stings an intruder while defending a hive, the worker sacrifices itself for its relatives, individuals who carry most of the same genes as that particular worker. The individual dies, but relatives survive to pass on their genes, including the gene for altruistic behavior. This concept is known as **kin selection** or **inclusive fitness**.

NATURAL SELECTION AND REPRODUCTIVE SUCCESS
Foraging Behavior: Cost-Benefit Analysis

Benefits are measured in terms of fitness enhancement—the more fit an animal is, the more likely it will get to pass more genes to the next generation. However, there must be a balance between the benefits of a behavior and the costs. Think about **foraging behavior** (all the behaviors involved in food gathering and eating). The benefits of eating are obvious. A robust, well-fed animal will have an advantage in competition for a mate. However, foraging has costs that might not be so obvious. Besides all the energy expended in foraging, it might be dangerous. For example, an animal might have to chase away another predator for the remains of a kill. Time spent foraging is also time lost from defending one's territory or from protecting one's young. *Ultimately, natural selection will favor behavior that minimizes the costs of foraging and maximizes its benefits.*

Mating Behavior and Mate Choice

Beyond the act of copulation and fertilization, mating behavior involves mating rituals, whether the animals are monogamous or polygamous, and the extent of parental care. Mating behavior even dictates morphological characteristics. Among **polygynous species**, such as elk, one male inseminates many females and the males are larger and more highly ornamented. This difference is called **sexual dimorphism**. In animal species where the female mates with more than one male, known as **polyandrous** species, the female is the showier of the two.

EVOLUTION OF BEHAVIOR

Some behaviors are simply controlled by a gene or a set of genes. Think of spiderwebs. Each web is characteristic of a particular species. Since the adult spider dies before the eggs hatch, a spider cannot learn to build a particular web. So the control of the behavior—spinning a web—must be genetic.

To understand the *ultimate cause* of a behavior, one must look at the behavior in terms of what selective advantage it confers on the individual or group. An interesting behavior that exemplifies this is altruism.

MULTIPLE-CHOICE QUESTIONS

Questions 1–4

For questions 1–4, choose from the list of scientists below.

(A) B. F. Skinner
(B) Karl von Frisch
(C) Niko Tinbergen
(D) Ivan Pavlov

1. Studied communication in bees

2. Classical conditioning

3. Operant conditioning

4. Fixed action pattern

Questions 5–8

For questions 5–8, choose categories of animal behavior from the list. You may use each one more than once or not at all.

(A) Fixed action pattern
(B) Associative learning
(C) Classical conditioning
(D) Imprinting

5. Pavlov's dogs

6. One stimulus becomes linked to another

7. Ducklings follow their mother

8. Innate, highly stereotypic behavior that must continue until it is completed

9. Pavlov's dogs learned to associate hearing a bell with food. Simply hearing a bell caused them to salivate. This is an example of

(A) habituation
(B) operant conditioning
(C) classical conditioning
(D) imprinting

10. Ethology is the study of

(A) endocrinology
(B) animal behavior and its relationship to its evolutionary history
(C) the brain and nervous system
(D) operant conditioning

11. "Mary had a little lamb; its fleece was white as snow. And everywhere that Mary went, the lamb was sure to go." The behavior of the lamb is best described as

(A) habituation
(B) imprinting
(C) operant conditioning
(D) classical conditioning

12. You want to train your puppy to wait at the curb until you tell him to cross the road. Your friend advises you to give your dog a treat every time he does as you ask. Your friend is advising that you train the dog using

(A) operant conditioning
(B) classical conditioning
(C) imprinting
(D) fixed action pattern

13. A sign stimulus that functions as a signal to trigger a certain behavior in another member of the same species is called

(A) a ritual
(B) a fixed action stimulus
(C) an inducer
(D) a releaser

14. To begin the mating dance, the male ostrich moves his head in a particular bobbing fashion. This initiates a specific response from the female, and the ritualized mating dance can begin. The male head bobbing is

(A) an imprinting stimulus
(B) a habituation
(C) a fixed action stimulus
(D) a releaser

15. Animals that help other animals are expected to be

    (A) stronger than other animals
    (B) related to the animals they help
    (C) male
    (D) female

16. Which of the following is related to altruistic behavior?

    (A) kin selection
    (B) fixed action pattern
    (C) a search image
    (D) imprinting

Answers to Multiple-Choice Questions

1. **(B)** Karl von Frisch studied and named the waggle dance in bees.

2. **(D)** Classical conditioning involves learning to associate an arbitrary stimulus with a reward or punishment. Ivan Pavlov "trained" dogs to salivate at the sound of a bell.

3. **(A)** Operant conditioning is also called trial and error learning. An animal learns to associate one of its own behaviors with a reward or punishment and then tends to avoid that behavior. The best known lab studies in operant conditioning were done by B. F. Skinner in the 1930s.

4. **(C)** A fixed action pattern enables an animal to engage in complex behavior automatically without having to "think" about it. An FAP is a sequence of behaviors that is unchangeable and usually carried out to completion once initiated. Niko Tinbergen is most associated with FAPs.

5. **(C)** An example of classical conditioning is how Pavlov trained his dogs to salivate at the sound of a bell.

6. **(B)** When one stimulus becomes associated with one response, it is associative learning.

7. **(D)** Konrad Lorenz imprinted his geese onto himself. They thought he was their mother and followed him everywhere.

8. **(A)** This is the definition of fixed action pattern.

9. **(C)** Pavlov is known for his work in classical conditioning by inducing dogs to salivate at the sound of a bell.

10. **(B)** Ethology originated in the 1930s with naturalists who were studying animals in their natural habitat. They studied animal behavior and how it connected to evolution and ecology.

11. **(B)** The reason the lamb followed Mary everywhere is because the lamb was imprinted on Mary. The lamb will continue to follow Mary until the animal is mature enough to live on its own and the sensitive period has passed.

12. **(A)** Operant conditioning is a type of learning that is the basis for most animal training. The trainer encourages a behavior by rewarding the animal. Eventually, the animal will perform the behavior without necessarily receiving a reward.

13. **(D)** A releaser is a sign stimulus that triggers a fixed action pattern among members of the same species.

14. **(D)** The releaser is a signal between two members of the same species that initiates a fixed action pattern. In this case, the releaser is the bobbing of the head. In the stickleback, it is the red color on the underbelly.

15. **(B)** Animals that help other animals are engaging in altruistic behavior. Altruistic behavior is seemingly selfless behavior that may save kin carrying genes similar to the individual that sacrificed itself.

16. **(A)** When an individual sacrifices itself for the family or group, it is sacrificing itself for relatives (the kin) that share similar genes. The kin are selected as the recipients of the altruistic behavior. They are saved and can pass on their genes. Altruism evolves because it increases the number of copies of a gene common to a related group.

FREE-RESPONSE QUESTION

> **Directions:** Answer all questions. You must answer the question in essay—**not** outline—form. You may use labeled diagrams to supplement your essay, but diagrams alone are *not* sufficient. Before you start to write, read each question carefully so that you understand what the question is asking.

Explain Darwin's theory of evolution by natural selection. Each of the following refers to one aspect of evolution. Discuss each term and explain it in terms of natural selection. For a. and b., see the chapter "Evolution."

a. Convergent evolution
b. Insecticide resistance
c. **Fixed action pattern**
d. **Imprinting**

Typical Free-Response Answer

Natural selection favors behavioral patterns that enhance survival and reproductive success. If a behavior does not increase reproductive success, it will be selected against and disappear from the gene pool. Fixed action pattern and imprinting are two behavioral patterns that exist because they favor reproductive success.

A **fixed action pattern (FAP)** is an innate, highly stereotypic behavior that once begun, continues to completion. FAPs are initiated by external stimuli called **sign stimuli**. When these stimuli are exchanged between members of the same species, they are known as **releasers**. An example of an FAP studied by **Tinbergen** involves the stickleback fish, which attacks other males that invade its territory. The releaser for the attack is the red belly of the intruder. The stickleback will not attack an invading male stickleback lacking a red underbelly, but it will readily attack a nonfishlike wooden model as long as a splash of red is visible. If a fixed action pattern enables an animal to survive long enough to reproduce, that animal will have a selective advantage over animals that do not automatically carry out the behavior.

Imprinting is learning that occurs during a sensitive or critical period in the early life of an individual and is irreversible for the length of that period. Konrad Lorenz studied this behavior pattern in geese. The hatchling responds to the first thing it sees that moves, becoming imprinted on the mother and following her everywhere. In species with parental care, mother-offspring bonding is critical to the survival of the offspring. If bonding does not occur, the parent will not initiate care and the offspring will die. If an animal hatches offspring that do not become imprinted on her, she has not reproduced successfully. Her genes will then be selected against and will be lost from the gene pool.

Investigations

17

INTRODUCTION

The new AP investigations are based on the former labs, but the approach has changed. There is no longer any "cookbook science," where the teacher or lab manual dictates what to investigate and how to conduct the investigation. In the new program, teachers will guide students. However, students will design experiments to test hypotheses, conduct investigations, analyze data, and communicate the results to the class. There are now 13 student-directed inquiry-based learning experiences. The minimum number of required investigations has been reduced from 12 to 8.

Besides being an opportunity to do real science, the investigations are important in another way. The College Board has shifted the required curriculum away from traditional teacher instruction onto the lab experience. For example, most topics from the plants unit have been removed from the required curriculum. However, you must learn about plant structure and the function of plants in order to do the photosynthesis and transpiration investigations.

GRAPHING

The purpose of showing data on a graph is to make it clear and easily understood. Here are some guidelines and reminders for constructing graphs.

Label Your Graph

a. Title it.

b. Use the *x*-axis for the **independent variable**, the value that you control, such as time.

c. Use the *y*-axis for the **dependent variable**, the value that changes as a result of changes in the independent variable. For example, if in the course of an experiment, you take a measurement every minute and the chunk of potato gets heavier and heavier, then time is independent and the mass of the potato is dependent.

Plot the Data Points, Then Draw a Best-Fit Line

If you are instructed to connect the dots, do so. Chances are, you will **not** be so instructed. You should draw a **best-fit line** (or curve). This may not be something you ever encountered in math class because the math teacher always provided you with values to graph that formed a perfectly straight line. In science, when you collect data, the numbers rarely fall into a straight line. However, if the line is not straight, you cannot make predictions by simply extending the line. Somehow, you must translate your rough data into a straight line. Here is how.

Plots your points as usual. Then, using a ruler (preferably a transparent one), draw a straight line that best shows the trend (slope) and that takes into account the location of all the data points. If one data point really differs from all the others, you may ignore it. Also, when drawing a best-fit line, your line does not have to pass through any data points and it does not have to pass through zero. Figure 17.1 shows a scatter plot graph with data points plotted and best-fit line drawn.

STUDY TIP

Can you draw a best fit line? Learn how.

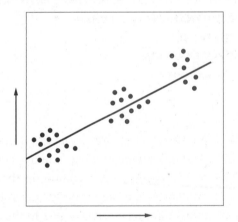

Figure 17.1 Best-Fit Line on Scatter Plot Graph

DESIGNING AN EXPERIMENT

The laboratory-based essay question may direct you to design an experiment. This makes many students nervous. However, if you follow these guidelines, you will have a good basis for devising a satisfactory experiment and writing a fine essay.

STUDY TIP

These guidelines for designing an experiment are very important.

1. **State a clear hypothesis**, what you expect to happen.

2. The **experiment must be reasonable to carry out and must work**. For example, having to set up 100 fish tanks is not reasonable.

3. A "controlled" experiment must **have a control**. The control must be exactly like the experimental except for the single factor you are testing. The control can be an organism left untreated.

4. Any experiment must have **only one variable**. For example, in an experiment where you are testing the effect of various light intensities on the growth of several plants, light intensity must be constant for each plant. To absorb the heat from the light, a heat sink must be placed between the light and the plant. Without something to absorb the heat from the light source, temperature would become a second variable.

5. **Have a large enough sample** to draw a reasonable conclusion. A sample of one organism is not acceptable. In Investigation 1, "Artificial Selection," the sample size is 150.

6. **Experimental organisms must be as similar as possible**. They must all be of the same variety, size, and/or mass, whichever is appropriate. State that fact.

7. **State that the experiment must be repeated**. This reduces the possibility that a change occurred by chance, some random factor, or individual variations in the experimental organism.

INVESTIGATION #1: ARTIFICIAL SELECTION
Introduction

In the classroom, students have few opportunities to study and measure natural selection in living multicellular organisms. Many lab investigations use computer simulations instead. However, this investigation is a good alternative for exploring selection and evolution in the classroom.

Objective

You will carry out an artificial selection with Wisconsin Fast Plants and try to answer the following question.

> **"Can extreme selection change the expression of a quantitative trait in a population in one generation?"**

Sprout some Wisconsin Fast Plants seeds. When they are 7–12 days old, examine them and decide as a class which single trait will be best to try to alter by artificial selection. The trait must be one that varies within a range, not an all-or-none trait. The most common trait studied in this experiment is the number of trichomes, which are hairlike structures on the leaves. (Trichomes deter predation by herbivores such as the butterfly.) Backlighting and using a magnifying glass will help you count the tiny trichomes. Do not count every hair. Instead, use a sampling technique agreed upon by the class. See Figure 17.2.

STUDY TIP

The most commonly asked questions on this topic concern:

- Imprinting
- Fixed action pattern
- Sign stimulus
- Releasers

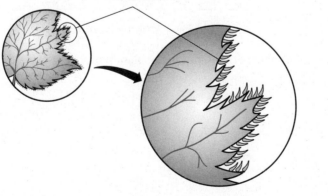

Figure 17.2

An appropriate sample size for the experiment is 150 plants. Figure 17.3 is a histogram that shows the number of trichomes in the first plant population.

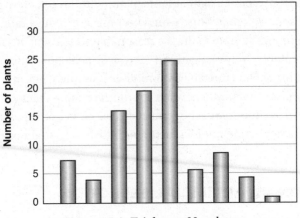

Figure 17.3 Trichome Numbers

Let's assume the class decides to increase the number of trichomes. Select the hairiest 10% of your original population of plants to use in your experiment. Perform one round of artificial selection by cross-pollinating the plants you have selected. After the F_1 plants germinate, mature, and flower, collect the seeds and grow them to produce another generation. As a control, pollinate an equal number of plants from the original population that were not part of your experimental group. Remember to keep the controls separate from the experimental plants. The histogram in Figure 17.4 shows the data from the F_2 experimental plants which were selected for having more trichomes.

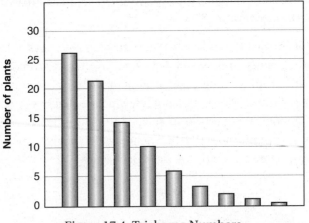

Figure 17.4 Trichome Numbers

Clearly, the average number of trichomes in the plant population increased in one generation.

INVESTIGATION #2: MATHEMATICAL MODELING: HARDY-WEINBERG

Introduction

One way to study evolution is to study how the frequency of alleles in a population changes over time. Mathematical models and computer simulations are tools used to explore evolution in this way.

Hardy-Weinberg theory describes the characteristics of a population that is stable or nonevolving. In a stable population, the allelic frequencies do not change from generation to generation. The Hardy-Weinberg equation enables us to calculate the frequencies of alleles in the population. The equation is:

$$p^2 + 2pq + q^2 = 1 \qquad \text{or} \qquad p + q = 1$$

Sample problems can be found in the chapter entitled "Evolution" earlier in this book or in the multiple-choice questions at the end of this chapter.

Objective

You will apply mathematical methods to data from real and simulated populations to predict what will happen to an evolving population.

You will use data from mathematical models based on the Hardy-Weinberg equilibrium to analyze genetic drift and the effect of selection in the evolution of specific populations.

You will build a spreadsheet that models how a hypothetical gene pool changes from one generation to the next. Any modern spreadsheet will work, including Microsoft Excel, Google's online Google Docs, and Zoho's online spreadsheet. This spreadsheet will let you explore how different factors, such as selection, mutation, and migration, affect allelic frequencies in a population.

You will be able to answer questions such as:

- Why do recessive alleles like cystic fibrosis stay in the human population?
- Why don't they gradually disappear?
- Why is the dominant trait polydactyly not a common trait in human populations?
- How do inheritance patterns or allelic frequencies change in a population?

INVESTIGATION #3: BLAST—COMPARING DNA SEQUENCES

Introduction

Bioinformatics is a field that combines statistics, mathematical modeling, and computer science to analyze biological data. An extremely powerful bioinformatics tool is BLAST, an acronym for **B**asic **L**ocal **A**lignment **S**earch **T**ool. **BLAST** is an algorithm that compares biological sequence information, such as the amino acid sequence in different proteins or the nucleotide sequence of different DNA sequences. It was designed by scientists at the National Institutes of Health (NIH) in 1990.

BLAST enables you to input a gene sequence of interest and search entire genomic libraries for identical or similar sequences in a matter of seconds. For example, by using BLAST, you might input a particular mouse gene and find that humans carry a very similar gene. You might instead discover, as happened a few years ago, that humans and neanderthals (*Homo neanderthalensis*) share much of the same DNA.

If you were to use BLAST to compare five DNA sequences from five different species, you would see something like the following on your computer screen:

Position:	1	2	3	4
Species *A*:	A C C G C T A C G A T T C G G C T A G C A T			
Species *B*:	A C C G C T G C G A T T C G G C C A G C A T			
Species *C*:	A C C G C T A C G T T T C G G C T A G C A T			
Species *D*:	A C C G C T G C G A T T C G G C T A G C A T			
Species *E*:	A G C G C T G C G A T T C G G C T A G C A T			

Objective

In this laboratory investigation, students will use BLAST to compare several genes and then use the information to construct a cladogram. A cladogram (or phylogenetic tree) is a visualization of the evolutionary relatedness of species. See Chapter 9 of this book for more information about cladograms. Figure 17.5 is a simple cladogram.

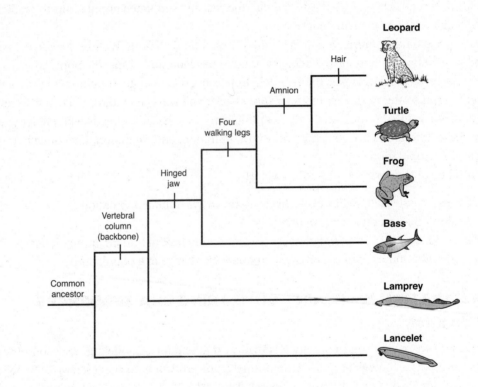

Figure 17.5

INVESTIGATION #4: DIFFUSION AND OSMOSIS

Introduction

Water potential (Ψ)

- Is measured in units of pressure called megapascals (MPa)
- Measures the relative tendency for water to move from one place to another
- Is the result of the combined effects of solute concentration and pressure

Water potential $=$ pressure potential $+$ solute potential

$$\Psi = \Psi_p + \Psi_s$$

- Water moves from high water potential to low water potential

This lab investigation consists of three parts.

- Part 1: Make artificial cells to study the relationship of surface area to volume.
- Part 2: Create models of living cells.
- Part 3: Observe osmosis in living cells.

Part 1: Make artificial cells to study the relationship of surface area to volume.

OBJECTIVE

Fashion "cells" from agar or gelatin stained with phenolphthalein. Explore the rates of diffusion in cells of different sizes and shapes.

Calculate the ratio of surface area to volume in both small and large cells. Predict which cell(s) might eliminate waste and take in nutrients faster by diffusion—a small cell or large cell. See Figure 17.6.

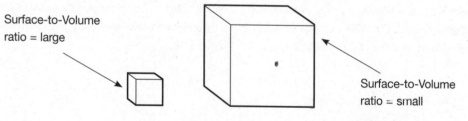

Figure 17.6

Part 2: Create models of living cells.

OBJECTIVE

Investigate the processes of diffusion, osmosis, and water potential in a model membrane system and in living cells. This section has three parts, A, B, and C.

PART 2A: OBSERVE DIFFUSION ACROSS A SEMIPERMEABLE MEMBRANE.

Fashion a length of semipermeable dialysis tubing into a bag, and fill it with two solutions: starch and glucose. Place the bag into a beaker containing Lugol's iodine solution, and allow the system to stand for 30 minutes. The contents of the bag will turn blue-black because iodine molecules are small enough to diffuse into the bag and react with the starch. However, the iodine solution in the beaker remains unchanged because the starch molecules are too large to diffuse out of the bag and mix with the iodine. See Figure 17.7.

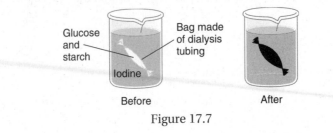

Figure 17.7

PART 2B: CALCULATE THE MOLARITY OF AN UNKNOWN SOLUTION BY OBSERVING OSMOSIS.

Fashion 6 lengths of semipermeable dialysis tubing into 6 bags. Fill 5 of them with sucrose solutions of varying molarities, 0.2M, 0.4M, 0.6M, 0.8M, and 1.0M. Fill 1 with distilled water. (The molecules of sucrose are too large to diffuse through the dialysis membrane.) Remove as much air from each bag as possible to allow room for expansion. Blot and weigh each bag. Place each bag into a beaker that contains distilled water. Allow them to sit for 30 minutes. Then remove the bags from the beakers. Blot and weigh the bags as before. Calculate the percent change in mass for each bag. Record and graph the data with the dependent variable (percent change in mass) on the *y*-axis and the various molarities (independent variables) on the *x*-axis. The bag containing distilled water did not change in mass because its contents are isotonic to the distilled water in the beaker and this bag is the control. *The mass of the other bags increased. The bag with the lowest molarity increased in mass the least, and the bag with the highest molarity increased the most.* See Figure 17.8.

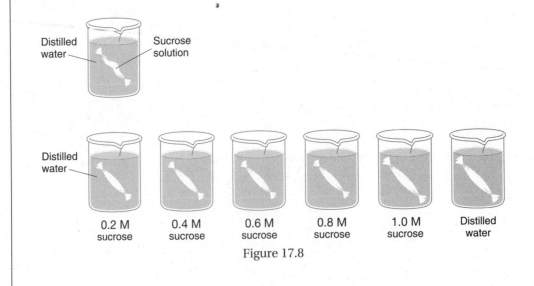

Figure 17.8

PART 2C: DETERMINE THE MOLARITY OF A LIVING (POTATO) CELL.

Cut identical-size small squares of potato and weigh them. Place each into a beaker covered with 300 mL of the following sucrose solutions: 0.2 M, 0.4 M, 0.6 M, 0.8 M, and 1.0 M and one with distilled water. Allow them to sit overnight, then weigh them again. Some pieces of potato will gain mass and some will lose mass. Calculate the percent change in mass for each. Plot a best-fit line on a graph, with the molarities of the solutions, on the x-axis and the percent change in mass on the y-axis. The point of the best-fit line, which crosses the x-axis, is the molarity where there was no change in mass of the potato and represents the molarity inside the potato cell; see Figures 17.9 and 17.10.

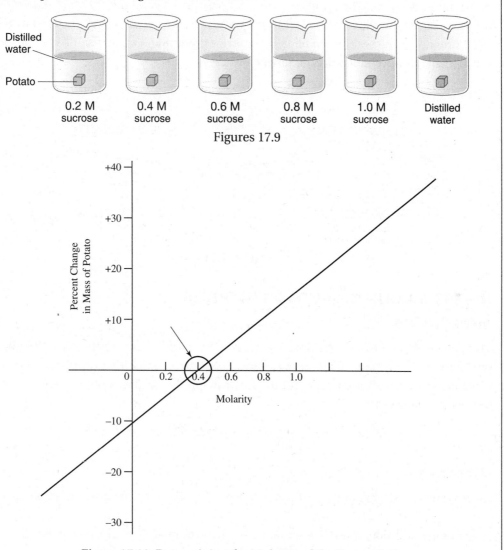

Figures 17.9

Figure 17.10 Determining the Molarity of the Potato Cell

Part 3: Observe osmosis in living cells.

OBJECTIVE

Observe plasmolysis in a living plant (elodea) cell

Tear a thin piece of healthy elodea tissue, and prepare a wet mount. Locate a single layer of cells under the light microscope at 40× with clearly visible chloroplasts. While observing the slide, expose the tissue to 5% (hypertonic) sodium chloride solution. Observe changes in the cytoplasm of the cell as water suddenly diffuses out of the cell from high water potential to low water potential. The cytoplasm shrinks (plasmolysis), and the chloroplasts condense into a small circle. If you wash away the salt water solution and rehydrate the cells with distilled water, you can see the cell fill up with water (turgor) and appear similar to the way it was initially. However, the cell is no longer alive; see Figure 17.11.

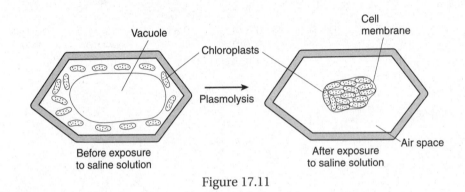

Figure 17.11

INVESTIGATION #5: PHOTOSYNTHESIS
Introduction

The process of photosynthesis provides oxygen for Earth's atmosphere. Like all enzyme-driven reactions, the rate of photosynthesis can be measured by either the disappearance of substrate or by the accumulation of product or by-products. The photosynthesis equation can be summarized as:

$$2H_2O + CO_2 + light \rightarrow carbohydrate\ (CH_2O) + O_2 + H_2O$$

Objective

Determine the rate of photosynthesis in living plant tissue using the floating leaf disks technique.

In this investigation, you will determine the rate of photosynthesis by measuring the accumulation of oxygen, a by-product of photosynthesis.

For consistency, use a hole punch to punch out small circles from a leaf. Place the leaf disks into a syringe along with a baking soda solution (bicarbonate). Create a vacuum by pulling the syringe out about 1 cm while holding your thumb over the opening of the syringe. The vacuum removes air from the air spaces in the spongy mesophyll in the leaf and replaces it with the solution of baking solution. See Figure 17.12.

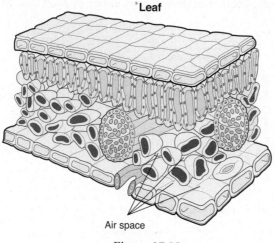

Leaf

Air space

Figure 17.12

The bicarbonate solution ($NaCO_3$) provides CO_2 that serves as a carbon source for photo-synthesis. With the air from the spongy mesophyll replaced by the denser baking soda solution, the leaf disks sink to the bottom of the syringe. Transfer them quickly to a beaker that also contains bicarbonate solution. Shine a light into the beaker so you can observe the leaf disks. As they carry out photosynthesis and begin to produce oxygen bubbles, the leaf disks begin to rise in the bicarbonate solution. The speed at which they rise is a function of the rate of photosynthesis.

Begin collecting data as soon as you shine a light onto the beaker. Stop timing when 50% of the disks have risen. See Figure 17.13.

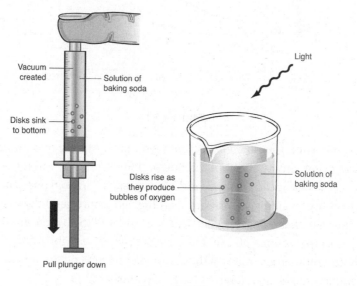

Vacuum created

Solution of baking soda

Disks sink to bottom

Light

Pull plunger down

Disks rise as they produce bubbles of oxygen

Solution of baking soda

Figure 17.13

Figure 17.14 is a sample graph that shows the number of disks floating versus time.

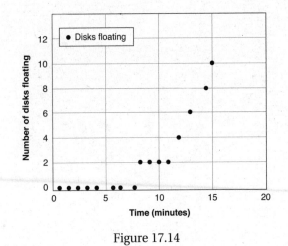

Figure 17.14

Figure 17.15 is a graph after you have converted the light intensity to the rate of photosynthesis.

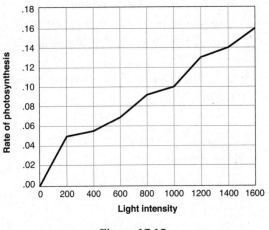

Figure 17.15

Remember that when you do an experiment, you must explore one variable at a time. If you are studying light and its effect on the rate of photosynthesis, you must be careful that heat from the lamp does not accidentally affect your results. To avoid this, place a **heat sink**—a clear glass beaker containing water—between the light source and the experimental container. Since the water has a high specific heat, it absorbs all the heat from the light source.

By using this simple setup, you can study many different variables and how each affects the rate of photosynthesis. Examples of these variables include different light intensities, different wavelengths of light, and different types of plants.

INVESTIGATION #6: CELL RESPIRATION
Introduction

Cell respiration is the process by which living cells produce energy by the oxidation of glucose with the formation of the waste products CO_2 and H_2O_2. Here is the equation for cellular respiration:

$$C_6H_{12}O_6 + 6O_2 \rightarrow 6CO_2 + 6H_2O + \frac{686 \text{ kcal of energy}}{\text{mole of glucose oxidized}}$$

The rate of cell respiration can be measured by the volume of oxygen used. The cells used to study respiration in this experiment are sprouted sweet pea seeds, and the control consists of sweet pea seeds that are dormant and not carrying out respiration. (This apparatus can also accommodate invertebrates, such as insects and earthworms.)

Objective

Construct a **respirometer** that will measure oxygen consumption of sprouted or unsprouted seeds. Observe the effects that different temperatures have on the rate of respiration; see Figure 17.16.

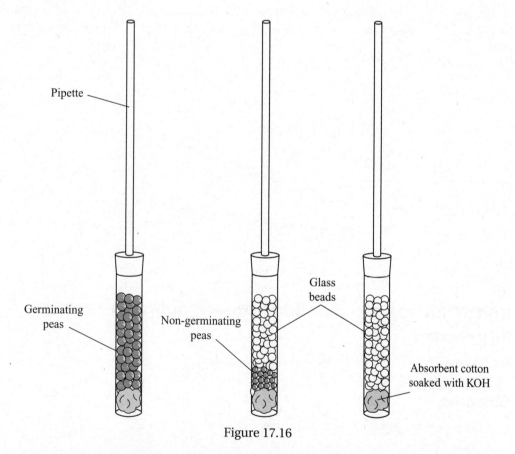

Figure 17.16

Since sprouted seeds have absorbed water and are larger than unsprouted seeds, glass beads are added to the respirometer holding the unsprouted seeds to equalize the volume of empty space in both vessels. Set up six respirometers as shown in Table 17.1.

Table 17.1

Effects of Temperature on Respiration Rate		
Respirometer	Temperature of Water	Contents
1	25°C	Germinating seeds
2	25°C	Dry seeds + beads
3	25°C	Glass beads alone
4	10°C	Germinating seeds
5	10°C	Dry seeds + beads
6	10°C	Glass beads alone

Place a thin layer of cotton at the bottom of each respirometer soaked with KOH to absorb CO_2 that is given off by respiration. This allows the change in volume inside the respirometer to measure *only* oxygen consumption.

Submerge the respirometers in water. As oxygen is used up inside the respirometer, water flows into the respirometer through the pipette and is a measure of oxygen consumption.

The results of the experiment are simply stated. The higher the temperature, the more oxygen is used and the faster the rate of respiration. Figure 17.17 shows a graph of the data.

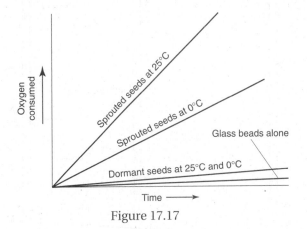

Figure 17.17

INVESTIGATION #7: CELL DIVISION—MITOSIS AND MEIOSIS

Introduction

See Chapter 6 for a review of mitosis and meiosis.

Objective

- Describe the events and regulation of the cell cycle.
- Explain how DNA is transmitted to the next generation by mitosis.
- Explain how DNA is transmitted to the next generation by meiosis followed by fertilization.
- Understand how meiosis and crossing-over lead to increased genetic diversity, which is necessary for evolution.

Part 1

Use models like pipe cleaners, beads, or "sockosomes" to learn about the stages of mitosis. Be able to answer the following questions:

- How does a chromosome replicate itself?
- Why do chromosomes condense before the nucleus divides?
- What happens if sister chromatids fail to separate properly?
- Can chemicals in the environment alter the process of mitosis in plants or animals? How would you design an experiment to prove your answer?

Part 2

Prepare a chromosome squash from an onion root tip. See Figure 17.18.

Count the cells in interphase and the cells in mitosis. Analyze the data.

Collect the class data for each group. Calculate the **mean** and **standard deviation** for each group.

Carry out an experiment comparing an onion root that has been treated with a chemical that alters the rate of mitosis with one that does not. Do a **chi-square analysis**. Determine if you can reject or fail to reject the **null hypothesis**.

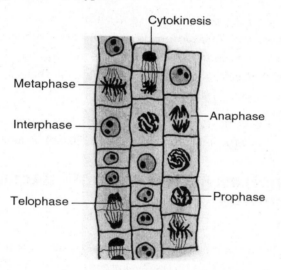

Figure 17.18

Part 3

Become familiar with normal karyotypes and abnormal ones that are characteristic of cancer.

Explore **HeLa cells** and **Philadelphia chromosomes**.

Part 4

Use models like pipe cleaners, beads, or "sockosomes" to learn about the stages of meiosis. Be able to answer the following questions:

- What is **crossing-over**?
- What is **independent assortment**?
- What happens if a pair of chromosomes fails to separate? How might this contribute to various genetic diseases?
- How are mitosis and meiosis fundamentally different?

Calculate the distance in map units between the gene for spore color and the centromere in the fungus *Sordaria fimicola*.

Sordaria fimicola is an **ascomycete** fungus that consists of fruiting bodies called **perithecia**, which hold many **asci**. Each ascus encapsulates eight **ascospores** or **spores**. When ascospores are mature, the ascus ruptures, releasing the ascospores. Each ascospore can develop into a new haploid fungus; see Figure 17.19.

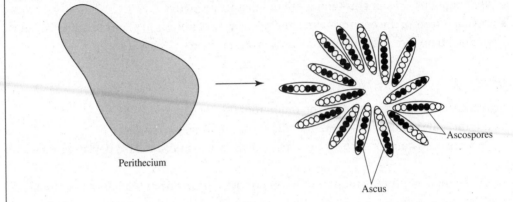

Perithecium

Ascospores

Ascus

Figure 17.19

By observing the arrangements of ascospores from a cross between wild-type (black) and a mutant tan type *Sordaria fimicola* fungus, you can estimate the percentage of crossing-over that occurs between the centromere and the gene that controls the tan spore color. Count 50–100 hybrid asci, and calculate the distance in map units between the gene for spore color and the centromere using this formula.

Divide the number of crossover asci (2:2:2:2 or 2:4:2)
by the total number of asci × 100 and divide by 2

INVESTIGATION #8: BIOTECHNOLOGY—BACTERIAL TRANSFORMATION

Introduction

Many years ago if a person was diabetic and needed insulin, the insulin was harvested from other mammals, such as pigs and cows, because their insulin was similar to human insulin. Today, human insulin is "manufactured" by bacteria in which scientists have inserted a gene that produces human insulin. This insulin is not similar to human insulin; it is human insulin. Many scientists took many years to figure out how to accomplish this. However, now it is done routinely.

Genes have been inserted in bacteria so the bacteria can "eat" oil and clean up major oil spills. Rice has been genetically modified to produce more protein. Cows have been genetically engineered to produce more milk. You may have seen the term GMO—genetically modified organism—on milk cartons or some other foods. The term means that the organism or food has been modified by altering its genes—by adding a foreign gene into its genome.

The science of inserting genes into a cell of an organism depends on small circles of exogenous DNA called **plasmids** that were first discovered in bacteria. Scientists insert a particular gene into a plasmid and then insert the plasmid into a cell. The host cell begins to produce the proteins encoded by the gene that was inserted into the plasmid. To force a bacterium to uptake a plasmid, the bacterium must be prepared in a special way. It must be made **competent**.

Objective

Cut lambda DNA with different restriction enzymes. Transform bacteria into an antibiotic-resistant form by inserting a plasmid.

To transform bacteria, first make them competent to uptake a plasmid. Use heat shock, a combination of alternating hot and cold, in the presence of calcium ions that disrupt the cell membrane. Once competent, the bacteria are incubated with plasmids that carry the resistance to a particular antibiotic. A control sample is run along with the experimental one, treated in exactly the same way except it does not get exposed to a plasmid. After the cells are allowed to rest, they are poured onto four petri dishes that contain Luria broth. Two petri dishes also contain antibiotic, two do not. Table 17.2 shows the setup.

Table 17.2

Experimental Data			
Test Tube	Plasmid	Petri Dish	Growth
1	+	+ Antibiotic	+
2	+	No Antibiotic	+
3 (Control)	–	+ Antibiotic	–
4 (Control)	–	No Antibiotic	+

Figure 17.20 is a sketch of the petri dishes showing the results of the experiment.

- Plate 1. These bacteria were incubated with plasmid. There are 7 colonies growing here. Each colony consists of bacteria that have been **transformed** and are resistant to the antibiotic because they absorbed the plasmid.
- Plate 2. This plate contains a lawn of bacteria. Any bacteria would grow on this plate; there is no antibiotic to prevent it. However, this culture did receive plasmid. We can assume that this plate contains bacteria resistant to the antibiotic as well as those susceptible to it.
- Plate 3. There is no growth on this dish. These bacteria were not incubated with plasmid and the plate contains no growth because the dish contains antibiotic that killed the bacteria.
- Plate 4. The bacteria were not incubated with plasmid and the plate also contains a lawn of growth because there is no antibiotic to kill the bacteria.

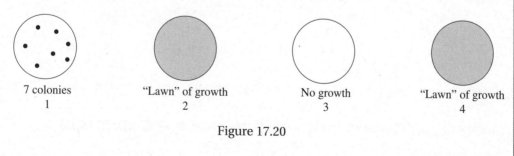

7 colonies
1

"Lawn" of growth
2

No growth
3

"Lawn" of growth
4

Figure 17.20

INVESTIGATION #9: BIOTECHNOLOGY— RESTRICTION ENZYME ANALYSIS OF DNA

Introduction

In this experiment, lambda DNA is incubated with two different restriction enzymes and then run through an electrophoresis gel. Place the lambda DNA into three test tubes. Incubate one sample with *Eco*R1, one with *Hind*III, and leave one untreated as a control. The two restriction enzymes digest the DNA of two samples, cutting them into characteristic pieces called **restriction fragments**. The three samples are placed into the wells on an agarose gel and drawn from the **cathode** to the **anode** by electric current. After staining, a pattern results on the gel that reveals the **restriction fragments**. The control sample of DNA remains uncut and can be seen as a large block of DNA near the well in lane 4. The two other samples were cut with restriction enzymes, and the restriction fragment or banding pattern is visible. By measuring the distance of each fragment from the well (similar to determining the R_f value in paper chromatography) and comparing this distance against a standard, the size of the fragment of DNA can be determined; see Figure 17.21.

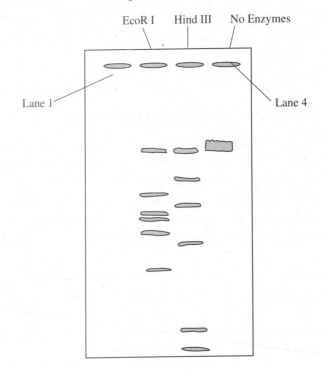

Figure 17.21

Objective

Use restriction enzymes and gel electrophoresis to identify and profile individuals.

INVESTIGATION #10: ENERGY DYNAMICS—FOOD CHAIN

Introduction

Almost all life on this planet is powered, either directly or indirectly, by sunlight. Producers capture solar energy and convert it to biomass. The net amount of energy captured and stored by the producers in a system is the system's net productivity. In contrast, the gross productivity is a measure of the total energy captured. Different plants have different strategies of energy allocation that reflect their role in different ecosystems. For example, annual weedy plants allocate more of their biomass production to reproduction and seeds than do slower-growing perennials. The producers are consumed or decomposed, and their stored chemical energy powers the other trophic levels of the biotic community. Energy dynamics in a biotic community is basic to understanding ecological interactions.

Objective

You will estimate the net primary productivity (NPP) of Wisconsin Fast Plants (the producers) growing under lights. You will also estimate the flow of energy from the plants to cabbage white butterfly larvae (the consumers) as the larvae eat cabbage family plants.

Figure 17.22 models how energy flows through a plant.

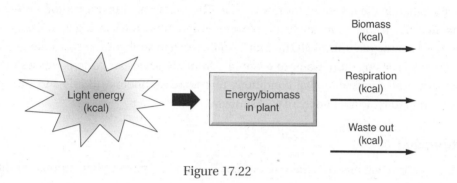

Figure 17.22

ESTIMATING NET PRIMARY PRODUCTIVITY (NPP) OF FAST PLANTS

Primary productivity is a rate. It is the energy captured by photosynthetic organisms in a given area per unit of time.

When light energy is converted to chemical energy in photosynthesis or transferred from one organism (a plant) to its consumer (insect), some energy will be lost as heat during the transfer.

In terrestrial ecosystems, productivity (or energy capture) is generally estimated by the change in biomass of plants produced over a specific time period. Measuring biomass or changes in biomass is simple. Take the mass of the organism(s) and record the mass over various time intervals.

Consider also that different organic compounds store different amounts of energy. Protein and carbohydrates store about 4 kcal/g dry weight. Fats store about 9 kcal/g of dry weight.

You must take this into consideration as you carry out your investigation.

Cabbage white butterfly larvae eat plants from the cabbage family. As with Wisconsin Fast Plants, measuring the energy flow into and out of these butterflies can be inferred from the biomass they gain and lose.

As butterfly larvae grow toward maturity, they pass through different developmental stages called instars. Use larvae that are already well along their 4th or 5th instar stage.

CONVERT BIOMASS MEASUREMENTS (GRAMS) TO ENERGY UNITS IN KILOCALORIES

For Wisconsin Fast Plants, assume that 1 gram (g) of dried biomass contains 4.35 kilocalories (kcal) of energy. Assume that the biomass of 4th-instar larvae is 40% of the wet mass.

Calculate the biomass of the larvae. For butterfly larvae, use an average value of 5.5 kcal/g of biomass to calculate energy of each larva.

To determine the energy content in the larval frass (droppings), use 4.76 kcal of energy/g of frass. Calculate the frass lost per individual larva.

To determine the energy content of the brussels sprouts eaten by each larva, convert the wet mass of the sprout to dry mass and multiply by 4.35 kcal/g.

Use the estimated percentage of biomass (dry mass) in fresh Wisconsin Fast Plants that you previously calculated.

Graph your results.

INVESTIGATION #11: TRANSPIRATION

Introduction

Transpiration is the loss of water from a leaf. The **theory of transpirational pull–cohesion tension** explains that as one molecule of water evaporates from a leaf, one molecule of water is pulled into the plant through the roots. An increase in sunlight increases the rate of transpiration by causing more water to evaporate from the leaves and also increases the rate of photosynthesis. Environmental conditions that increase photosynthesis and transpiration are low humidity, wind, and increased sunlight.

Objective

In this lab, you will measure the rate of transpiration in a plant under various conditions in a controlled experiment.

Insert two-week old bean seedlings (*Phaseolus vulgaris*) into a **potometer**, and measure the amount of water used by the plants under varying conditions of humidity (misting the leaves), light (a strong lightbulb shining on the plant), and wind (a fan). Set up several potometers: one for each variable and one with only ambient conditions as the control; see Figure 17.23.

Here is what you can expect to happen.

- Increasing humidity, by placing a bag over or misting the leaves, will decrease transpiration.
- Increasing light will increase transpiration.
- Increasing wind with a fan will lower the humidity around the leaves and thereby increase transpiration.

When you increase the light intensity, make sure that you also do not increase the temperature surrounding the plant, thus, introducing another variable. To accomplish this, place a **heat sink**, a large bowl or beaker of water, between the light and the plant to absorb the heat while allowing the light to pass through.

Since each plant does not have the same leaf surface area, your results are meaningless unless you can demonstrate consistency. You must measure the area of the leaves on the plants and calculate the water loss in terms of square meters. To measure the surface area of the leaf, trace all the leaves from each plant onto graph paper with squares of known size. Construct a graph with "water loss (mL/m²)" on the y-axis and "time (min)" on the x-axis.

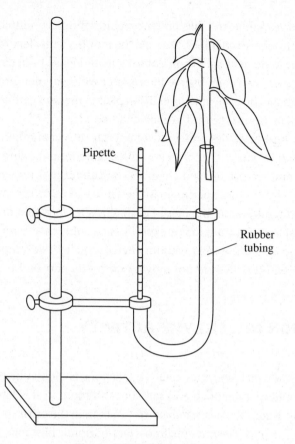

Figure 17.23 Potometer

INVESTIGATION #12: FRUIT FLY BEHAVIOR
Introduction

Drosophila melanogaster, the common fruit fly, has been studied by scientists since 1907. It lives throughout the world and feeds on rotting fruit. *Drosophila* is a model research organism. Students have been doing genetics experiments with these flies for generations. However, most students today can accomplish the classic genetic experiments using computer simulations instead of real fruit flies. In this experiment, you will use live fruit flies to study animal behavior.

Objective

You will investigate the relationship between a model organism, *Drosophila*, and its response to different environmental conditions

You will build a **choice chamber** to study which environment the flies prefer.

Choose a large enough sample of fruit flies. For example, 30 flies is a good sized sample. A sample of 10 flies is too small. However, 100 may be too difficult to manage and might result in errors.

Build a choice chamber. Use two graduated cylinders or clear plastic water bottles with the necks cut off.

To test the hypothesis that fruit flies prefer specific lighting conditions, put them to sleep. Place the flies into the choice chamber, and seal the mouths with clear tape. Cover one side of the choice chamber to prevent light from entering it, and leave both chambers undisturbed. When the flies wake up, you can observe them. After what you determine is an appropriate amount of time, remove the cover and tape. Then place a piece of cardboard over each chamber to keep the flies in. Count the flies in each chamber.

To test the hypothesis that flies prefer one temperature over another, warm one half of the choice chamber. Before you put the flies into the chamber, make sure that you can control and measure the temperature of each chamber. Be careful that the temperature is not too warm or too cold. Carry out the experiment in the same way as before. After the same amount as time as before, carefully count the flies in each chamber to gather quantitative data.

You must always include a control in any experiment. In this case, your control is a chamber just like the experimental one except without altering the light or temperature. After all, you have to make sure that the flies are not moving from one part of the chamber to the other solely by chance.

INVESTIGATION #13: ENZYME ACTIVITY
Introduction

In general, the rate at which enzymes catalyze reactions is altered by environmental factors such as pH, temperature, inhibitors, and salt concentration. In this lab, the enzyme used is one that is found in all aerobic cells, **catalase**. Its function in the cell is to decompose hydrogen peroxide, a by-product of cell respiration, into a less-harmful substance. Here is the reaction:

$$2 H_2O_2 \rightarrow 2 H_2O + O_2\uparrow.$$

In this lab, the enzyme catalase is added to peroxide of known concentration and bubbling is observed. The bubbles are oxygen gas. (The test for the presence of oxygen gas is a glowing splint. It will burst into flames in the presence of O_2.) The quantity of bubbling is proportional to the activity of the catalase.

Objective

To observe the effects that acid and high heat have on enzyme function.

For this experiment, extract fresh catalase from liver tissue by homogenizing it in a blender and filtering the liquid. Keep in on ice at all times. In addition, you must establish how much active peroxide is in each peroxide sample because it decomposes readily on its own. This is done by titrating the hydrogen peroxide to which H_2SO_4 has been added with $KMnO_4$. When a persistent pink color remains, the titration is complete and the concentration of H_2O_2 is known because *the concentration of $KMnO_4$ is proportional to the concentration of H_2O_2.* The less H_2O_2 is left in the beaker, the less $KMnO_4$ will be needed to turn the liquid in the beaker pink or light brown.

Here is the equation that is the basis for the experiment.

$$5 H_2O_2 + 2 KMnO_4 + 3 H_2SO_4 \rightarrow K_2SO_4 + 2 MnSO_4 + 8 H_2O + 5 O_2$$

To observe the reaction to be studied, take three samples of fresh catalase. Leave one unchanged as a control, boil one, and acidify the other. Compare the effectiveness of decomposing H_2O_2 by observing the bubbling. To quantify how much H_2O_2 actually remains in each sample, run a titration. See Figure 17.24.

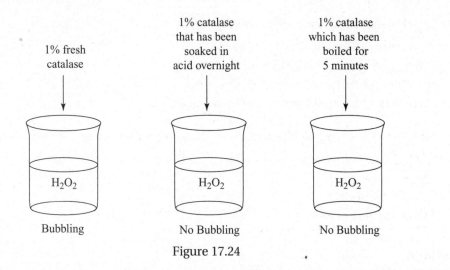

1% fresh
catalase

1% catalase
that has been
soaked in
acid overnight

1% catalase
which has been
boiled for
5 minutes

H$_2$O$_2$

H$_2$O$_2$

H$_2$O$_2$

Bubbling

No Bubbling

No Bubbling

Figure 17.24

The results will demonstrate that high heat and strong acid denature the enzyme catalase. There will be no bubbling in those beakers.

MULTIPLE-CHOICE QUESTIONS

1. If a piece of potato is allowed to sit out on the counter and dry out, the water potential of the potato cells would

    (A) increase
    (B) decrease
    (C) remain the same
    (D) increase, then decrease

2. Which of the following correctly indicates the gradient of water potential from the highest to the lowest?

    (A) root, soil, stem, leaf, air
    (B) soil, root, stem, leaf, air
    (C) air, leaf, stem, root, soil
    (D) root, leaf, stem, soil, air

3. In a diffusion lab with 6 bags of dialysis tubing, why do you use the percent change in mass rather than simply change in mass?

    (A) Doing so was arbitrary and applies to these instructions only; it really does not matter.
    (B) Doing so is the convention.
    (C) The percent change in mass is not as accurate.
    (D) The bags did not all weigh exactly the same mass at the start.

4. Which is the test for the presence of oxygen?

   (A) Limewater turns from clear to cloudy.
   (B) A glowing splint bursts into flames.
   (C) Phenolphthalein turns from clear to pink.
   (D) A lighted splint will make a popping noise.

5. Which is NOT an important event that occurs during or as a result of meiosis?

   (A) independent assortment of alleles
   (B) crossing-over
   (C) cloning of cells
   (D) production of gametes

Question 6

The list below contains choices for question 6.

   (A) KOH
   (B) the vial containing glass beads
   (C) unsprouted (dormant seeds)

6. Which component of the experiment that uses respirometers is the control?

   (A) A and B
   (B) both B and C
   (C) C only
   (D) all of the above

7. If two bands on an electrophoresis gel have nearly the same base pair size, how might you best separate them?

   (A) Run them in a gel that is more concentrated.
   (B) Run them in a gel that is less concentrated.
   (C) Turn up the current and run them through the gel faster.
   (D) Stain them.

8. Which is true of the genetics of fruit flies?

   (A) The correct order of development is egg $\rightarrow$ pupa $\rightarrow$ larva.
   (B) There are four instar stages in fruit flies.
   (C) The males are larger than the females.
   (D) White-eyed trait is sex-linked recessive.

9. The allele for the hair pattern called widow's peak is dominant over the allele for no widow's peak. In a population of 1,000 individuals, 841 show the dominant phenotype. How many individuals would you expect to be hybrid for widow's peak?

   (A) 16%
   (B) 24%
   (C) 48%
   (D) 100%

10. In a certain population, the dominant phenotype of a certain condition occurs 91% of the time. What is the frequency of the dominant allele?

    (A) 0.3
    (B) 3%
    (C) 0.7
    (D) 7%

11. If a population is in Hardy-Weinberg equilibrium and the frequency of the recessive phenotype for a particular trait appears in 16% of the population, what will the frequency for the **hybrid condition** be in 500 years?

    (A) 16%
    (B) 24%
    (C) 48%
    (D) 64%

Answers to Multiple-Choice Questions

1. **(B)** As tissue drives out, its water potential decreases. Water flows from higher water potential and osmotic potential to lower. Water would tend to flow into the dehydrated cells.

2. **(B)** Water moves from the highest water potential to the lowest. In a plant, water moves from the soil to the roots to the stems and leaves and into the air spaces in the spongy mesophyll and out to the environment.

3. **(D)** Since the bags did not all weigh the same mass, the only way to control for the variation is to show change in mass as a percentage.

4. **(B)** A lighted splint in the presence of oxygen is explosive. Limewater is the test for CO_2. Phenolphthalein is the test for a base. Cobalt chloride tests for water or moisture. A lighted splint pops in the presence of hydrogen.

5. **(C)** A clone is an identical copy. The cloning of cells is characteristic of mitosis, not meiosis. The purpose of meiosis is to produce gametes that are different from the parent cells. All the other events are important parts of meiosis.

6. **(B)** The unsprouted seeds are used as a comparison for the sprouted seeds. However, the glass beads that were introduced to equalize the volume in the various containers, must be included to demonstrate that they have no effect on the results in vials #2 and #5.

7. **(A)** If the gel is more concentrated, it will present more of an impediment and separate the two bands more effectively. Another method would be to lower the current and allow them to run very slowly and overnight. One must simply try both techniques to see which works better.

8. **(D)** White-eyed trait is sex-linked recessive. The correct order of development is egg → larva or instar → pupa. There are 3 instar stages in the fruit fly's development. The female is larger than the male fruit fly. Females remain virgin for about 10 hours.

9. **(C)** If approximately 84% of the population shows the dominant trait, then 16% show the recessive phenotype and are homozygous recessive. Therefore, the frequency of the recessive allele is 0.4 and of the dominant allele is 0.6. The frequency of the hybrid in the Hardy-Weinberg equation is $2pq$: $2(0.4 \times 0.6) = .48 = 48\%$.

10. **(C)** The frequency of a dominant allele is q; the frequency of a recessive allele is p. If the dominant phenotype appears 91% of the time, then the recessive phenotype appears 9% of the time. If $q^2 = 0.9$, then $q = 0.3$ and $p = 0.7$.

11. **(C)** The population is in equilibrium, it will not change. If $q^2 = .16$, then $q = .4$ and $p = .6$. Therefore, the frequency of the hybrid is 48%.

Five Themes to Help You Write a Great Essay

18

→ **EVOLUTION**
→ **ENERGY TRANSFER**
→ **RELATIONSHIP OF STRUCTURE TO FUNCTION**
→ **REGULATION**
→ **INTERDEPENDENCE OF NATURE**

In preparing for the AP Exam, you should be aware that there are major concepts that unify the study of biology and that appear over and over, particularly in Part II questions. You should be prepared to discuss these ideas and to give examples of each at the **molecular**, **cellular**, **organ**, **organism**, and **population** level. Below is a list of the major themes and a review of the ideas in each.

Evolution
Energy transfer
Relationship of structure to function
Regulation
Interdependence of nature

EVOLUTION

At the Molecular/Cellular Level

- The theory of endosymbiosis states that millions of years ago, small prokaryotic cells were engulfed by larger prokaryotic cells. Those small cells became the chloroplasts and mitochondria that are critical for life as it is known.
- Structural and functional modifications for dry environments include:

    C4 and CAM photosynthesis
    Kranz anatomy and the Hatch-Slack pathway
    Seeds

- Countercurrent exchange is a mechanism that evolved to enhance diffusion rates where necessary.

At the Organ/Organism/Population Level

- Mutations and genetic recombination generate heritable variation that is subject to natural selection.

STUDY TIP

Studying the material in this chapter will really help you with the essay part of the test.

- The peppered moth population changed drastically from white to dark in less than 50 years in response to the industrialization in England. The phenomenon is referred to as *industrial melanism.*
- Bacteria become resistant to antibiotics quickly after exposure to antibiotics. After a course of antibiotics, bacteria that are not resistant to the antibiotic are killed, while those that happen to be resistant to the antibiotic reproduce rapidly with no competition. This results in a new population that is resistant to the antibiotic.
- Cladograms and phylogenetic trees are diagrammatic reconstructions of evolutionary history. They reflect phylogenetic relationships among organisms.
- Plants and animals have evolved many modifications to move from sea to land. Plants have evolved a waxy cuticle, supporting tissue like collenchyma and sclerenchyma, roots, and vascular vessels. Greatly reduced leaf size in conifers reduces water loss.
- Animals have evolved limbs for support and movement. Scales minimize water loss, and lungs allow breathing. Nephrons or nephridia produce hypertonic urine to remove nitrogenous waste and to conserve water.
- Social behavior is any kind of interaction among two or more animals, usually of the same species. Types of social behaviors include cooperation, agonistic, dominance hierarchies, territoriality, and altruism.

ENERGY TRANSFER
At the Molecular/Cellular Level

- During the process of photosynthesis, plants convert solar energy to chemical bond energy. This is an energy conversion upon which all life depends.
- During the electron transport chain of cellular respiration, the exergonic flow of electrons is coupled with the energy-absorbing formation of a proton gradient.
- A proton gradient across membranes in mitochondria and chloroplasts powers the synthesis of ATP through a process called oxidative phosphorylation.
- Energy is transferred from ATP to energy-absorbing reactions. For example, ATP provides the energy to pump sodium and potassium ions across the cell membrane of the axon of a neuron.

At the Organ/Organism/Population Level

- Plants transfer light energy into chemical bond energy.
- Energy is transferred from producers to animals through the food chain. Food chains are never more than four or five trophic levels because energy is lost along the chain.
- Energy transfer is about 10% from one trophic level to the next.

RELATIONSHIP OF STRUCTURE TO FUNCTION
At the Molecular/Cellular Level

- The structure of the cell membrane (fluid-mosaic model) explains how various substances can cross the membrane, into or out of the cell.
- The structure of DNA, two strands connected by hydrogen bonds, enables scientists to understand how DNA replicates itself.
- The winding internal membrane (cristae) of the mitochondria allows for vastly increased ATP production.

- The many layers of thylakoid membranes that make up the grana of chloroplasts are responsible for increased absorption of light by thousands of photosystems.
- A nerve cell has a special shape to accommodate its function of transmitting impulses long distances. Some nerve cells in humans are more than 1 foot (3 m) long.
- Lipid-storing cells consist almost entirely of a vacuole that stores fat. The nucleus is pushed to the edge, and there is very little cytoplasm.
- The particular conformation of an enzyme determines what substrate it will act upon. If the enzyme is denatured (changes its shape), it will not function at all.
- When hemoglobin combines with one substrate molecule (oxygen), the binding triggers a slight conformational change in the hemoglobin that amplifies hemoglobin's ability to bind with more oxygen.

At the Organ/Organism/Population Level

- The long small intestine in humans provides increased surface area for enhanced digestion and absorption.
- The type of teeth an animal has reflects its dietary habits. Rodents that are vegetarians have sharp, chisellike incisors that facilitate gnawing and no canines. Predatory carnivores, like the large cats, have large canines that can kill prey and rip an animal apart. Humans are omnivores and have three different types of teeth suited to eating a variety of foods.
- Neurons have long bodies to transmit an impulse a long distance.

REGULATION

At the Molecular/Cellular Level

- The operon is a major regulatory mechanism of gene transcription in bacteria.
- Examples of cells communicating are gap junctions, tight junctions, and desmosomes.
- The plasma membrane, by virtue of its structure, controls what enters and leaves a cell.
- Hox genes control the expression of other genes and the development of a growing embryo.
- Enzymes control the rate of reactions by speeding them up or slowing them down. They speed up reactions by lowering the activation energy. They slow down or inhibit reactions through competitive, noncompetitive, and allosteric inhibitions.
- The cell cycle and cell division are under strict control by a built-in system of checkpoints or restriction points in G_1, G_2, and M. This timing is regulated by cyclins and cyclin-dependent kinases (CDKs).
- Receptors on the surface of the cell membrane help regulate cell and population growth. If cells become too crowded, they stop dividing. This phenomenon is called contact inhibition.

At the Organ/Organism/Population Level

- The nervous and endocrine systems regulate the body overall.
- Negative feedback mechanisms in the endocrine system maintain homeostasis.
- Body temperature, breathing rate, and heart rate are under the control of the brain and nervous system.

- Hormones like ecdysone and juvenile hormone control the development of a metamorphosing insect.
- Plant hormones, such as auxins, ethylene, gibberellins, abscisic acid, and cytokinins, help to maintain homoeostasis in plants.
- Countercurrent heat exchange maintains stable body temperatures in polar animals.
- Apoptosis is programmed cell death. This built-in suicide mechanism that is controlled by genes is essential to the normal development of the nervous and immune systems in animals.

INTERDEPENDENCE OF NATURE

At the Molecular/Cellular Level

- Plants and animals are mutually dependent. Photosynthesis requires CO_2 and produces O_2 while respiration requires O_2 and releases CO_2.

At the Organ/Organism/Population Level

- All life depends on the ability of plants as producers. The solar energy that plants convert to chemical energy is passed along the food chain from consumer to consumer.
- Organisms depend on other organisms through various symbiotic relationships: mutualism, commensalism and parasitism.
- Decomposers, such as bacteria and fungi, recycle nutrients within an ecosystem. Without them, life would not exist.

Learn How to Grade an Essay

19

→ **SAMPLE ESSAY A**

→ **ANALYSIS OF ESSAY A**

→ **SAMPLE ESSAY B**

→ **ANALYSIS OF ESSAY B**

INTRODUCTION

If you want to write a good essay, you have to know how it is graded.

Here are two sample essays similar to those written by students. The essays have already been graded. Each time a student gives the correct terminology or a correct explanation of a concept, a point is given in the margin. Below each answer is an explanation of what the student did or did not receive credit for. For sample essay B, cover the numbers in the margin on the right, read the essay, then try to grade it yourself.

Question: *Explain, in detail, how enzymes speed up and slow down reactions.*

SAMPLE ESSAY A

Enzymes, which are proteins, help a reaction speed up. It does this by using a catalyst. The function of enzymes relates to the structure of the enzyme. Certain enzymes can help only those that are right for it. They work by "lock and key." Enzymes speed up reactions by lowering the reaction barrier. They slow down reactions by raising the reaction barrier. Enzymes attach to a specific substrate. They can also inhibit reactions through competitive and noncompetitive inhibition. There is also denaturization. Enzymes are affected badly by heat and strong pH. In the wrong environment, they will be denatured, their shape will change, and they will not work in any reaction.

1

1

1
—
Total
3 pts.

PRACTICE

How many points would you give this essay?

Analysis of Essay A

The numbers on the right side of the essay represent points the student earned. Here is what the student received credit for:

1 pt.   Enzymes are proteins.

1 pt.   Enzymes attach to a specific substrate.

1 pt.   Enzymes can be denatured (their shape altered) by strong heat or pH. Because they are denatured, they no longer function.

3 pts.   Total

Here is what the student seemed to be trying to say but for which he/she received no credit.

In the first sentence, the student states that enzymes speed up reactions. That is true. However, that information was given in the question.

Enzymes *are* catalysts; they do not work by *using* a catalyst.

Enzymes do not work by "lock and key." They work by "induced fit."

Enzymes speed up reactions by lowering the energy of activation and facilitating more effective collisions.

Although it is true that enzymes slow reactions by competitive and non-competitive inhibition, these terms were only mentioned, not explained.

SAMPLE ESSAY B

Enzymes are organic catalysts. They can speed up a reaction by lowering the energy   1
of activation, but they cannot make a reaction happen that wouldn't normally   1
happen. Enzymes are globular proteins that exhibit tertiary structure due to vari-   1
ous intramolecular attractions, including Van der Waals forces, hydrophobic and   1
hydrophilic interactions, and disulfide bridges. They have a particular conformation
or 3-D shape that determines how they function. Since enzymes are not used up in
a reaction, an organism has a limited number of enzymes. Enzymes have *specificity*,   1
meaning that one enzyme catalyzes one specific chemical reaction. The molecule
an enzyme bonds to is called a substrate, and when the two connect an enzyme-   1
substrate complex is formed. An enzyme bonds to a substrate at an active site. The   1
process by which enzymes actually catalyze reactions can be explained using the
induced-fit model. Since all reactions need effective collisions for them to occur, the
angle at which particles collide with each other is important. Enzymes, by binding
with the substrate, increase the angle of collisions. Because the collisions between
particles become more effective, less energy is needed to activate the reaction. An   1
example of an enzyme that catalyzes a reaction is catalase. Catalase assists H_2O_2 into
breaking down into H_2O and O_2 gas. The catalase binds with the substrate H_2O_2 at   1
the active site. Enzymes are named after the substrate they work on plus the suffix   1
"ase." For example, maltase is the name of the enzyme that breaks apart maltose into
glucose molecules.   1

In addition to speeding up reactions, enzymes can also slow them down or inhibit them competitively, noncompetitively, or allosterically. In competitive inhibition, the enzyme has one active site but two substrates. If there is a higher concentration of substrate *A* than substrate *B*, the enzyme will catalyze a reaction involving substrate *A*. In noncompetitive inhibition, the enzyme has two active sites and two substrates. When one substrate goes into its active site, it blocks the other active site, preventing the other substrate from fitting into it. In allosteric regulation, a modifier is needed to bind with the enzyme at the allosteric site, to change its conformation so it can bind with its substrate. If the modifier is not there, the substrate's reaction will be impeded.

1

1

1

I have explained the structure of enzymes and how they can speed up and slow down reactions.

**Total
14 pts.**

Analysis of Essay B

The numbers on the right side of the essay represent points the student earned. Although this is not a beautifully written essay, there is a substantial amount of correct information given, and it would earn full credit, 10 points.

REMEMBER

Explain or define all scientific terms you use.

1 pt.	Enzymes are organic catalysts.
1 pt.	They speed up reactions by lowering the energy of activation.
1 pt.	They cannot make a reaction happen that would not normally happen.
1 pt.	Enzymes are globular proteins that exhibit tertiary structure due to—
1 pt.	Various intramolecular attractions, as listed.
1 pt.	Form/conformation dictates function.
1 pt.	Enzymes are not used up and are needed in limited quantity.
1 pt.	Enzymes are specific.
2 pts.	Enzymes bond to a substrate at an active site, forming an enzyme-substrate complex.
1 pt.	More effective collisions result.
1 pt.	Example: catalase
1 pt.	Competitive inhibition is correctly explained
1 pt.	Noncompetitive inhibition is correctly explained.

What the Student Did Not Get Credit For

The student did not receive credit for mentioning "induced fit," because the concept was not explained.

Allosteric inhibition is not clearly explained. The following should be added. "There are two active sites, one for the substrate and one for the allosteric inhibitor."

The student states, "Since enzymes are not used up . . . an organism has a limited number of enzymes." This would have been better stated as "Since enzymes are not used up, very small amounts are needed by the body."

Students can receive points for discussing something in depth. The mention of "catalase" as an example of an enzyme would give the student a point for "depth of understanding." The instructions did not state to include an example; but doing so was a good idea.

Notice, there is no mention of denaturing as in the first essay. **You do not have to discuss every single point to get full credit.** You gain points as you state correct information using related terminology. **Points are never taken away on an essay if information is incorrect.**

MODEL
TESTS

ANSWER SHEET
Model Test 1

1. (A) (B) (C) (D)	17. (A) (B) (C) (D)	33. (A) (B) (C) (D)	49. (A) (B) (C) (D)
2. (A) (B) (C) (D)	18. (A) (B) (C) (D)	34. (A) (B) (C) (D)	50. (A) (B) (C) (D)
3. (A) (B) (C) (D)	19. (A) (B) (C) (D)	35. (A) (B) (C) (D)	51. (A) (B) (C) (D)
4. (A) (B) (C) (D)	20. (A) (B) (C) (D)	36. (A) (B) (C) (D)	52. (A) (B) (C) (D)
5. (A) (B) (C) (D)	21. (A) (B) (C) (D)	37. (A) (B) (C) (D)	53. (A) (B) (C) (D)
6. (A) (B) (C) (D)	22. (A) (B) (C) (D)	38. (A) (B) (C) (D)	54. (A) (B) (C) (D)
7. (A) (B) (C) (D)	23. (A) (B) (C) (D)	39. (A) (B) (C) (D)	55. (A) (B) (C) (D)
8. (A) (B) (C) (D)	24. (A) (B) (C) (D)	40. (A) (B) (C) (D)	56. (A) (B) (C) (D)
9. (A) (B) (C) (D)	25. (A) (B) (C) (D)	41. (A) (B) (C) (D)	57. (A) (B) (C) (D)
10. (A) (B) (C) (D)	26. (A) (B) (C) (D)	42. (A) (B) (C) (D)	58. (A) (B) (C) (D)
11. (A) (B) (C) (D)	27. (A) (B) (C) (D)	43. (A) (B) (C) (D)	59. (A) (B) (C) (D)
12. (A) (B) (C) (D)	28. (A) (B) (C) (D)	44. (A) (B) (C) (D)	60. (A) (B) (C) (D)
13. (A) (B) (C) (D)	29. (A) (B) (C) (D)	45. (A) (B) (C) (D)	61. (A) (B) (C) (D)
14. (A) (B) (C) (D)	30. (A) (B) (C) (D)	46. (A) (B) (C) (D)	62. (A) (B) (C) (D)
15. (A) (B) (C) (D)	31. (A) (B) (C) (D)	47. (A) (B) (C) (D)	63. (A) (B) (C) (D)
16. (A) (B) (C) (D)	32. (A) (B) (C) (D)	48. (A) (B) (C) (D)	

SECTION I

Time: 90 minutes

PART A: 63 MULTIPLE-CHOICE QUESTIONS

Directions: For each question choose the best answer choice. After you have completed Part A, move on to Part B.

1. Here is a sketch of a molecule of the sex hormone, testosterone, that is derived from cholesterol.

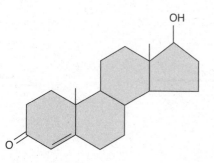

Which of the following statements best describes the action of this hormone on cells of the human gonads?

(A) The hormone acts as the first messenger when it binds to and activates the G protein-coupled receptor on the surface of cells in the testes. This activates the mobile G protein located inside the cell.

(B) The hormone enters cells in the testes by first binding with a membrane receptor, which causes a channel to open in the membrane, allowing the testosterone to flood into the cell.

(C) The hormone readily passes through the cell membrane and binds to a receptor in the cytoplasm. The hormone and receptor then enter the nucleus and act as a transcription factor that turns on one or more genes.

(D) The hormone binds with cAMP on the surface of the cell. Once attached to cAMP, the hormone enters the cell and initiates a signal transduction pathway.

2. Mice are mammals. As such, they are endotherms. They maintain internal heat metabolically. A lizard is an ectotherm and gains its body heat from the environment. Which of the following graphs most accurately depicts this situation?

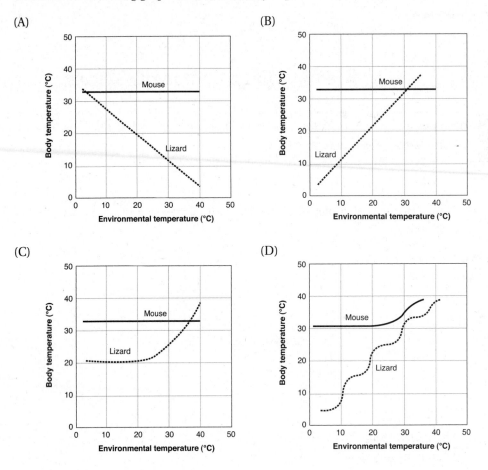

(A)

(B)

(C)

(D)

3. Which of the following is a direct result of depolarizing the presynaptic membrane of an axon?

(A) The postsynaptic cell produces an action potential.
(B) Synaptic vesicles fuse with the presynaptic membrane.
(C) Voltage-gated calcium channels in the membrane open.
(D) Neurotransmitter is released into the synapse.

4. There are two types of enzyme inhibition: competitive and noncompetitive. Competitive inhibitors compete with the substrate for one active site, while noncompetitive inhibitors bind to another part of the enzyme and alter its shape. Malonate is an inhibitor of the respiratory enzyme succinate dehydrogenase in the following reaction.

$$\text{Succinate} \xrightarrow[\substack{\text{Malonate}\\ \text{(inhibitor)}}]{\substack{\text{Succinate dehydrogenase}\\ \text{(enzyme)}}} \text{Fumarate}$$

Which of the following choices best describes the best way to determine whether malonate is a competitive or a noncompetitive inhibitor of succinate dehydrogenase?

(A) In the presence of malonate, increase the concentration of succinate, the substrate. If the rate of reaction increases, then malonate is a competitive inhibitor.

(B) In the presence of malonate, increase the concentration of succinate, the substrate. If the rate of reaction increases, then malonate is a noncompetitive inhibitor.

(C) Malonate can alternate between a competitive and noncompetitive inhibitor depending on what is required.

(D) A coenzyme must be affecting the rate of the reaction.

5. Water climate diagrams summarize the climate for a region. Average temperature is shown on the left axis, precipitation on the right axis, and months of the year on the x-axis. Assuming that a major limiting factor for plant growth is availability of water, choose the graph below that depicts the region with the best conditions for plant growth.

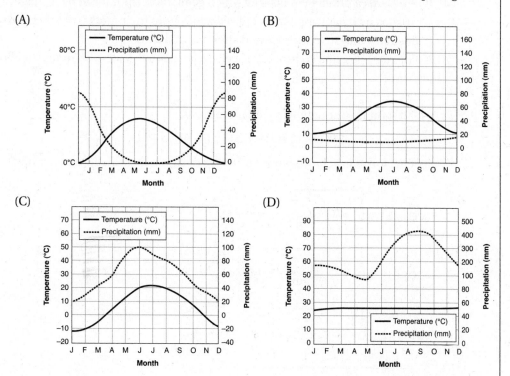

6. Here is a graph showing the rate of photosynthesis of a green plant plotted against light intensity.

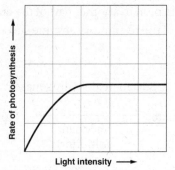

Which of the following statements accurately accounts for the shape of the graph?

(A) The rate of photosynthesis stops increasing because chlorophyll molecules begin to decompose due to increased heat, not increased light.

(B) The rate of photosynthesis slows as chlorophyll molecules, which are proteins, denature.

(C) Photosynthesis ceases when protons are no longer released by the photolysis of water.

(D) The rate of photosynthesis increases until the light-harvesting apparatus in the thylakoid membranes become saturated and cannot make use of additional light.

7. The lac operon is found only in prokaryotes and consists of structural genes and a promoter and operator. Which of the following statements explains why we study the lac operon?

(A) It represents a principal means by which gene transcription can be regulated.

(B) It explains how baby mammals utilize lactose from their mothers as they nurse.

(C) It illustrates how RNA is processed after it is transcribed.

(D) It illustrates possible control of the cell cycle and may lead to an understanding of cancer.

8. Here is the simplified equation for photosynthesis:

$$6CO_2 + 12\,H_2O \rightarrow C_6H_{12}O_6 + 6H_2O + 6O_2$$

Which of the following choices correctly traces the atom identified with an arrow?

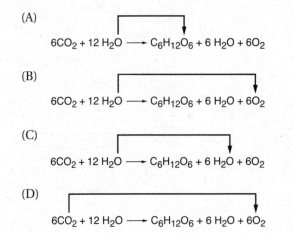

(A)

$$6CO_2 + 12\,H_2O \longrightarrow C_6H_{12}O_6 + 6\,H_2O + 6O_2$$

(B)

$$6CO_2 + 12\,H_2O \longrightarrow C_6H_{12}O_6 + 6\,H_2O + 6O_2$$

(C)

$$6CO_2 + 12\,H_2O \longrightarrow C_6H_{12}O_6 + 6\,H_2O + 6O_2$$

(D)

$$6CO_2 + 12\,H_2O \longrightarrow C_6H_{12}O_6 + 6\,H_2O + 6O_2$$

9. The plasma membrane is selectively permeable and regulates what enters and leaves a cell. It consists of a phospholipid bilayer embedded with proteins. Which of the following statements about the membrane is correct?

(A) Carbon dioxide is a polar molecule and readily diffuses through the hydrophilic layers of the membrane.

(B) Starch readily diffuses across the membrane into the liver, where it is stored as glycogen.

(C) Aquaporins are special water channels in the plasma membrane that facilitate the uptake of large amounts of water without the expenditure of energy.

(D) Oxygen passes through the cristae of mitochondria mainly through ATP synthase channels.

10. Active transport involves the movement of a substance across a membrane against its concentration or electrochemical gradient. Active transport is mediated by specific transport proteins and requires the expenditure of energy. Which of the following statements describes an example of active transport?

(A) Glucose is transported across some membranes by carrier proteins down a concentration gradient.

(B) Freshwater protists such as amoeba and paramecia have contractile vacuoles that pump out excess water.

(C) Countercurrent exchange in the capillaries of fish gills enables fish to absorb large amounts of oxygen from the surrounding water where oxygen levels are low.

(D) A red blood cell will lyse (burst) if placed into a hypotonic solution because large amounts of water flood into the cell.

11. Here is a sketch of an animal cell.

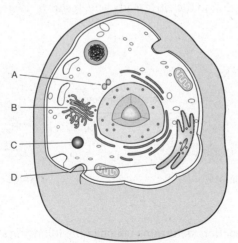

Which of the following statements is correct?

(A) Structure *A* detoxifies poisons in the cell.

(B) Structure *B* packages proteins for export.

(C) Structure *C* synthesizes RNA.

(D) Structure *D* consists of cytoskeleton.

12. Chronically high levels of glucocorticoids can result in obesity, muscle weakness, and depression. This looks like several diseases but actually is only one, Cushing syndrome. Excessive activity of either the pituitary or the adrenal gland can cause the disease. To determine which gland has abnormal activity in a particular patient, doctors use the drug dexamethasone, a synthetic glucocorticoid that blocks ACTH (adrenocorticotropic hormone) release.

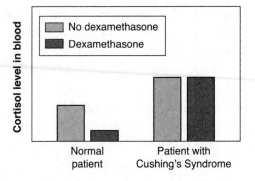

Based on the graph, which gland is affected in the patient with Cushing syndrome and what is the reasoning behind your answer?

(A) The pituitary, because although ACTH is blocked, the pituitary is still sending a signal to the adrenal glands.

(B) The pituitary, because blocking ACTH has no effect on cortisol levels.

(C) The adrenal gland, because the pituitary is prevented from stimulating the adrenal glands, and yet cortisol levels are still high.

(D) The adrenal gland, because the pituitary is sending a signal to the adrenal gland and the adrenal glands have stopped producing cortisol.

13. Here is the final reaction in the citric acid cycle. It shows the regeneration of oxaloacetate.

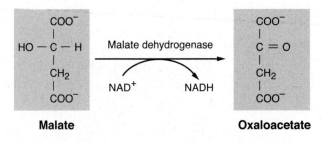

After studying the reaction, determine which of the following statements is correct.

(A) The enzyme malate dehydrogenase is allosteric.

(B) The reaction is exergonic; the released energy is absorbed by NAD$^+$.

(C) The reaction is a reduction reaction.

(D) NAD$^+$ is oxidized into NADH.

14. Which of the following statements about the DNA in one of your liver cells is correct?

    (A) Your liver cells contain the same DNA as your white blood cells.
    (B) Most of the DNA in your liver cells will be transcribed but not translated.
    (C) Most of the DNA in your liver cells will be transcribed in the nucleus and translated in the cytoplasm.
    (D) DNA in your liver cells contains genes unique to storing glycogen, while other cells do not contain those genes.

15. Two genes (*B* and *E*) determine coat color in Labrador retrievers. Alleles *B* and *b* code for how much melanin is present. The dominant allele *B* codes for black fur, the recessive allele *b* codes for brown hair. A second gene consisting of two alleles, *E* and *e*, codes for the deposition of pigment in the hair. In this case, the alleles for pigment deposition (*E* or *e*) are said to be epistatic to the gene that codes for black or brown pigment. If the dominant allele (*E*) is not present, regardless of the genotype at the black/brown locus, the animal's fur will be yellow.

    Which of the following statements is correct about the genetics of coat color in Labradors?

    (A) *BBEe* will be brown.
    (B) *BbEe* will be brown.
    (C) *Bbee* will be brown.
    (D) *BBee* will be yellow.

16. Within a cell, the amount of protein synthesized using a given mRNA molecule depends in part on which of the following?

    (A) DNA methylation suppresses the expression of genes.
    (B) Transcription factors mediate the binding of RNA polymerase and the initiation of transcription, which will determine how much protein is manufactured.
    (C) The speed with which mRNA is degraded will determine how much protein is synthesized.
    (D) The location and number of ribosomes in a cell will solely determine how much protein is synthesized.

17. Oxygen is carried in the blood by the respiratory pigment hemoglobin, which can combine loosely with four oxygen molecules, forming the molecule oxyhemoglobin. To function properly, hemoglobin must bind to oxygen in the lungs and drop it off at body cells. The more easily the hemoglobin binds to oxygen in the lungs, the more difficult the oxygen is to unload at the body cells. Here is a graph showing two different saturation-dissociation curves for one type of hemoglobin at two different pH levels.

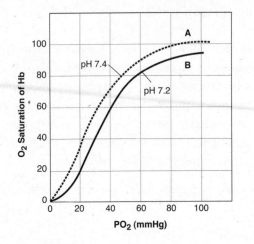

Based on your knowledge of biology and the information in this graph, which statement about the hemoglobin curves is correct?

(A) Hemoglobin *B* has a greater affinity for oxygen and therefore binds more easily to oxygen in the lungs.
(B) Hemoglobin *A* is characteristic of a mammal that evolved at sea level where oxygen levels are high.
(C) In an acidic environment, hemoglobin drops off oxygen more easily at body cells.
(D) Hemoglobin *A* is found in mammals with a higher metabolism.

18. Banana plants, which are triploid, are seedless and therefore sterile. Which is the most likely explanation for this phenomenon?

(A) Because they are triploid, bananas cannot generate gametes because homologous pairs cannot line up during meiosis.
(B) Because they are triploid, there are no male or female banana plants.
(C) Because they are triploid, bananas cannot generate homologous pairs during mitosis.
(D) Because they are triploid, bananas cannot carry out crossing-over.

19. Animals maintain a minimum metabolic rate for basic functions such as breathing, heart rate, and maintaining body temperature. The minimum metabolic rate for an animal at rest is the basal metabolic rate (BMR). The BMR is affected by many factors, including whether an animal is an ectotherm or endotherm; its age and sex; and size and body mass. Here is a graph that shows the relationship of BMR per kilogram of body mass to body size for a group of mammals.

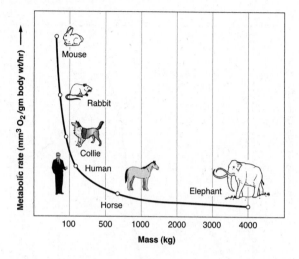

Which statement correctly describes the relationship between BMR and body mass?

(A) The relationship between BMR and body mass is a direct one. The larger the body mass the higher the BMR is and the greater the breathing rate is.

(B) The relationship between BMR and body mass is a direct one. The larger the body mass the higher the BMR is and the lower the breathing rate is.

(C) The relationship between BMR and body mass is inversely proportional. The larger the body mass, the lower the BMR is and the lower the breathing rate is.

(D) The larger the animal, the faster the heart rate and breathing rate both are.

20. Here is a graph that depicts a first exposure to antigen *A* on Day 1 with a subsequent, primary immune response. A second exposure to antigen *A* on Day 30 results in a secondary immune response due to the presence of circulating memory cells that release antibodies against antigen *A*. There is also a first exposure to antigen *B* on Day 30.

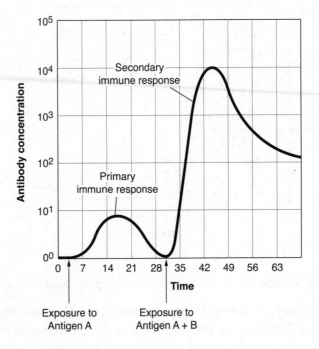

Which of the following graphs accurately depicts the immune response to antigen *B* and the reason for it?

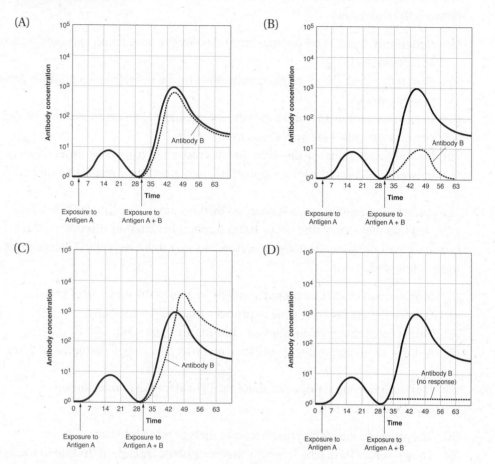

(A) Graph A. The primary response to antigen *B* is almost as fast and large as the secondary immune response to antigen *A* because the entire immune system was activated by the first exposure to antigen *A*.

(B) Graph B. Immune responses are specific. The fact that there was an earlier exposure to antigen *A* has no bearing on the response to antigen *B* on Day 30.

(C) The response to antigen *B* on Day 30 is larger than the secondary immune response to antigen *A* because the immune system has already been activated and all new responses are heightened.

(D) There is almost no immune response to antigen *B* because the immune system is fully engaged in a secondary response to antigen *A*.

21. Incomplete dominance and epistasis are two terms that define genetic relationships. Which of the following statements best describes the differences or similarities between the two terms?

    (A) Both terms describe inheritance controlled by the interaction of multiple alleles at different loci.

    (B) Both terms describe inheritance controlled by the interaction of multiple genes at different loci.

    (C) Incomplete dominance describes the interaction of two genes at different loci; epistasis describes the interaction of two alleles.

    (D) Incomplete dominance describes the interaction of two alleles on one gene; epistasis describes the interaction of two genes and their respective alleles.

22. Before the human genome was sequenced by the Human Genome Project, scientists expected that they would find about 100,000 genes. In fact, they discovered that humans have only about 24,000 genes. How can we exhibit so many different traits from so few genes?

    (A) Modification of histone proteins usually increases the function of genes.

    (B) Epigenetics enables one gene to produce many different traits.

    (C) This is proof that pseudogenes and introns are expressed.

    (D) A single gene can produce more than one trait because of alternative splicing

23. Which of the following statements about the light-dependent reactions of photosynthesis is correct?

    (A) They provide the carbon that becomes incorporated into sugar.

    (B) They produce PGA, which is converted to glucose by carbon fixation in the light-independent reactions.

    (C) Water is split apart, providing hydrogen ions and electrons to NADP for temporary storage.

    (D) They occur in the stroma of chloroplasts.

24. The acrosome and cortical reactions that occur immediately after fertilization ensure which of the following?

    (A) The sperm will fertilize the ovum.

    (B) The zygote will begin to divide normally.

    (C) One and only one sperm will fertilize the egg.

    (D) The zygote will not divide into two embryos, producing twins.

25. Which aspects of cell structure best reveals the unity of all life?

    (A) All cells are surrounded by a plasma membrane.

    (B) All cells have at least one nucleus.

    (C) All cells carry out cellular respiration in mitochondria.

    (D) The surface-to-volume ratio of all cells is the same.

26. Refer to the codon table for the following question.

		SECOND BASE			
		U	C	A	G	

SECOND BASE

FIRST BASE (5' end) / THIRD BASE (3' end)

	U	C	A	G	
U	UUU ⎤ Phe UUC ⎦ UUA ⎤ Leu UUG ⎦	UCU ⎤ UCC ⎥ Ser UCA ⎥ UCG ⎦	UAU ⎤ Tyr UAC ⎦ UAA Stop UAG Stop	UGU ⎤ Cys UGC ⎦ UGA Stop UGG Trp	U C A G
C	CUU ⎤ CUC ⎥ Leu CUA ⎥ CUG ⎦	CCU ⎤ CCC ⎥ Pro CCA ⎥ CCG ⎦	CAU ⎤ His CAC ⎦ CAA ⎤ Gln CAG ⎦	CGU ⎤ CGC ⎥ Arg CGA ⎥ CGG ⎦	U C A G
A	AUU ⎤ AUC ⎥ Ile AUA ⎦ AUG Met or start	ACU ⎤ ACC ⎥ Thr ACA ⎥ ACG ⎦	AAU ⎤ Asn AAC ⎦ AAA ⎤ Lys AAG ⎦	AGU ⎤ Ser AGC ⎦ AGA ⎤ Arg AGG ⎦	U C A G
G	GUU ⎤ GUC ⎥ Val GUA ⎥ GUG ⎦	GCU ⎤ GCC ⎥ Ala GCA ⎥ GCG ⎦	GAU ⎤ Asp GAC ⎦ GAA ⎤ Glu GAG ⎦	GGU ⎤ GGC ⎥ Gly GGA ⎥ GGG ⎦	U C A G

Here is a small stretch of mRNA that would be translated at the ribosome:

...AUG CUG **U**AA UCAGGG...

Suppose a spontaneous mutation altered the boldface A and changed it to a U. What effect, if any, would this have on the protein formed?

(A) Because of redundancy in the code, there would be no change in the protein formed.

(B) The amino acid sequence formed from this stretch of DNA would be Met–Leu–Lys–Ser–Gly.

(C) The polypeptide would not form because translation would stop at UAA.

(D) Translation would continue and a polypeptide would form because AUG codes for start as well as for methionine.

27. Stretching out from the equator is a wide belt of tropical rain forests that are being cut down to provide exotic woods for export to the U.S. and to make land available to graze beef cattle. Which of the following statements is a negative consequence of clear-cutting the tropical rain forests?

(A) Indigenous populations will have access to modern conveniences.

(B) There will be less precipitation in those clear-cut areas.

(C) U.S. markets will have access to less expensive beef.

(D) Deforestation will allow more sunlight to penetrate areas that were kept dark by dense vegetation.

28. Mosquitoes resistant to the pesticide DDT first appeared in India in 1959 within 15 years of widespread spraying of the insecticide. Which of the following statement best explains how the resistant mosquitoes arose?

(A) Some mosquitoes experienced a mutation after being exposed to DDT that made them resistant to the insecticide. Then their population expanded because these moquitoes had no competition.

(B) Some mosquitoes were already resistant to DDT when DDT was first sprayed. Then their population expanded because all the susceptible mosquitoes had been exterminated.

(C) DDT is generally a very effective insecticide. One can only conclude that it was manufactured improperly.

(D) Although DDT is effective against a wide range of insects, it is not effective against mosquitoes.

29. DNA sequences in many human genes are very similar to the corresponding sequences in chimpanzees. Which statement gives the most likely explanation of this fact?

(A) Humans evolved from chimpanzees millions of year ago.

(B) Chimpanzees evolved from humans millions of years ago.

(C) Humans and chimpanzees evolved from a recent common ancestor—about 6 million years ago.

(D) Humans and chimpanzees evolved from a distant common ancestor about 4 billion years ago.

Questions 30 and 31

Two ecologists, Peter and Rosemary Grant, spent thirty years observing, tagging, and measuring finches (a type of bird) in the Galápagos Islands. They made their observations on Daphne Major—one of the most desolate of the Galápagos Islands. It is an uninhabited volcanic cone where only low to the ground cacti and shrubs grow. During 1977, there was a severe drought and seeds of all kinds became scarce. The small, soft seeds were quickly eaten by the birds, leaving mainly large, tough seeds that the finches normally ignore. The year after the drought, the Grants discovered that the average width of the finches' beak had increased.

Before   After

30. Which of the following statements best explains the increase in beak size?

(A) Finches with bigger and stronger beaks were able to attack and kill the finches with smaller beaks.

(B) Finches with bigger beaks were larger animals with stronger wings that could fly to other islands and gather a wide variety of seeds.

(C) During the drought, the finches' beaks grew larger to accommodate the need to eat tougher seeds.

(D) Finches with larger beaks were able to eat the tougher seeds and were healthier and reproduced more offspring that inherited the trait for wider beaks.

31. Which of the following statements best explains the mechanism behind the change in beak size?

(A) A new allele appeared in the finch population as a result of a mutation.

(B) A change in the frequency of a gene was due to selective pressure from the environment.

(C) A new trait appeared in the population because of recombination of alleles.

(D) A new trait appeared in the population because of genetic drift.

Questions 32 and 33

Answer the following two questions based on this pedigree for the biochemical disorder known as alkaptonuria. Affected individuals are unable to break down a substance called alkapton, which colors the urine black and stains body tissues. Otherwise, the condition is of no consequence. Males are shown as squares, females as circles. Afflicted individuals are shown in black. If there is a carrier condition, it is not displayed.

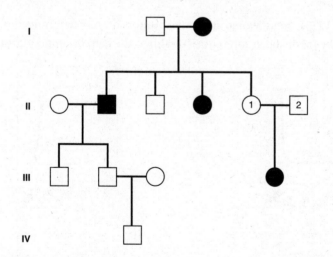

32. Which of the following best states the pattern of inheritance shown in the pedigree?

(A) The trait is sex-linked dominant.

(B) The trait is sex-linked recessive.

(C) The trait is autosomal dominant.

(D) The trait is autosomal recessive.

33. Which of the following statement is supported by the information given in the pedigree?

(A) The P generation mother is X-X.
(B) The P generation father is X-Y.
(C) If parents 1 and 2 in row II have another child, the chance that the child would be afflicted with alkaptonuria is 25%.
(D) The genotype of woman 1 is X-X.

___

34. Sea otters in the North Pacific are a keystone species. That means they are not abundant in a community. However, they do exert major control over other species in the community. Here is a food chain in which the sea otter is a keystone species.

Kelp $\rightarrow$ Sea urchin $\rightarrow$ Sea otter $\rightarrow$ Orca

Here are two graphs showing the populations of sea otters and sea urchins from 1970 to 2000.

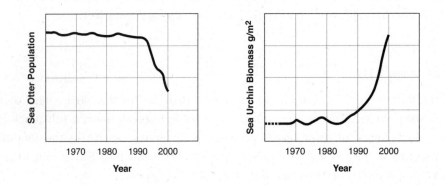

In the late 1990s, orcas, which are active hunters of sea otter, moved into the area. Which of the following graphs correctly shows the kelp population from 1970 to 2000?

(A)

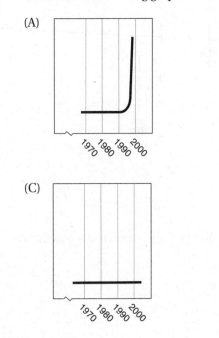

(B)

(C)

(D)

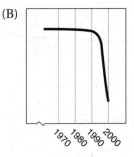

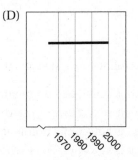

Questions 35–38

Four vials were set up to investigate bacterial transformation. Vials 1 and 2 each contained *E. coli* bacteria that had been made competent and had then been mixed with a plasmid containing the gene for ampicillin (pAMP) resistance. Vials 3 and 4 both contained *E. coli* that had also been made competent but had not been mixed with a plasmid. Each vial of bacteria was poured onto a nutrient agar plate. Vials 1 and 3 were poured onto plates that contained the antibiotic ampicillin. Vials 2 and 4 were poured onto plates that did not contain ampicillin.

This figure shows what the nutrient agar plates looked like. The shaded plates represent extensive growth, and the dots represent individual bacterial colonies.

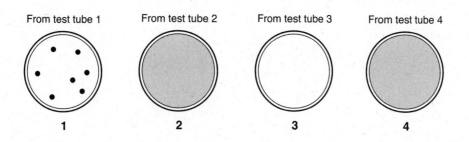

35. Plates that have only ampicillin-resistant bacteria growing on them include which of the following?

   (A) Plate 1 only
   (B) Plate 2 only
   (C) Plate 3 only
   (D) Plates 1 and 3 only

36. Which of the following statements explains why there was no growth on Plate 3?

   (A) The heat shock technique to make the *E.coli* competent killed the bacteria.
   (B) Those particular *E.coli* bacteria were inhibited from growing by the nutrient in the agar.
   (C) Those bacteria were not transformed.
   (D) The bacteria culture died because they have a short life span.

37. Which of the following statements best explains why there were fewer colonies on Plate 1 than on Plate 2?

   (A) The bacteria on Plate 2 did not transform.
   (B) There was no antibiotic in the agar in Plate 2 that would have restricted growth of bacteria.
   (C) The transformation of bacteria on Plate 2 was more successful.
   (D) The bacteria on Plates 1 and 2 were not taken from the same culture.

38. In a variant of this experiment, the plasmid contained GFP (green fluorescent protein) in addition to ampicillin resistance. Which of the following plates would have the highest percentage of bacteria that would fluoresce?

(A) Plate 1 only
(B) Plate 2 only
(C) Plate 3 only
(D) Plate 4 only

39. Here is a sketch of a neuromuscular junction in a patient with an autoimmune disease. Acetylcholine (ACh) is the stimulatory neurotransmitter.

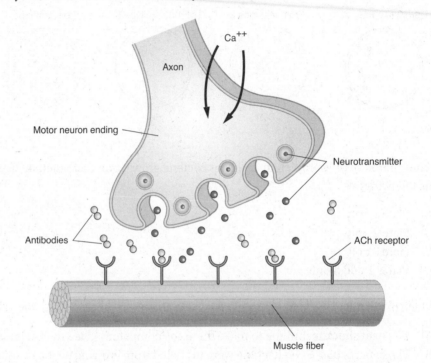

Which of the following predicts what will happen in the continued presence of the antibodies?

(A) Ca⁺⁺ ions will flood into the motor neuron ending, increasing the release of more ACh.
(B) The amount of neurotransmitter being released will decrease.
(C) The number of action potentials in the motor neuron will decrease.
(D) Antibodies will destroy the postsynaptic receptors, and the muscle response will diminish.

40. Which cell type is most likely involved in storage?

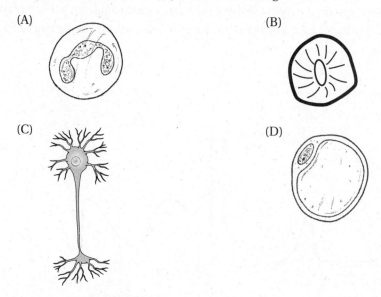

(A)

(B)

(C)

(D)

41. In 1953, Stanley Miller, while working under the guidance of Harold Urey at the University of Chicago, carried out a series of experiments with substances that mimicked those of early Earth. His experimental setup yielded a variety of amino acids that are found in organisms alive on Earth today. The purpose of these experiments best supports which of the following hypotheses?

(A) The basic building blocks of life originated in outer space and came to Earth carried by comets or meteorites.
(B) The molecules necessary for life to develop were located in deep-sea vents.
(C) The molecules necessary for life to develop could have formed under the conditions of the early Earth.
(D) The molecules necessary for life on Earth were self-replicating proteins, just like the ones produced in Miller's experiments.

42. The graph on the left (*A*) shows an absorption spectrum for chlorophyll *a* extracted from a plant. The graph on the right (*B*) shows an action spectrum from a living plant, with wave lengths of light plotted against the rate of photosynthesis as measured by release of oxygen.

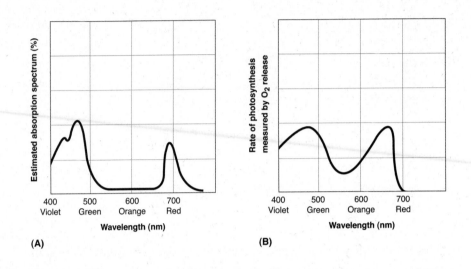

(A)   **(B)**

Which statement best explains the difference between the two spectra?

(A) Graph *A* plots the absorption spectrum of a red plant; graph *B* plots an absorption spectrum for a green plant.

(B) The chlorophyll from Graph *A* cannot carry out the light-dependent reactions; but the chlorophyll in graph *B* can.

(C) The data from Graph *A* characterize several photosynthetic pigments that reflect almost no light; the data from Graph *B* characterize chlorophyll *a*, which reflects only green light.

(D) Graph *A* shows an absorption spectrum for an unusual type of chlorophyll *a*.

Questions 43–44

Cystic fibrosis is the most common inherited disease in the U.S. Among people of European descent, 4% are carriers of the recessive cystic fibrosis allele. The most common mutation in individuals with cystic fibrosis is a mutation in the CFTR protein that functions in the transport of chloride ions between certain cells and extracellular fluid. These chloride transport channels are defective or absent in the plasma membranes of people with the disorder. The result is abnormally high concentration of extracellular chloride that causes the mucus that coats certain cells to become thicker and stickier than normal. Mucus builds up in the pancreas, lungs, digestive tract, and other organs. This leads to multiple effects, including poor absorption of nutrients from the intestines, chronic bronchitis, and recurrent bacterial infections.

43. Scientific work has been carried out to measure where the relative amounts of CFTR protein are localized in the affected cells.

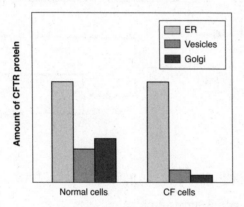

   After studying the graph, which of the following statements about CFTR protein is correct?

   (A) Transcription is not occurring.
   (B) Translation is not occurring.
   (C) CFTR protein does not fold properly after it is synthesized.
   (D   CFTR protein is not being packaged in the cytoplasm.

44. From the description of cystic fibrosis above, which of the following statements is correct?

   (A) The disease is similar to cancer in that it has both an environmental and a genetic cause.
   (B) Cystic fibrosis is an example of a genetic disease caused by polygenic inheritance because several genes must be mutated in order for the disease to occur.
   (C) Cystic fibrosis is an example of pleiotropy because one mutated gene causes multiple effects.
   (D) A person who has one cystic fibrosis allele will have the disease.

45. Referring to the simplified reaction of cellular respiration shown here, which of the following statements is correct?

$$C_6H_{12}O_6 \ + \ 6O_2 \ \rightarrow \ 6CO_2 \ + \ 6H_2O \ + \ energy$$

(A) $C_6H_{12}O_6$ is reduced.

(B) O_2 is reduced.

(C) CO_2 is reduced.

(D) H_2O is reduced.

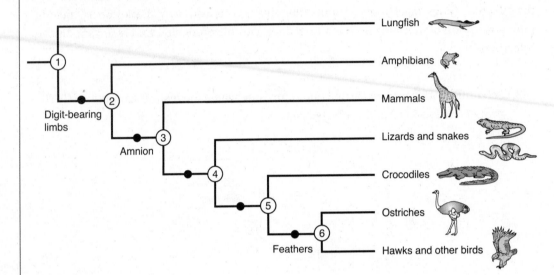

46. Based on the cladogram, which of the following is NOT true?

(A) Only some tetrapods have amnions.

(B) Mammals and amphibians are more closely related than mammals and birds.

(C) Ancestor 2 lived before ancestor 3; but we do not know when that was.

(D) Lungfishes and amphibians share a common ancestor.

47. On Andros Islands in the Bahamas, populations of mosquito fish, *Gambusia hubbsi*, colonized a series of ponds. These ponds are no longer connected. However, the environments are very similar except that some ponds contain predatory fish, while others do not. In high predation ponds, selection has favored the evolution of a body shape that enables rapid bursts of speed. In low predation ponds, another body type is favored—one that is well-suited to swim for long periods of time.

When scientists brought together sample mosquito fish from the two types of ponds, they found that the females mated only with males that had the same body type as their own. Which of the following statements best describes what happened to the mosquito fish as evidenced by the mating choice in the female fish?

(A) Reproductive isolation caused geographic isolation.

(B) Reproductive isolation was not complete.

(C) Allopatric isolation brought about reproductive isolation.

(D) Sympatric isolation brought about reproductive isolation.

48. From 1972 to 2004, researchers studying the greater prairie chicken observed that a population collapse mirrored a reduction in fertility as measured by the hatching rate of eggs. Comparison of DNA samples from the Jasper County, Illinois, population with DNA from feathers in a museum collection showed that genetic variation had declined in the studied population. The researchers translocated prairie chickens from Minnesota, Kansas, and Nebraska into the Illinois prairie chicken population in 1992 and found that the hatching rate in Illinois prairie chickens changed. Here is a graph of the data.

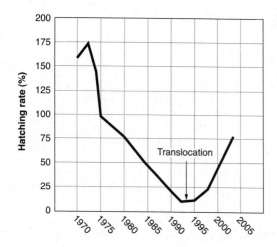

Which of the following statements most accurately explains what happened to the studied population of prairie chickens after translocation and why?

(A) The hatching rate increased because genetic variation declined.
(B) The hatching rate increased because genetic variation increased.
(C) The hatching rate decreased because genetic variation decreased.
(D) The hatching rate decreased because the translocated animals were invasive and grew to dominate the population.

49. Which of the following statements about the immune system is NOT correct?

(A) A lymphocyte has receptors for many different antigens.
(B) An antigen can have different epitopes.
(C) Dendritic cells are antigen-presenting cells.
(D) T cells attack infected body cells.

50. In a study of dusty salamanders, *Desmognathus ochrophaeus*, scientists brought individuals from different populations into the laboratory and tested their ability to mate and produce viable, fertile offspring. Here is the graph of the data.

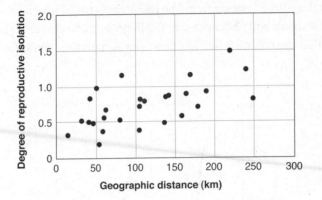

The degree of reproductive isolation is represented by an index ranging from 0 (no isolation) to 2 (complete isolation).

Which of the following statements best describes the evolutionary history of dusty salamanders?

(A) Mutation alone caused the two populations to diverge.

(B) Both mutation and genetic drift cause the two populations to diverge.

(C) Both mutation and genetic drift caused the two populations to become separate species after they were separated by great distance.

(D) Reproductive isolation between the two populations increases as the distance between them increases.

51. Which of the following statements best describes what a Barr body is and its significance?

(A) It is an inactivated Y chromosome and results in a man being sterile.

(B) It is an inactivated Y chromosome, and the person who has it appears female.

(C) It is an inactivated X chromosome and results in females with half their cells having one X inactivated and the other half of their cells having the other X inactivated.

(D) It is an inactivated X chromosome and results in females who are sterile.

52. Scientists carried out a series of experiments to study innate immunity in fruit flies. They began with a mutant fly strain in which a pathogen is recognized but the signaling that would normally trigger an innate response is blocked. As a result, the flies did not make any antimicrobial peptides (AMP). The researchers then genetically engineered some of the mutant flies to express significant amounts of a single AMP, either defensin or drosomycin. They then infected the flies with a fungus, *Neurospora crassa*, by shaking anesthetized flies for 30 seconds in a Petri dish containing a sporulating fungal culture. The 6-day survival rate was monitored and recorded.

Here is a graph displaying some data from the experiment.

KEY:
A Wild type
B Mutant + drosomycin
C Mutant + defensin
D Mutant

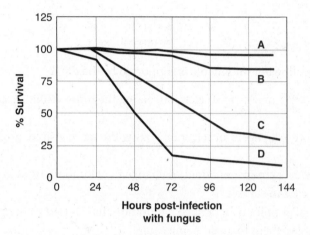

Which of the following statements is supported by the data?

(A) Each AMP provided minimal immunity against fungal infection.
(B) Both AMPs provided better immunity than did either one alone.
(C) Drosomycin provided immunity against the fungal infection.
(D) There was no control in this experiment.

Questions 53–54

An experiment was carried out with guppies, which are brightly colored, popular, aquarium fish. Three hundred guppies were added to 12 large pools. Cichlids, a voracious predator, were added to 4 of the pools. Killifish that rarely eat guppies were added to 4 other pools. No other fish were added to the last 4 pools. After 16 months, a time period that represents 10 generations for guppies, all the guppies were analyzed for size and coloration. Here are the data.

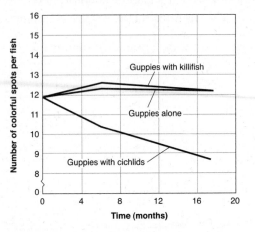

53. Which of the following statements is supported by the data?

(A) Cichlids were an agent of selection, eating the more visible, brightly colored guppies.
(B) Killifish were an agent of selection, eating the more visible, brightly colored guppies.
(C) Guppies, as a group, experienced change in coloration, from brightly colored to dull in order to survive.
(D) Individual guppies experienced a mutation that caused a change in coloration and enabled them to avoid being eaten.

54. Which of the following statements best expresses the point of this experiment?

(A) Mutations can be brought about by the environment.
(B) Mutations occur randomly.
(C) Agents of selection are not always readily apparent.
(D) Evolution does not always require millions of years.

55. In the Tularosa Basin of New Mexico are black lava formations surrounded by light-colored sandy desert. Pocket mice inhabit both areas. Dark-colored ones inhabit the lava formations, while light-colored mice inhabit the desert. Which of the following statements is correct about this scenario?

(A) The two varieties of mice descended from a recent common ancestor.
(B) The mouse population was originally one population that diverged into two species because of mutations.
(C) Selection favors some phenotypes over others.
(D) Originally the mice were all dark colored. As the lava decomposed into sand, the color of some mice changed because that color was favored in that environment.

56. What is the best explanation for the fact that tuna, sharks, and dolphins all have a similar streamlined appearance?

(A) They all share a recent common ancestor.
(B) They all sustained the same set of mutations.
(C) They acquire a streamlined body type to survive in their particular environment.
(D) The streamlined body has a selective advantage in that environment.

57. Which of the following statements is **NOT** correct about lipids?

(A) Lipids consist of fatty acids and glycerol.
(B) Steroids are examples of lipids.
(C) The molecules of a saturated fatty acid are packed so close together that they form a solid at room temperature.
(D) The head of a phospholipid is hydrophobic, and the tails are hydrophilic.

Questions 58–59

Here is a sketch of prokaryotic DNA undergoing replication and transcription simultaneously.

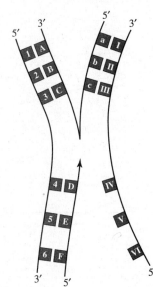

58. If 1 is thymine, what is *a*?

(A) Adenine
(B) Thymine
(C) Cytosine
(D) Uracil

59. If 4 is adenine, what is *D*?

(A) Adenine
(B) Thymine
(C) Cytosine
(D) Uracil

60. Yeast cells carry out both aerobic and anaerobic respiration. A yeast cell that is fed on glucose is moved from an aerobic to an anaerobic environment. Which of the following statements is correct and gives the correct reason for it?

    (A) The cell dies because it cannot make ATP.
    (B) The cell would need to consume glucose at a much greater rate because aerobic respiration is much more efficient as compared with anaerobic respiration.
    (C) The cell would need to consume another food source other than glucose because it will not be able to make adequate ATP with only glucose.
    (D) The cell will begin to divide rapidly because larger cells require more oxygen and glucose than smaller ones.

Questions 61–62

Intact chloroplasts are isolated from dark-green leaves by low-speed centrifugation and are placed into six tubes containing cold buffer. A blue dye, DPIP, which turns clear when reduced, is also added to all the tubes. Then each tube is exposed to different wavelengths of light. A measurement of the amount of decolorization is made, and the data are plotted on a graph. Although the wavelengths of light vary, the light intensity is the same.

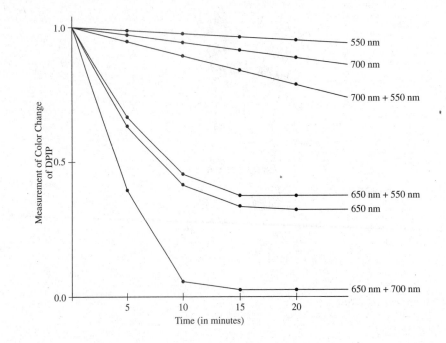

61. Which statement below best describes the results of the experiment?

    (A) The lower the wavelength of light, the greater the rate of photosynthesis.
    (B) The highest wavelengths of light provide the fastest rate of photosynthesis.
    (C) The highest rate of photosynthesis results from exposure to two different wavelengths of light.
    (D) The greatest reduction in DPIP occurs at 550 nm light intensity.

62. What is the best explanation for the results in Question 61?

(A) 650 nm and 700 nm of light have the greatest penetrating power.
(B) The photosynthetic pigments in the experiment do not absorb light in the 650 nm and 700 nm range.
(C) Only wavelengths of 650 nm and 700 nm are reflected by the chlorophyll.
(D) The chloroplasts have two photosystems that absorb light in different wavelengths.

---

63. A group of scientists wished to learn about the energy efficiency of different modes of locomotion. They studied the literature that was based on accurate measurements and produced this graph based on the data they analyzed.

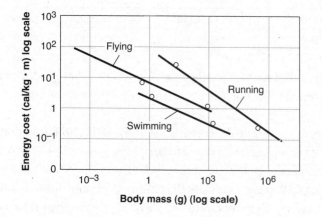

Which of the following statement accurately describes what the scientists discovered?

(A) An animal with a large body mass expends more energy per kilogram of body mass than a small animal, regardless of the type of locomotion.
(B) Swimming has to overcome drag as well as gravity.
(C) Running is the least energy-efficient means of locomotion.
(D) The best parameter to measure in this type of experiment is CO_2 consumption.

PART B: 6 GRID-IN QUESTIONS

> **Directions:** Read each of the 6 questions, then provide a numeric answer for each. Units are not required.

1. A female fruit fly hybrid for both gray body (*Gg*) and normal wings (*Nn*) is crossed with a male with black body (*gg*) and vestigial wings (*nn*): *GgNn* × *ggnn*

   The F1 results of the cross are shown in the chart below.

A	B	C	D
Gray normal	Black vestigial	Gray vestigial	Black normal
969	941	190	184

   Given the data above, calculate the rate of cross-over that produced the gray vestigial and black normal offspring. Give your answer to the nearest tenth.

2. A student crossed a purple dahlia with a yellow dahlia and found that all the offspring produced purple-flowers. She then crossed those purple-flowered offspring (F1) and produced 350 plants. Of their offspring (F2), 101 produced yellow flowers and 249 produced purple flowers. Calculate the chi-squared value for the null hypothesis that the purple-flowered F1 plants were all hybrid. Give your answer to the nearest tenth.

3. Here is a graph that shows a growth curve for a population of bacteria.

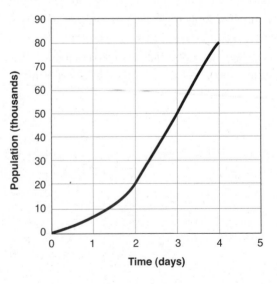

   Calculate the mean rate of population growth between day 2 and day 4. Give your answer to the nearest whole number.

4. In a certain population of birds, the allele for a crown on the head (*C*) is dominant to the allele for no crown on the head (*c*). A particularly cold and long winter favored the birds with no crown. When spring came, researchers determined that the population was currently in Hardy-Weinberg equilibrium and that the occurrence of the birds with no crown was up to 24%. What will the frequency of the no-crown allele be in 10 years? Give your answer to the nearest hundredth.

5. Here is the volume of the world's water supply ($\times$ 1,000 km³). Our drinking water comes from the freshwater located in groundwater, lakes, ice sheets, glaciers, and rivers.

Source	Volume
Oceans, seas, and bays	1,338,000.
Ice sheets and glaciers	24,064.
Groundwater 　Fresh 　Saline	23,400. (10,530.) (12,870.)
Soil moisture	16.5
Ground ice and Permafrost	300.
Lakes 　Fresh 　Saline	176.4 (91.0) (85.4)
Atmosphere	12.9
Swampwater	11.47
Rivers	2.12
Biological water	1.12
Total	**1,385,984**

What percentage of all the water on Earth is freshwater? Give your answer in percent to the tenths place.

6. Here is a sketch that shows Earth's water cycle including the oceans and continents. Assuming that total precipitation equals total evaporation, what is the value of evaporation from the land to the atmosphere?

Earth's Water Cycle

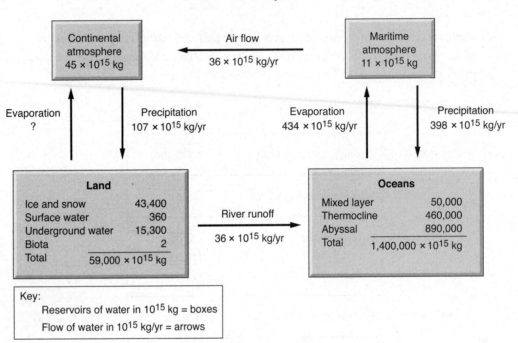

SECTION II

8 Free-Response Questions

Time: 90 minutes

8 FREE-RESPONSE QUESTIONS

Directions: The first two questions are long, free-response questions. Spend about 20 minutes answering each. Questions 3 through 8 are short, free response questions. Take about 6 minutes answering each. Write your response on a separate piece of paper and ask your teacher to "grade" it.

1. Cells communicate with each other using a complex process that involves signals, receptors, and pathways. Discuss the different ways in which cells send signals to each other. Include the following:

   (a) Different types of signals based on distance from a signal's origin
   (b) Different types of receptors
   (c) What constitutes a cellular response
   (d) What the signal transduction pathway says about the evolution of cell communication

2. Terrestrial biomes are diverse. They are controlled by the climate.

   (a) Describe the difference between climate and weather.
   (b) Describe one biome. Give its location on Earth and characterize the organisms that live there.
   (c) Describe how a disturbance might affect this biome and how it might reconstitute itself.
   (d) Give an example of how a geographic cline might affect the evolution of an organism.

3. Name two properties of a eutrophic lake.

4. Many organisms live in estuaries and experience both freshwater and saltwater conditions every day as the water rises and falls with the tides. Explain how changing salinity levels challenge the survival of the organisms.

5. It is now known that very little of the human genome actually codes for proteins. Briefly explain how microRNAs, also called miRNA, relate to this.

6. Explain and give one example of stabilizing selection.

7. Give two differences between innate and adaptive immunity.

8. Explain why a poison that inhibits an enzyme in the Calvin cycle would also inhibit the light-dependent reactions.

ANSWER KEY

Section I Part A: Multiple-Choice Answers

1. **(C)**	12. **(C)**	23. **(C)**	34. **(B)**	45. **(B)**	56. **(D)**
2. **(B)**	13. **(B)**	24. **(C)**	35. **(A)**	46. **(C)**	57. **(D)**
3. **(C)**	14. **(A)**	25. **(A)**	36. **(C)**	47. **(C)**	58. **(B)**
4. **(A)**	15. **(D)**	26. **(C)**	37. **(B)**	48. **(B)**	59. **(D)**
5. **(D)**	16. **(C)**	27. **(B)**	38. **(A)**	49. **(A)**	60. **(B)**
6. **(D)**	17. **(C)**	28. **(B)**	39. **(D)**	50. **(D)**	61. **(C)**
7. **(A)**	18. **(A)**	29. **(C)**	40. **(D)**	51. **(C)**	62. **(D)**
8. **(B)**	19. **(C)**	30. **(D)**	41. **(C)**	52. **(C)**	63. **(C)**
9. **(C)**	20. **(B)**	31. **(B)**	42. **(B)**	53. **(A)**	
10. **(B)**	21. **(D)**	32. **(D)**	43. **(D)**	54. **(D)**	
11. **(B)**	22. **(D)**	33. **(C)**	44. **(C)**	55. **(C)**	

Section I Part B: 6 Grid-In Answers

1. 16.4%
2. 2.99
3. 30,000
4. .49
5. 2.5%
6. 71×10^{15} kg

Section II Free-Response Answers

See Answer Explanations.

ANSWER EXPLANATIONS

Section I Part A: Multiple-Choice Answers

1. **(C)** The sex hormone testosterone is lipid soluble (hydrophobic) and dissolves directly through the cell membrane. Choices A and B correctly describe the action of a hydrophilic signal. Cyclic AMP is a secondary messenger found in the cytoplasm, not on the surface of a cell.

2. **(B)** The body temperature of an ectotherm increases with the ambient temperature. A homeotherm or an endotherm can raise and lower its body temperature to maintain a steady body temperature.

3. **(C)** This question focuses on the order of activity at the synapse. An electrical impulse travels down the axon, causing voltage-gated calcium channels in the presynaptic membrane to open. Calcium ions flood into the terminal branch. This causes presynaptic vesicles to fuse with the membrane and release their contents, neurotransmitter, into the synapse.

4. **(A)** If malonate is a competitive inhibitor, it competes with the enzyme (succinate dehydrogenase) for succinate. If malonate is a noncompetitive inhibitor, increasing succinate will not increase the rate of reaction because the enzyme is being inhibited by malonate.

5. **(D)** This region has the most consistently warm temperature and the highest precipitation.

6. **(D)** The graph flattens out but doesn't decrease. Therefore, photosynthesis is still going on. That indicates that choices A, B, and C cannot be correct.

7. **(A)** Remember that operons are found in prokaryotes, not higher organisms. However, higher animals may utilize a similar mechanism. For the same reason, you must eliminate Choice B. Operons are a means of controlling transcription of genes, whereas Choice C focuses on what happens after transcription. Operons have nothing to do with the cell cycle.

8. **(B)** The process of hydrolysis releases oxygen from water into the air.

9. **(C)** Carbon dioxide is a balanced and nonpolar molecule that can readily diffuse through the membrane. Starch is too large to diffuse through a membrane. Only protons pass through mitochondrial membranes.

10. **(B)** Whenever a pump is involved, the mechanism is active transport. Glucose is always transported by either simple diffusion or facilitated diffusion. However, neither process requires the expenditure of energy. Choices C and D refer to passive transport.

11. **(B)** Structure *B* is the Golgi apparatus, which packages and secretes proteins made by the ribosomes in the E.R. The centrioles shown as structure *A* consist of microtubules. Structure *C* is a lysosome. Strucure *D* consists of ribosomes on the E.R.

12. **(C)** The patient is still producing too much cortisol even though the pituitary is not sending a signal to the adrenal glands. ACTH, which would normally stimulate the adrenal to release cortisol, is blocked by the dexamethasone. If the adrenals were healthy, they should not be producing anything. Unfortunately, they are. Compare the patient with the normal person.

13. **(B)** Energy is released, and NAD^+ binds to hydrogen released by malate.

14. **(A)** All cells in your body contain the same genes; the difference is which ones are turned on and which are turned off. Also, less than 3% of all a human cell's DNA is transcribed and translated.

15. **(D)** The trait for deposition of pigment (*E*) is epistatic to the presence of color (*B/b*). If there is no dominant trait for deposition, the animal will be yellow or golden. An animal with the genotypes *BBEe* or *BbEe* is black. An animal with the genotypes *bbEE* or *bbEe* is brown. However, the genotype *BBee* results in a yellow dog.

16. **(C)** Choices A and B refer to transcription of DNA. However, the question asks about protein manufactured from an RNA molecule.

17. **(C)** The line to the right shows a lower affinity for oxygen. CO_2 dissolved in water produces carbonic acid. What this graph demonstrates is that the blood near respiring cells, which release carbon dioxide, make the surrounding area acidic. This acidic environment causes hemoglobin to release its oxygen to the cells where it is needed.

18. **(A)** No evidence is provided in the question about bananas not having separate sexes. Meiosis, not mitosis, produces gametes. It is true that there can be no crossing-over if there is no synapsis (when homologous pairs line up to exchange DNA), but gametes could form. There would not be any variation in the gametes.

19. **(C)** The energy to maintain each gram of body mass is inversely related to body size. Each gram of mouse requires 20 times as many calories as a gram of elephant even though the whole elephant uses more calories than a whole mouse. The higher the metabolic rate, the higher the oxygen and food requirement per unit of body mass.

20. **(B)** This may seem tricky, but Choice B is correct. Immune responses are specific. Each challenge with a new antigen results only in a primary immune response for that antigen.

21. **(D)** This is a fact.

22. **(D)** Modification of histone proteins, as in methylation, silences genes. Epigenetics refers to the expression of a trait due to non-DNA factors. Pseudogenes and introns are not expressed.

23. **(C)** All the other choices refer to the light-independent reactions.

24. **(C)** The acrosome reaction enables the sperm to penetrate the coat surrounding the ovum. The cortical reaction ensures that a fertilization membrane forms and prevents another sperm from penetrating the egg. The cortical reaction is responsible for a slow block to polyspermy.

25. **(A)** Prokaryotes do not have internal membranes. They do not have a defined nucleus. Instead, they have DNA in a nucleoid region. All cells carry out cellular respiration, but they do not necessarily do it in mitochondria. The surface-to-volume ratio of cells varies with the cell's size. Smaller cells have a greater surface-to-volume ratio.

26. **(C)** UAA codes for a stop sequence. Translation would cease at that point in the process.

27. **(B)** Rain forests are responsible for the cycle of rain, absorption of water by roots, transpiration from leaves, cloud formation, and rain again. Without trees to absorb and evaporate water, the region will dry up and perhaps become a desert under a hot sun.

28. **(B)** Resistance to a certain chemical arises because exposure to the chemical kills all the insects except the ones that are already resistant to it. The resistant population is selected for in this way and reproduces a new population in which all the individuals are resistant.

29. **(C)** Humans and chimps share an ancestor that walked upright about 6–7 million years ago.

30. **(D)** This is classic Darwinian evolutionary theory—the survival of the best adapted.

31. **(B)** Choices A, C, and D all have the same problem. A new trait may have appeared by a single mutation, by recombination of alleles, or by genetic drift. However, that does not explain how the birds, on average, came to have longer beaks.

32. **(D)** Look at individuals 1 and 2. Neither has the condition, but they have a daughter with the condition. Sex-linked dominant and autosomal dominant are eliminated because neither parent has the condition. Sex-linked recessive is also eliminated because if the father passed the trait to his daughter, he would have to have the

condition. For the daughter to have the condition, she would have had to inherit two affected X chromosomes—one from each parent.

33. **(C)** Since the trait is inherited as autosomal recessive, individuals 1 and 2 are each hybird. The chance that any child of theirs will have the condition is 25%.

34. **(B)** When orcas moved into the area, the population of sea otters declined. This caused the population of sea urchins to increase and, ultimately, the kelp population to decline.

35. **(A)** The few colonies growing on Plate 1 consist of bacteria that have taken up the plasmid and are resistant to ampicillin. Plate 2 consists of both antibiotic-resistant bacteria as well as nonresistant bacteria. There is no antibiotic in the agar on Plate 2 to distinguish the two bacteria.

36. **(C)** The bacteria on Plate 3 were not transformed. That means they did not uptake a plasmid and were not resistant to antibiotic. There is antibiotic in the agar on that plate, which killed the nontransformed bacteria.

37. **(B)** The few colonies growing on Plate 1 consist of bacteria that have taken up the plasmid and are resistant to ampicillin. Plate 2 consists of both antibiotic-resistant bacteria as well as nonresistant bacteria. There is no antibiotic in the agar on Plate 2 to distinguish the two bacteria.

38. **(A)** There would be fluorescent bacteria on both Plates 1 and 2. However, 100% of the bacteria on Plate 1 would fluoresce while only a percentage of the bacteria on Plate 2 would do so.

39. **(D)** In an autoimmune disease, the immune system mistakenly attacks its own body structures. The antibodies attack the postsynaptic muscle receptors, not the neuron function.

40. **(D)** This happens to be a fat-storing cell from an animal. It has the obvious trait of a large vacuole. Choice A is a white blood cell. Choice B is a sclerenchyma cell from a plant. Choice C is a neuron.

41. **(C)** Some scientists believe that some complex molecules necessary for life on Earth might have come from outer space. That might be true. However, the purpose of Miller's experiment was to demonstrate that the molecules necessary for life to develop could have formed under the conditions of the early Earth. He did not succeed in demonstrating that the first molecules were self-replicating.

42. **(B)** Graph *A* shows the absorption spectrum for chlorophyll that was extracted from a living plant. Once separated from the grana and stroma, chlorophyll by itself cannot carry out photosynthesis.

43. **(D)** Ribosomes produce protein for export. Once synthesized, the protein is packaged into vesicles in the Golgi for secretion. The graphs show that the amount of CFTR is high where synthesis occurs in the E.R. and low in the vesicles and Golgi. So the protein was synthesized in the ribosomes but not packaged.

44. **(C)** Cystic fibrosis is solely an inherited disease. No environmental component causes the disease. It is an autosomal recessive disease. In order to have the disease, a person must have inherited two mutated genes, one from each parent. Polygenic

inheritance involves the inheritance of several genes. An example is height, skin, or hair color—any trait where a tremendous variety occurs in a population.

45. **(B)** This equation shows cellular respiration. The definition of reduction is the gain of electrons or hydrogen. Free oxygen on the left side of the equation exists by itself, but on the right side it has combined with hydrogen. Also, remember that during cellular respiration, oxygen combines with protons and electrons during oxidative phosphorylation.

46. **(C)** This cladogram does not have a time scale attached to it. All we can say about any clade or lineage is that one evolved earlier or later than another. A cladogram is constructed to show the development of different traits and the lineage of related organisms that evolved with that trait.

47. **(C)** Allopatric isolation refers to isolation due to geographic separation. Sympatric isolation refers to isolation caused by something other than geography, such as polyploidy. These fish clearly exhibited allopatric isolation because the two populations lived in different ponds with different selective factors. The pressure for selection came from the predatory fish. After years of being separated, the two fish populations could no longer mate. Therefore, allopatric isolation brought about reproductive isolation.

48. **(B)** Prior to translocation, the chickens had become inbred, diversity had decreased, and fertility had declined.

49. **(A)** Notice the question says NOT. A lymphocyte has receptors for only one type of antigen. The other statements are correct.

50. **(D)** The question does not reveal anything about mutation or genetic drift. The point here is that in this case, divergence is a function of geographic distance.

51. **(C)** Every cell in the mammalian females has two X chromosomes, but one of them will be deactivated in every cell. Which one of the two becomes inactive occurs randomly and independently in each embryonic cell. So, females are considered to be mosaics because they contain two types of cells, in terms of what genes a cell contains. This does not apply to men because men only have one X chromosome in each cell.

52. **(C)** The survival of mutant flies + drosomycin was greater than mutant flies + defensin. Two controls, *A* and *D*, were included in this experiment.

53. **(A)** Killifish do not eat guppies, but cichlids do. So cichilds were the agent of selection. No organism or population changes in order to survive. If individuals are not adapted, they die. No individual changes during its lifetime. Rather, the frequency of a particular trait in a population may change.

54. **(D)** The evolution in this population—the change in frequency of the trait for brightly colored spots—occurred rapidly, within 16 months. Evolution does not always require millions of years. The rate of change is a function of the pressure from the environment to change.

55. **(C)** The question does not give you enough background to determine anything more than to say that some traits are favored more than others in a particular environment.

56. **(D)** Once again, the environment provides the direction for evolutionary change. Tuna and sharks are fish; dolphins are mammals. Fish and mammals do not share a recent common ancestor.

57. **(D)** The head of a phospholipid is hydrophilic, and the tails are hydrophobic. The phosphate head is charged and therefore polar.

58. **(B)** First orient yourself in the drawing. The process at the top of the sketch is replication; the bottom is transcription. So if 1 is thymine, then its complement, I, is adenine. Therefore, if I is adenine, then its base pair is thymine.

59. **(D)** The bottom of the diagram represents transcription. Therefore, if 4 is adenine, then *D* is its complement in RNA, which is uracil.

60. **(B)** Anaerobic respiration is much less efficient at making ATP than is aerobic respiration. Glycolysis produces only 2 net ATPs from each molecule of glucose. The yeast can survive on glucose alone.

61. **(C)** The greatest reduction (change in color of DPIP) occurs at 650 nm + 700 nm of light. The steeper the decline of the graph, the greater the reduction of DPIP.

62. **(D)** The shorter the wavelength of light, the greater the penetrating power. Therefore, light with a wavelength of 550 nm has the greatest penetrating power. Light that is reflected is not absorbed.

63. **(C)** According to the data, an animal with a small body mass expends more energy per kilogram of body mass regardless of the type of locomotion. Swimming is the most efficient means of locomotion. Swimming does not have to overcome gravity; the swimmer is supported by the water.

Section I Part B: 6 Grid-In Answers

1. The answer is 16.4%.

   Divide the total number of recombinations (190 + 184 = 374) by the total number of offspring (2,284)

   $374 \div 2,284 = 0.1637 = 16.4\%$

2. The answer is 2.99.

   The hypothesis is that the purple-flowered F1 plants are hybrid. The expected ratio for the F2 plants is 3:1, purple to yellow. Out of 350 F2 plants, 249 had purple flowers and 101 had yellow flowers. So fill in the chart as shown.

Phenotype	Observed (*o*)	Expected (*e*)	(*o* − *e*)	(*o* − *e*)²	(*o* − *e*)² ÷ *e*
Purple Flowers	249	263	249 − 263 = −14	196	196 ÷ 263 = 0.74
Yellow Flowers	101	87	101 − 87 = 14	196	196 ÷ 87 = 2.25

The chi-squared formula found on the reference table is $\chi^2 = \Sigma(o - e)^2 \div e$. When applied to this cross, we get $0.74 + 2.25 = 2.99$. For the null hypothesis, the critical value for $p = 0.05$ with 1 degree of freedom is 3.84 (from the reference table). Since our value is less than that, we fail to reject the null hypothesis.

3. The answer is 30,000.

Use the formula from the reference table to determine the per capita rate increase: $\Delta N \div \Delta T$, where ΔN = change in population size, and ΔT = time interval. In this example, ΔN = 20,000 to 80,000 = 60,000 and ΔT = 2 days.

$\Delta N \div \Delta T = 30,000/\text{day}$

4. The answer is .49.

According to Hardy-Weinberg protocol, q^2 = homozygous recessive, q = the recessive allele; p^2 = homozygous dominant, and p = the dominant allele. The question asks about the frequency of the recessive allele in 10 years. The frequency of the allele is currently 24% = q^2, so $q = 4.89$. Since the population is in Hardy-Weinberg equilibrium, the frequency will not change in 10 years.

5. The answer is 2.5%.

Add all the freshwater sources, and then divide by the total volume of all water on Earth.

$24,064. + 10,530. + 300. + 91.0 + 2.12 = 34,987.12$
$34,987.12 \div 1,385,984 = 0.025 = 2.5\%$

6. The answer is 71×10^{15} kg.

Add up all the precipitation from the maritime and continental atmospheres. Subtract that value from the evaporation from the oceans.

$(398 \times 10^{15}) + (107 \times 10^{15}) = 505 \times 10^{15}$
$(505 \times 10^{15}) - (434 \times 10^{15}) = 71 \times 10^{15}$

Section II: Free-Response Answers

1. Important terms are highlighted in this essay.

   (a) Cells must communicate with each other and with their environment, and they do it in a wide variety of ways. Simple bacterial cells secrete molecules that enable them to respond to changes in their population density by a phenomenon called **quorum sensing**. In plant cells, **plasmodesmata** (plant cell junctions) allow chemical signals to pass between adjacent cells. In animal cells, signals can diffuse locally or can travel long distances. **Autocrine signals** diffuse from one part of a cell to another part of the same cell. **Synaptic signaling** in the nervous system employs neurotransmitters between adjacent cells. **Paracrine signals** send a message to nearby cells by diffusion, while **endocrine signals** (hormones) can travel anywhere in the blood to reach their target cells.

(b) There are two types of signaling molecules, hydrophilic and hyrdrophobic. Hydrophilic molecules bind to receptors on the cell surface. Hydrophobic molecules diffuse through the membrane and bind to a receptor once inside the cell. Receptors on the cell surface are all similar. They span the entire thickness of the membrane and are in contact with both the extracellular environment as well as the cytoplasm. The binding of a ligand to a receptor causes a change in the shape of the cytoplasmic side of the same receptor. Thus the signal is transmitted (transduced) from the membrane surface into the cytoplasm, but the signaling molecule never enters the cell. Cell surface receptors are so prevalent that they make up about 30% of all human proteins. Three examples of cell surface receptors are ion channel receptors, G-protein coupled receptors, and protein kinase receptors.

Ion-channel receptors open and shut a channel or a gate in a membrane, allowing an influx of ions, such as Na^+. When a ligand binds to the extracellular domain of a **G-protein coupled receptor (GPCR)** shape change occurs in the cytoplasmic side of the receptor. This shape change activates the mobile intermediary G protein located inside the cell. G protein, in turn, activates an enzyme, causing a cellular response. In effect, the G protein shuttles back and forth between the receptor and the enzyme, acting like an on-off switch. **Protein Kinase Receptors (PKRs)** belong to a class of plasma membrane receptors that exhibit enzymatic activity. One particular kinase receptor called **receptor tyrosine kinase (RTK)** influences many processes: the cell cycle, cell migration, cell metabolism, and cell proliferation. Once the cell surface receptor binds to a ligand, the receptor activates a second messenger, such as cyclic AMP (cAMP), inside the cell. This second messenger carries the message to the nucleus and genes.

In contrast, hydrophobic ligands diffuse directly through the plasma membrane and bind to receptors in the cytoplasm. Once inside the cell, the form a ligand–receptor complex that makes its way to the nucleus, where it switches genes on or off. Such hydrophobic chemical messengers include **steroid hormones** like testosterone and estrogen as well as thyroid hormones and the gas NO, nitric oxide.

(c) Regardless of where the receptor is located, the message arrives in the cytoplasm or nucleus, in order to stimulate a response. An activated receptor acts as a **transcription factor** that turns on one or more genes. By controlling the synthesis of an enzyme, a signal can control an entire chemical pathway. Sometimes instead of regulating the synthesis of a protein, an activated receptor can regulate protein activity. For example, a signal may open or shut an ion channel in the plasma membrane or may increase or decrease the metabolism in the cell.

(d) Many signal transduction pathways have been identified and studied extensively across several kingdoms: bacteria, yeast, animal cells, and plants. The amazing similarity in all of these pathways and in all these diverse organisms suggests that signal transduction pathways evolved hundreds of millions of years ago in a common ancestor. In addition, signal transduction pathways are highly specific, highly regulated, and highly efficient. A single signal molecule amplifies a signal, stimulating the release of thousands of molecules of product in what is called the **cascade effect**.

2. Notice that you should not include an introduction and conclusion. Just dive in and answer the question. Including the number and letter of the question you are answering will make your answer very clear.

(a) Climate is the long-term weather conditions that prevail at a given place. Weather is the kind of conditions one experiences each day.

(b) The northern coniferous forest is spread across northern North America and Eurasia and all the way to the arctic tundra. Winters are usually cold, and summers may be hot. This biome is dominated by cone-bearing trees, such as pine, fir, spruce, and hemlock, that depend on fire to regenerate. Animals include many diverse species including birds, moose, brown bears, chipmunks, and squirrels.

(c) Fires, both naturally occurring and those started by people, kill thousands of acres of trees and other plants. Fires can destroy a forest leaving just the soil intact. If this occurs, the forest will regenerate by secondary ecological succession. The result may be a more diverse population. Periodic outbreaks of insects that feed on the dominant trees can permanently destroy thousands of acres of trees.

(d) Rabbits of a single species may exhibit different coloring and ear size depending on their geographic cline. In the cold north, rabbits have evolved to be white to blend in with the snow and have short ears to conserve heat in the cold winter. In the south, rabbits are mottled brown and have long ears to radiate heat away from their bodies.

3. Humans have disrupted freshwater ecosystems, causing a process called eutrophication. Runoff from sewage and manure from pastures increases nutrients in lakes and causes excessive growth of algae and other nuisance plants. Shallow areas become choked with weeds, causing swimming and boating to become impossible. As large populations of photosynthetic organisms crowd out each other and die, two things happen. First, organic material accumulates on the lake bottom, reducing the depth of the lake. Second, detritivores use up oxygen as they decompose the dead organic matter. Lower oxygen levels make it impossible for some fish to live. As fish die, decomposers expand their activity and oxygen levels decrease even more. The cycle of death, accumulation of organic matter, and decrease in oxygen levels continues until the lake fills in and disappears.

4. Water readily diffuses into or out of an organism depending on water potential. A freshwater organism inundated with saltwater will lose tremendous amounts of water and must have a mechanism to take in more water to compensate for the loss. In contrast, a saltwater organism inundated with freshwater will tend to take in excess water and must have a means of pumping out excess water.

5 The human genome consists of 6 billion bases of DNA (3 billion base pairs) and about 24,000 genes. However, about 98% of human DNA does not code for protein. These noncoding regions consist of regulatory sequences that control gene expression. For example, there are introns located between genes, and different types of repetitive sequences that never get transcribed or translated. Researchers also have discovered different types of RNA that play a role in gene expression. One type of RNA is microRNA, or miRNA, that can prevent translation of a strand of mRNA by binding to it. In effect, miRNA can inhibit or silence a coding region of DNA.

6. Stabilizing selection eliminates the extremes and favors the more common intermediate forms in a population. Human babies born with a very high or a very low birth weight do not fare as well as babies who are born with a birth weight in the range of 6–8 pounds.

7. Innate immunity is nonspecific. It involves defenses to keep invaders out of the body, such as skin, stomach acid, and mucous membranes that trap antimicrobial substances. It also involves a second line of defense called an inflammatory response, which limits the spread of microbes that evade the first line of defense. In the second line of defense, histamine triggers vasodilation (enlargement of blood vessels), increasing blood supply to the area and bringing in more phagocytes to destroy the pathogens.

In contrast to the nonspecific innate immune system, adaptive immunity is specific. It includes antigen-presenting cells, APC. They identify specific antigens or epitopes and then trigger special cells of the adaptive immune system. B cells secrete antibodies against specific epitopes. T cells destroy body cells infected by viruses and other agents.

8. The Calvin cycle generates NADP from NADPH and ADP from ATP to be used in the light-dependent reactions. NADP picks up protons to form NADPH and ADP becomes phosphorylated to form ATP in the light-dependent reactions.

ANSWER SHEET
Model Test 2

1. Ⓐ Ⓑ Ⓒ Ⓓ
2. Ⓐ Ⓑ Ⓒ Ⓓ
3. Ⓐ Ⓑ Ⓒ Ⓓ
4. Ⓐ Ⓑ Ⓒ Ⓓ
5. Ⓐ Ⓑ Ⓒ Ⓓ
6. Ⓐ Ⓑ Ⓒ Ⓓ
7. Ⓐ Ⓑ Ⓒ Ⓓ
8. Ⓐ Ⓑ Ⓒ Ⓓ
9. Ⓐ Ⓑ Ⓒ Ⓓ
10. Ⓐ Ⓑ Ⓒ Ⓓ
11. Ⓐ Ⓑ Ⓒ Ⓓ
12. Ⓐ Ⓑ Ⓒ Ⓓ
13. Ⓐ Ⓑ Ⓒ Ⓓ
14. Ⓐ Ⓑ Ⓒ Ⓓ
15. Ⓐ Ⓑ Ⓒ Ⓓ
16. Ⓐ Ⓑ Ⓒ Ⓓ

17. Ⓐ Ⓑ Ⓒ Ⓓ
18. Ⓐ Ⓑ Ⓒ Ⓓ
19. Ⓐ Ⓑ Ⓒ Ⓓ
20. Ⓐ Ⓑ Ⓒ Ⓓ
21. Ⓐ Ⓑ Ⓒ Ⓓ
22. Ⓐ Ⓑ Ⓒ Ⓓ
23. Ⓐ Ⓑ Ⓒ Ⓓ
24. Ⓐ Ⓑ Ⓒ Ⓓ
25. Ⓐ Ⓑ Ⓒ Ⓓ
26. Ⓐ Ⓑ Ⓒ Ⓓ
27. Ⓐ Ⓑ Ⓒ Ⓓ
28. Ⓐ Ⓑ Ⓒ Ⓓ
29. Ⓐ Ⓑ Ⓒ Ⓓ
30. Ⓐ Ⓑ Ⓒ Ⓓ
31. Ⓐ Ⓑ Ⓒ Ⓓ
32. Ⓐ Ⓑ Ⓒ Ⓓ

33. Ⓐ Ⓑ Ⓒ Ⓓ
34. Ⓐ Ⓑ Ⓒ Ⓓ
35. Ⓐ Ⓑ Ⓒ Ⓓ
36. Ⓐ Ⓑ Ⓒ Ⓓ
37. Ⓐ Ⓑ Ⓒ Ⓓ
38. Ⓐ Ⓑ Ⓒ Ⓓ
39. Ⓐ Ⓑ Ⓒ Ⓓ
40. Ⓐ Ⓑ Ⓒ Ⓓ
41. Ⓐ Ⓑ Ⓒ Ⓓ
42. Ⓐ Ⓑ Ⓒ Ⓓ
43. Ⓐ Ⓑ Ⓒ Ⓓ
44. Ⓐ Ⓑ Ⓒ Ⓓ
45. Ⓐ Ⓑ Ⓒ Ⓓ
46. Ⓐ Ⓑ Ⓒ Ⓓ
47. Ⓐ Ⓑ Ⓒ Ⓓ
48. Ⓐ Ⓑ Ⓒ Ⓓ

49. Ⓐ Ⓑ Ⓒ Ⓓ
50. Ⓐ Ⓑ Ⓒ Ⓓ
51. Ⓐ Ⓑ Ⓒ Ⓓ
52. Ⓐ Ⓑ Ⓒ Ⓓ
53. Ⓐ Ⓑ Ⓒ Ⓓ
54. Ⓐ Ⓑ Ⓒ Ⓓ
55. Ⓐ Ⓑ Ⓒ Ⓓ
56. Ⓐ Ⓑ Ⓒ Ⓓ
57. Ⓐ Ⓑ Ⓒ Ⓓ
58. Ⓐ Ⓑ Ⓒ Ⓓ
59. Ⓐ Ⓑ Ⓒ Ⓓ
60. Ⓐ Ⓑ Ⓒ Ⓓ
61. Ⓐ Ⓑ Ⓒ Ⓓ
62. Ⓐ Ⓑ Ⓒ Ⓓ
63. Ⓐ Ⓑ Ⓒ Ⓓ

SECTION I

Time: 90 minutes

PART A: 63 MULTIPLE-CHOICE QUESTIONS

> **Directions:** For each question choose the best answer choice. After you have completed Part A, move on to Part B.

1. Oxygen is carried in the blood by the respiratory pigment hemoglobin, which can combine loosely with four oxygen molecules to form the molecule oxyhemoglobin. To function properly, hemoglobin must bind to oxygen in the lungs and drop it off at body cells. The more easily the hemoglobin bonds to oxygen in the lungs, the more difficult for the hemoglobin to unload the oxygen at the body cells. Here is a graph showing two different saturation-dissociation curves for one type of hemoglobin at two different pH levels.

    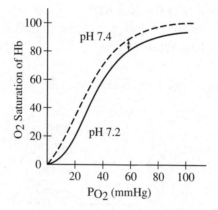

    Based on your knowledge of biology and the information in this graph, which statement about the hemoglobin curves is correct?

    (A) Hemoglobin *B* has a greater affinity for oxygen and will therefore bind more easily to oxygen in the lungs.
    (B) Hemoglobin *B* is characteristic of a mammal that evolved at a high elevation where oxygen is rare.
    (C) When CO_2 levels in the blood are high, as shown in hemoglobin *B*, hemoglobin releases oxygen more readily.
    (D) Hemoglobin *A* is the type found in mammals with a higher metabolism. Hemoglobin *B* is characteristic of mammals with a lower metabolism.

2. The geneticist Mary Lyon hypothesized the existence of structures visible just under the nuclear membrane in mammals, which were later named Barr bodies. Which of the following statement is NOT correct about Barr bodies?

(A) In the early female embryo, one copy of the X chromosome becomes inactivated in every body cell.
(B) The same chromosome in every cell of a normal female is inactivated.
(C) A male with the XXY genotype will have one Barr body.
(D) Barr bodies consist of highly methylated DNA.

3. Energy is harvested during cellular respiration in stages. Which of the following correctly states which phase of cellular respiration harvests the most energy and the correct explanation why?

(A) The most energy is released during the Krebs cycle because it is here that pyruvate is completely broken down into CO_2.
(B) The most energy is released during the Krebs cycle because in addition to the production of ATP, both $FADH_2$ and NADH are produced. Each of those molecules will release 2 ATPs and 3 ATPs, respectively.
(C) The most energy is released during oxidative phosphorylation because in addition to the phosphorylation of ADP into ATP, all the potential energy held in NADH and FADH is transferred to ATP.
(D) The most energy is released during oxidative phosphorylation because H_2O is completely broken down into H^+ and O_2.

4. The following reaction occurs in the citric acid cycle. Study the reaction as it is shown here.

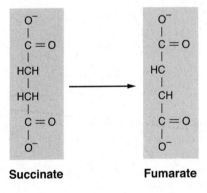

Succinate   Fumarate

Which of the following statements is correct?

(A) This is part of the reaction that produces pyruvate.
(B) The reaction is an example of negative feedback.
(C) The reaction is an oxidation reaction.
(D) The reaction occurs in the cristae membrane.

5. Which of the following statements is correct about a man who has hemophilia and whose wife does not have the disease and who does not have any relatives with the disease?

(A) All his daughters will have the disease.
(B) All his sons will have the disease.
(C) All his sons will be carriers.
(D) All his daughters will be carriers.

6. Identify the phase of meiosis when recombination occurs, and state the reason for your decision.

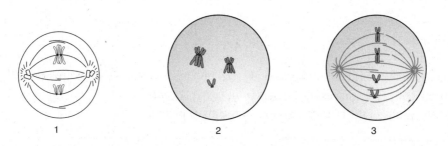

(A) Picture 1, because sister chromosomes are lined up on the metaphase plate.
(B) Picture 1, because homologous pairs are lined up on the metaphase plate.
(C) Picture 2, because homologues are paired up in synapsis.
(D) Picture 3, because homologous pairs are exchanging genetic material.

7. A rooster with gray feathers and a hen with the same phenotype produce 17 gray, 7 black, and 8 white chicks. What is the simplest explanation for the inheritance of these colors in chickens?

(A) The trait for gray is epistatic over the trait for black.
(B) The trait for gray is epistatic over the trait for white.
(C) The traits for black and white color demonstrate incomplete dominance.
(D) The traits for black and white color demonstrate codominance.

8. Hydrangea flowers have one gene for flower color. Plants of the same genetic variety have flowers that range in color from blue to pink with the color varying due to the type of soil in which they are grown. Which of the following statement best explains this phenomenon?

(A) The alleles for flower color show incomplete dominance where neither trait is dominant; expression of the genes shows a blending of traits.

(B) The alleles for flower color are codominant; both traits show depending on the environment.

(C) In this case, the environment alters the expression of a trait.

(D) The genes for flower color show polygenic inheritance.

Questions 9–10

A female fruit fly hybrid for both gray body (*Gg*) and normal wings (*Nn*) is crossed with a male with black body (*gg*) and vestigial wings (*nn*): *GgNn* × *ggnn*.

The F1 results of the cross are shown in the chart below.

A	B	C	D
Gray normal	Black vestigial	Gray vestigial	Black normal
969	941	190	184

9. Which of the following statements best explains the results?

(A) The two alleles for body color and wing structure are located on separate chromosomes and assort independently.

(B) The two alleles for body color and wing structure are located on separate chromosomes and assorted independently, but there was some crossing-over.

(C) The two alleles are linked, located on the same chromosome, immediately next to one another, and inherited together.

(D) The two alleles are linked and located on the same chromosome but are far apart and experienced some crossing-over.

10. Which of the following sketches depicts the most likely location of the alleles for body color and wing shape?

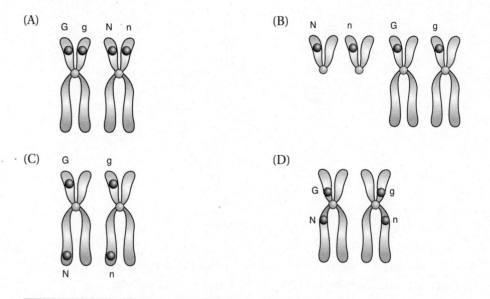

(A)   G   g   N   n

(B)   N   n   G   g

(C)   G   g
      N   n

(D)   G   g
      N   n

11. Which of the following is an example of a coupling of an exergonic reaction with an endergonic one?

(A) Unicellular organisms that live in freshwater, such as amoeba, must pump out excess water using their contractile vacuoles.

(B) The enzyme lactase binds with lactose to produce molecules of glucose and galactose.

(C) Electrons escaping from chlorophyll *a* are replaced by those released by the hydrolysis of water.

(D) The flow of electrons down an electron transport chain in mitochondria powers the pumping of protons against a gradient into the outer compartment.

12. Which of the following statements best describes what this cell will accomplish as part of the immune system?

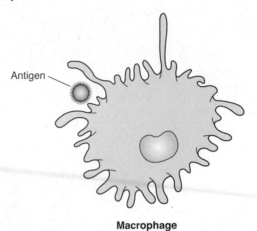

Macrophage

(A) It will secrete a chemical that will form holes in the pathogen.
(B) It will produce antibodies against the antigen it is engulfing.
(C) It will circulate in the bloodstream for a lifetime and will be able to identify the antigen in the future.
(D) It will present pieces of the antigen on its surface membrane.

13. A researcher is experimenting with nerve transmission using neurons from a giant squid. At the beginning of the experiment, the axon is at rest and has a threshold measured at –70 mV at 37°C. The axon receives one stimulus that triggers one action potential in the axon. Subsequent stimuli are much stronger. Which of the following statements explains what happens when several strong stimuli are experienced by the axon?

(A) The axon becomes hyperpolarized to –90 mV.
(B) The axon becomes hyperpolarized to –50 mV.
(C) The strength of the action potential becomes greater.
(D) The frequency of the action potential increases.

14. The frequency of a particular allele in a population of 1,000 birds in Hardy-Weinberg equilibrium is 0.3. If the population remains in equilibrium, what would be the expected frequency of that allele after 500 years?

(A) It will increase if the allele is favorable or will decrease if it is unfavorable for individuals in that population.
(B) Not enough information is provided to determine if the frequency of the allele will remain the same or change.
(C) The frequency of the allele will remain at 0.3 because the population is at Hardy-Weinberg equilibrium.
(D) The frequency of the allele will remain the same; this is an example of the bottleneck effect.

Questions 15–16

You carry out an experiment to study transpiration in plants using a two-week-old bean plant. You set up a potometer by cutting the stem of a plant and securing it into clear flexible tubing that has been tightly connected to a calibrated pipette. Water fills the potometer from the plant stem to the tip of the pipette.

    You measure water loss from your potometer at 10-minute intervals and plot your data on the graph as Line *B*.

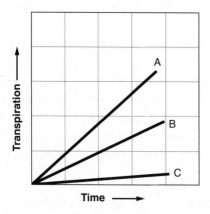

15. You then explore what happens if you expose the plant to different conditions. Which of the following statements is correct based on proposed changes to the potometer?

    (A) Increasing the green wavelengths of light will increase the rate of transpiration and account for Line *A* on the graph.

    (B) Placing a plastic bag over the plant will increase the rate of transpiration and account for Line *A*.

    (C) Placing a fan near the plant will decrease the rate of transpiration and account for Line *C*.

    (D) Painting the upper leaf surface of most of the leaves with clear nail polish will decrease the rate of transpiration and account for Line *C*.

16. Which of the following is NOT correct about transpiration in plants?

    (A) Water moves upward from the roots to the shoots in the xylem by active transport from an area of high osmotic potential to an area of low osmotic potential.

    (B) Plants lose water through stomates in their leaves.

    (C) The movement of water through a plant is facilitated by the physical and chemical properties of the water itself.

    (D) Placing a container of water between the light source and the plant is necessary to prevent the introduction of heat as a second variable in the experiment.

17. When the first tiny prokaryotic cell took up residence inside a larger prokaryotic cell, it heralded the advent of the eukaryotic cell and led to an explosion of new life on Earth. Since then, most cells on Earth have internal organelles. Which of the following best summarizes an advantage of having internal membranes and organelles?

    (A) DNA can reproduce more efficiently.
    (B) Even though prokaryotes do not have mitochondria, they contain structures that carry out the same function.
    (C) Organelles separate specific reactions in the cell and increase metabolic efficiency.
    (D) Compartmentalization enables prokaryotes to reproduce more quickly.

18. Many different fermentation pathways occur in different organisms in nature. For example, skeletal muscle cells convert pyruvate into lactic acid when no oxygen is present. Yeast cells can produce alcohol and carbon dioxide under the same circumstances. However, regardless of the specific reactions, the purpose of glycolysis is an important one. Which statement best describes the importance of glycolysis?

    (A) It produces large amounts of ATP by substrate level phosphorylation.
    (B) It reoxidizes NADH so that glycolysis can continue.
    (C) It produces pyruvate, which is the raw material for oxidative phosphorylation.
    (D) It occurs in the cytoplasm.

19. Human DNA and the DNA of a banana are 50% identical. Which statement best explains this fact in terms of what you know about the evolution of life on Earth?

    (A) Humans evolved from an organism that resembled a banana more than a billion years ago.
    (B) All organisms on Earth evolved from an ancestral eukaryotic cell 2 billion years ago.
    (C) Humans and bananas evolved from a recent common ancestor within the last 6 million years.
    (D) Bananas and humans have experienced thousands of mutations that have made us appear so very different from each other.

20. Which of the following principles is NOT part of Darwin's theory of evolution by natural selection?

(A) Evolution is a gradual process that occurs over a long period of time.
(B) Every population has tremendous variation.
(C) Mutations are the main source of all variation in a population.
(D) Organisms will overpopulate area, giving rise to competition.

21. Which of the following statements is supported by the information presented in the graph, which shows a comparison of two regions in Oregon, one deforested and the other undisturbed?

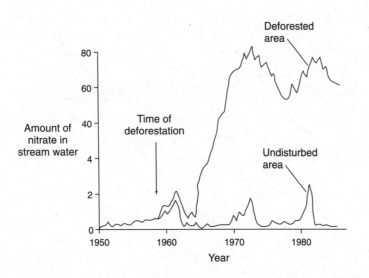

(A) Roots of plants are efficient at absorbing nitrates from the soil.
(B) Variation in the amount of nitrate in stream water is due to periods of intense rain.
(C) Replanting trees after an area has been clear-cut can prevent nitrate runoff into rivers.
(D) The presence of trees in an area causes an increase of nitrates in the soil.

22. Here is a food web for a habitat that is threatened by developers who will remove three-fourths of the grasses in the area on which the mice and rabbits feed.

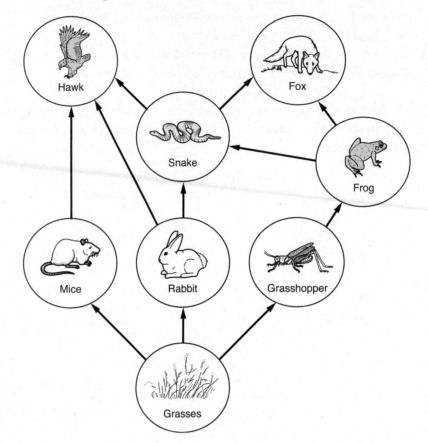

Which of the following statements describes what will most likely happen to the wildlife in the area?

(A) The hawk will begin to eat frogs instead of snakes and mice.
(B) Since three trophic levels are shown in the food web, 75% of the organisms in this food web will die.
(C) Based on the number of trophic levels, about 25% of the mice and rabbits will die.
(D) The hawk population will decrease.

23. The pH of two lakes is measured. Lake *A* has a pH of 8.0; Lake *B* has a pH of 6.0. Which of the following statements is correct about these lakes?

(A) Lake *B* is alkaline.
(B) Lake *B* is 100 times more alkaline than Lake *A*.
(C) The hydrogen ion concentration of Lake *B* is 10^4 M.
(D) The pH of both lakes will decrease in response to acid rain.

24. The *BCL-2* gene codes for a protein that normally inhibits apoptosis. In some cases, the *BCL-2* gene is mutated and becomes activated inappropriately. Cells with this mutated, overexpressed gene fail to undergo apoptosis; they continue to divide and form cancerous tumors. The mutated gene causes many cancers, including lymphoma and breast, colon, and prostate cancers. Oblimersen is a drug now in clinical trials that is an antisense RNA. It binds to *BCL-2* mRNA and inactivates it. Which of the following sketches accurately demonstrates the action of the antisense drug Oblimersen?

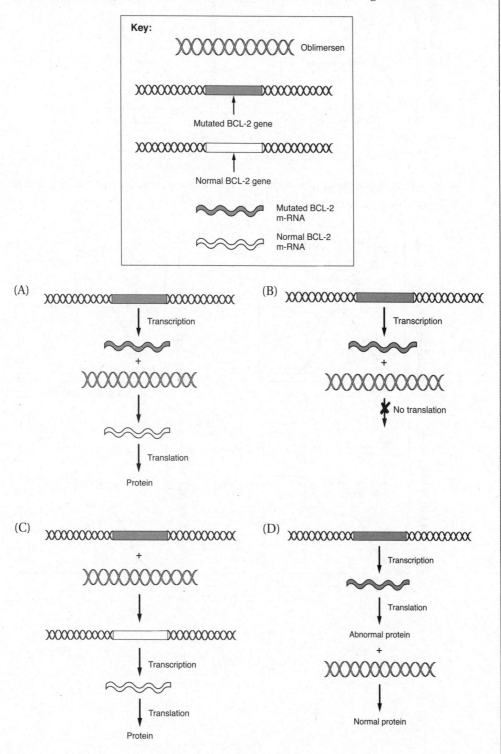

25. Figure 1 shows the growth of the filamentous green algae *Spirogyra* in a flask of sterilized pond water that contains nitrate. If phosphate is added to the flask as shown, the growth curve changes to that shown in Figure 2.

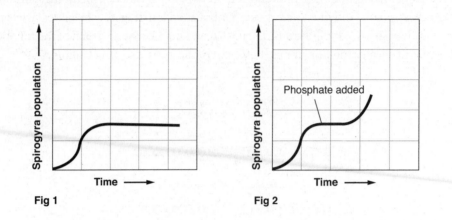

Fig 1   Fig 2

Which of the following graphs is the best prediction of the algae growth if more nitrate (NO$_3$) is added along with the phosphate?

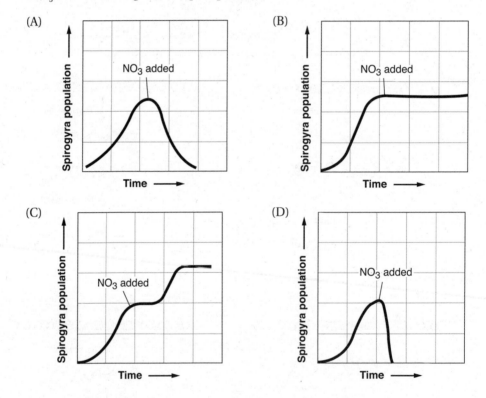

(A)

(B)

(C)

(D)

26. Which of the following statements correctly characterizes an autoimmune disease?

(A) T cells attack healthy body cells.

(B) An aneurysm is a weakness in the wall of a blood vessel that causes it to balloon out. The vessel can rupture and cause a hemorrhage.

(C) Several factors cause hypertension: smoking, being overweight, lack of physical activity, and too much salt in the diet.

(D) Alzheimer's disease is caused by the buildup of plaque within the brain, causing brain cells to die.

27. Apoptosis, which is programmed cell death, is a necessary process in living organisms. Which of the following is NOT correct about apoptosis?

(A) It occurs randomly.

(B) A particular cell dies when it is no longer needed by the organism.

(C) A cell carries out apoptosis when too many mutations have accumulated.

(D) Plant cells carry out apoptosis as a defense measure in cells that have been infected by parasites.

28. African lungfish often live in small, stagnant, freshwater pools. With reduced gills, they differ from other fish in that they breathe air. They also differ in that they release the nitrogen waste urea instead of ammonia. Which of the following statements correctly explains how this is an advantage for lungfish?

(A) Urea is insoluble in water and sinks to the bottom of the small pools. Thus the poisonous substance is isolated from the fish.

(B) Ammonia, which is highly toxic, would endanger lungfish that inhabit the small pools.

(C) Other fish in the small pools can use urea as an energy source.

(D) The highly toxic urea makes the pool uninhabitable to competitors.

Questions 29–30

29. During oxidative phosphorylation, the energy needed to drive ATP synthesis comes from one source. Which of the following statements states the immediate source of that energy?

(A) Electrons flow down an electron transport chain as they are attracted to oxygen.

(B) The transfer of a phosphate group to ADP is exergonic.

(C) The bonding of electrons to oxygen at the end of the chain releases energy.

(D) The proton gradient across the membrane where the ATP synthase is embedded represents potential energy.

30. Which of the following statements about cellular respiration is correct?

(A) Most CO_2 produced during cellular respiration is released from glycolysis.

(B) Protons are pumped through ATP synthase by active transport.

(C) The final electron acceptor of the electron transport chain is NAD^+.

(D) ATP is formed because an endergonic reaction is coupled with an exergonic reaction.

---

31. The light reactions of photosynthesis supply the Calvin cycle with which of the following?

(A) The light reactions provide oxygen for the light-independent reactions.

(B) ATP and NADPH provide the power and raw materials for the Calvin cycle.

(C) Water entering the plant through the roots provides hydrogen directly to the Calvin cycle.

(D) CO_2 released by the light-dependent reactions provides the raw material for the Calvin cycle.

32. Here is a sketch showing cyclic photophosphorylation in the grana of plant cells.

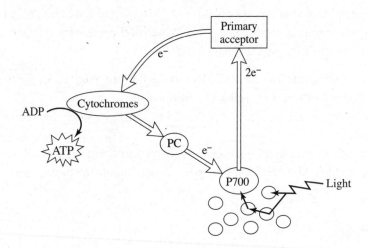

Which of the following statements about this process is correct, and what is the reason for it?

(A) It is similar to oxidative phosphorylation in cellular respiration because its function is to generate ATP.

(B) It is similar to glycolysis because it involves an electron transport chain.

(C) It is the opposite of cellular respiration because it releases oxygen rather than utilizes it.

(D) It is the opposite of the citric acid cycle because ATP is utilized, not produced.

33. Which of the following statements about lipids is correct?

  (A) Unsaturated fatty acids are linked to heart disease.
  (B) Lipids make up most cell surface receptors.
  (C) Phospholipids are water soluble.
  (D) Steroids are lipids that consist of glycerol and fatty acids.

34. Unlike large populations, small populations are vulnerable to various processes that draw populations down an extinction vortex toward smaller and smaller populations until no individuals survive. Which of the following statements correctly identifies the factors that endanger a population?

  (A) Inbreeding and loss of genetic variation threaten a population.
  (B) Migration of new individuals into the population threatens a population.
  (C) Mutation reduces the health of a population.
  (D) Breeding with individuals from a different population may cause the extinction of the first population due to a decrease in diversity.

35. Which of the following statements about the immune system is NOT correct?

  (A) Innate immunity, also known as the adaptive immune response, relies on circulating phagocytes engulfing foreign substances.
  (B) Adaptive immunity is a slower response than innate immunity.
  (C) Innate immunity activates a humoral response.
  (D) Dendritic cells attack infected body cells.

36. The disaccharide lactose is available to *E. coli* in the human colon if the host drinks milk. Digestion of lactose into glucose and galactose begins with the hydrolysis of lactose by the enzyme β-galactosidase, which is encoded by gene *Z* in the *lac* operon. Which of these diagrams correctly depicts the *lac* operon when lactose is being utilized?

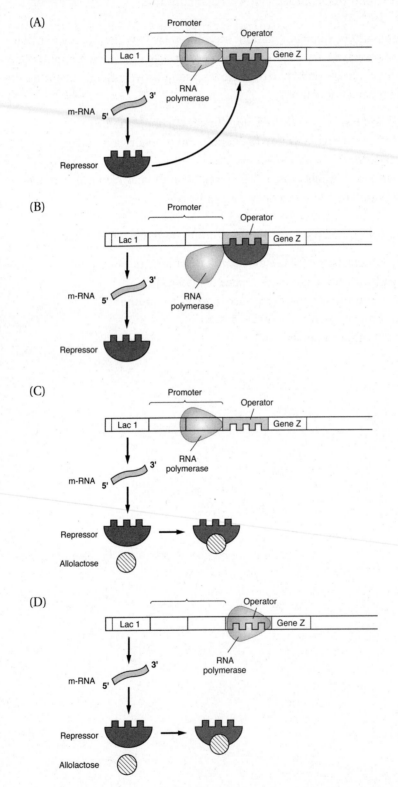

37. Which of the following statements explains how a point mutation can have no negative impact in the protein formed from a sequence in DNA?

(A) The first codon of a gene is always AUG, a leader sequence; a mutation in that sequence will have no effect on the protein produced.
(B) Several codons code for the same amino acid.
(C) DNA polymerase and DNA ligase carry out excision repair before transcription.
(D) RNA processing will repair point mutations before mRNA leaves the nucleus.

38. When DNA replicates, each strand of the original DNA molecule is used as a template for the synthesis of a second, complementary strand. Which of the following sketches most accurately illustrates the synthesis of a new DNA strand at the replication fork?

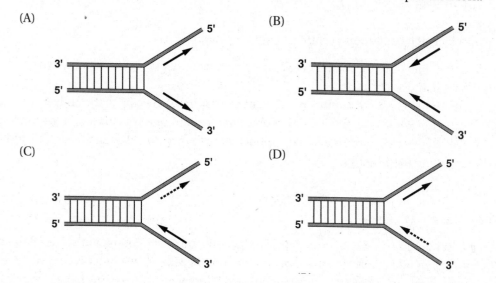

(A)

(B)

(C)

(D)

39. Which of the following statements correctly describes what happens to the preRNA during RNA processing?

(A) Introns are removed in the nucleus. A 5′ cap and the poly-A tail are added after the preRNA leaves the nucleus.
(B) Introns are removed and both a 5′ cap and a poly-A tail are added before the preRNA leaves the nucleus.
(C) Exons are removed and a 5′ cap is added before the preRNA leaves the nucleus. After leaving the nucleus, the poly-A tail is added.
(D) Point mutations are repaired, exons are removed, and both a 5′ cap and a poly-A tail are added before the preRNA leaves the nucleus.

40. Which of the following statements most accurately explains what *Eco*R1 is?

   (A) It is a bacterium that lives in the human large intestine.
   (B) It is the best-studied bacteriophage, a virus that attacks *E. coli*.
   (C) It is a restriction enzyme extracted from *E. coli*.
   (D) It is a type of DNA used extensively in research.

41. snRNPs are most closely associated with which of the following?

   (A) Cell division
   (B) Apoptosis
   (C) Replication of DNA
   (D) RNA processing

42. Which of the following is correct about vaccination?

   (A) It increases the number of different receptors that recognize a particular pathogen.
   (B) It increases the number of different macrophages specific to the antigen.
   (C) It increases the number of epitopes that the immune system can recognize.
   (D) It increases the number of lymphocytes with receptors that can recognize and bind to the antigen.

Questions 43–44

A scientist wanted to understand the role of the gonads and their hormones in the development of the sex of an individual. He carried out an experiment with rabbit embryos while they were still in the mothers' uterus at a stage before sex differences were observable. He surgically removed the portion of each embryo that would develop into ovaries or testes. When the babies were born, he made note of their chromosome sex and the sexual differentiation of the genitals. Here are his data.

Sex Chromosome of Individual	Appearance of Genitals	
	No Surgery	Embryonic Gonad Removed
XY	Male	Female
XX	Female	Female

43. Which of the following statements is a correct conclusion to draw from this experiment?

(A) In rabbits, female development requires a hormonal signal from the female gonad.
(B) In rabbits, male development requires a hormonal signal from the male gonad.
(C) In mammals, gonad development is triggered by sex hormones.
(D) In mammals, all embryos develop male gonads without signals to do otherwise.

44. What is the correct prediction if the researcher were to replace the surgically removed gonad with a crystal of testosterone?

(A) The chromosomal male embryo would develop testes.
(B) The chromosomal female embryo would develop testes.
(C) Both chromosomal male and female embryos would develop male gonads.
(D) Neither chromosomal male nor female embryos would develop male gonads.

45. Which of the following statements about meiosis is correct?

(A) The daughter cells are genetically identical to the parent cells.
(B) Homologues pair during prophase II.
(C) DNA replication occurs before meiosis I and meiosis II.
(D) The number of chromosomes is reduced.

46. Which of the following is correct about organisms that are the first to colonize a habitat undergoing ecological succession?

(A) They are the fiercest competitors in the area.
(B) They maintain the habitat as it is for their own kind.
(C) They change the habitat in a way that makes it more habitable for other organisms.
(D) They are small invertebrates.

47. In the microscopic world of a pond, paramecia are ferocious predators that prey on smaller protists. In a classic experiment, two species of paramecia were grown separately in culture. The species in culture *A* was *P. caudatum*. The species in culture *B* was *P. bursaria*. Then the two species were combined in one culture dish (culture *C*).

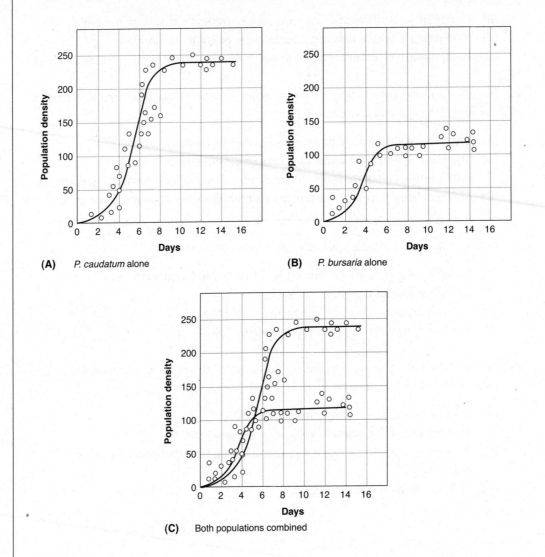

Which of the following is the most likely explanation for the growth pattern of the two populations combined in culture *C*?

(A) *P. caudatum* is driving *P. bursaria* to extinction because *P. caudatum* is the fitter species.

(B) *P. caudatum* and *P. bursaria* share a niche.

(C) *P. caudatum* and *P. bursaria* occupy different niches.

(D) *P. caudatum* is feeding on *P. bursaria* but only to a limited degree.

48. When driving from east to west in the U.S. from a forest ecosystem to grasslands, trees gradually give way to grasses. Which of the following is the critical factor responsible for this shift?

(A) Amount of sunlight
(B) Availability of oxygen
(C) Availability of water
(D) Length of growing season

49. Which of the following is an example of a cline?

(A) The hybrid tomato plant is stronger and produces better fruit than the pure genotype.
(B) There are two distinct varieties in one population of snail that inhabits an island in the Pacific Ocean.
(C) Males of some species have long antlers to fight other males for females.
(D) In one species of rabbit, the ones that evolved in the cold, snowy north are white, while the ones that evolved in the south are brown.

50. Which of the following statements is NOT correct about ancient Earth?

(A) Earth was formed about 4½ billion years ago.
(B) Amino acid synthesis could have been stimulated by volcanic eruption.
(C) The first cells formed about 3½ billion years ago, according to the fossil record.
(D) The presence of free oxygen facilitated the rapid formation of organic molecules.

51. Intact chloroplasts are isolated from dark-green leaves by low-speed centrifugation and placed into six tubes containing cold buffer. A blue dye, DPIP, which turns clear when reduced, is also added to all the tubes. The amount of decolorization is a function of how much reduction and, therefore, how much photosynthesis occurs. Each tube is exposed to different wavelengths of light. A measurement of the amount of decolorization is made, and the data are plotted on a graph. Although the wavelengths of light vary, the light intensity in each tube is the same.

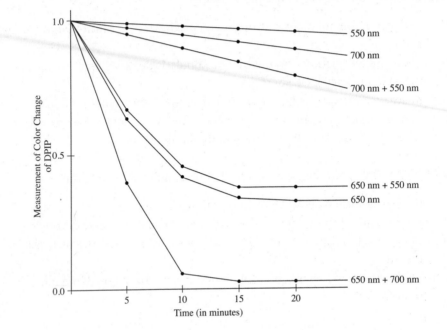

Which statement below best explains the results of this experiment and the reason for it?

(A) The rate of photosynthesis is highest when a tube is exposed to light in the 550 nm range because that wavelength of light contains the greatest amount of energy.

(B) The rate of photosynthesis is highest when exposed to light in the 550 nm range because that wavelength of light contains the least amount of energy.

(C) The rate of photosynthesis is highest when exposed to combined light in the 650 nm and 700 nm ranges because the combination of the two wavelengths of light contains the greatest amount of energy.

(D) The rate of photosynthesis is highest when exposed to combined light in the 650 nm and 700 nm ranges because there are two photosystems in chloroplasts that absorb light in two different wavelengths.

52. Today, two distinctly different beak sizes occur in a single population of finch in an isolated region of West Africa. This finch, the black-bellied seedcracker, is considered a delicacy. The oldest residents of the region remember that all black-bellied seedcrackers used to appear identical. The change in population is shown in the graph.

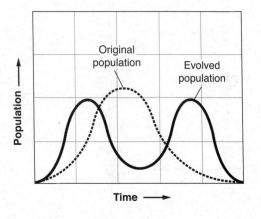

Which of the following statements best identifies the change in the population of finches and the most likely reason for it?

(A) Convergent evolution; two distinct varieties evolved into one variety because only one type of seed now exists.

(B) Diversifying selection; one population of finch divided into two populations because at least two types of seeds are now available.

(C) Directional selection; one original variety was replaced by another because one food source was replaced by another.

(D) Stabilizing selection; the original population died out, leaving only individuals with either long or short beak size.

53. Here is a figure showing the change in one ancestral population of birds over time.

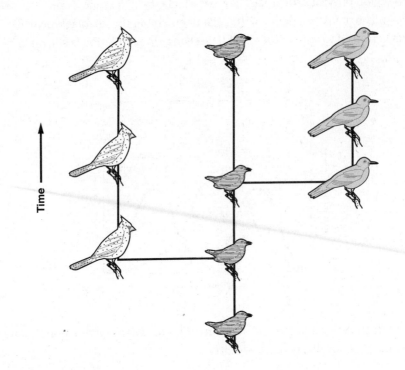

Which of the following statements best supports the evidence presented in the picture?

(A) Evolution was gradual and regular with small changes over a long period of time.
(B) The environment stayed the same for long periods of time.
(C) These birds evolved from a single species that went extinct.
(D) These birds evolved mainly as a result of genetic drift.

54. Which of the following statements is correct about the DNA content of a particular diploid cell just prior to mitosis if the DNA content of the same diploid cell in G1 is X?

(A) The DNA content of the cell in metaphase I is 0.5X
(B) The DNA content of the cell in metaphase I is X.
(C) The DNA content of the cell in metaphase I is 2X.
(D) The DNA content of the cell in metaphase I is 4X.

55. Which of the following sketches shows the correct way in which two nucleotides pair up in a molecule of DNA?

(A)

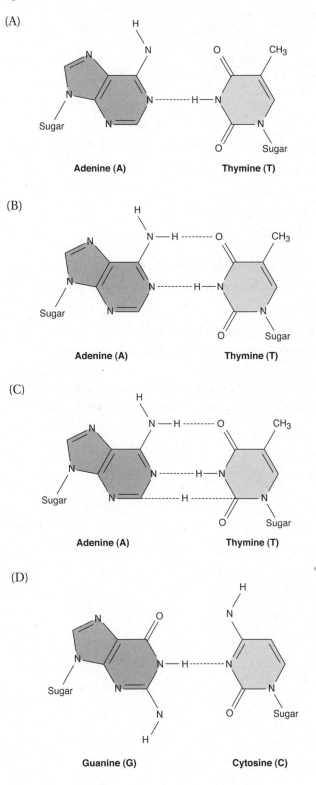

Adenine (A)   Thymine (T)

(B)

Adenine (A)   Thymine (T)

(C)

Adenine (A)   Thymine (T)

(D)

Guanine (G)   Cytosine (C)

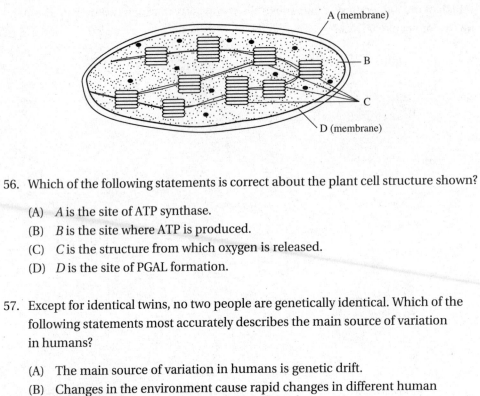

56. Which of the following statements is correct about the plant cell structure shown?

(A) *A* is the site of ATP synthase.
(B) *B* is the site where ATP is produced.
(C) *C* is the structure from which oxygen is released.
(D) *D* is the site of PGAL formation.

57. Except for identical twins, no two people are genetically identical. Which of the following statements most accurately describes the main source of variation in humans?

(A) The main source of variation in humans is genetic drift.
(B) Changes in the environment cause rapid changes in different human populations.
(C) Shuffling of alleles is responsible for the greatest variation in humans.
(D) Mutations are responsible for the greatest changes in different human populations.

MODEL TEST 2

58. β-thalassemia is a disease that strikes mainly people of Mediterranean and Asian descent. It involves a flaw in the gene that codes for β-globin, one of the protein chains that make up hemoglobin (Hb). A trial was conducted. The patient in this trial was typical of individuals with β-thalassemia. He needed monthly blood transfusions and daily treatments to lower blood iron levels. In 2007, when the patient was 19 years old, scientists removed some of his bone marrow cells and treated them with a modified virus that was engineered to carry a good copy of the β-globin gene. They infused the repaired cells into the patient. After several months, the scientists assessed the patient. Here is a graph of the results.

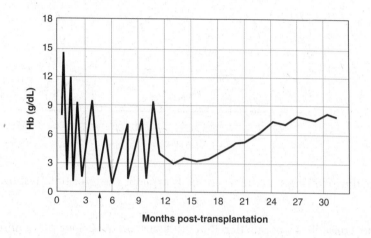

Which of the following statements is supported by the information provided?

(A) The gene therapy trial failed.

(B) The gene therapy trial was a success, but the patient still requires blood transfusions.

(C) Researchers transplanted a gene directly into the patient's red blood cells.

(D) Researchers inserted a gene into a virus vector that carried the normal hemoglobin gene into the patient's cells.

59. This graph represents three idealized survivorship curves for different populations.

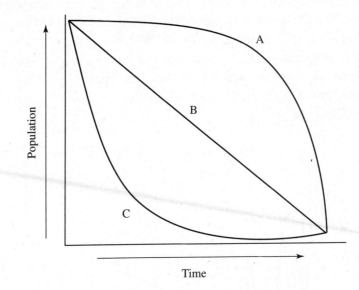

Which of the following statements correctly explains the information shown on this graph?

(A) Line *A* describes a population that produces few offspring and requires much parental care.

(B) Line *B* describes a population like sea stars that exhibit external fertilization.

(C) Line *C* describes a population that exhibits high survival of the young with high death rates of older individuals in the population.

(D) Line *A* describes survivorship patterns of insects.

60. The agave, or century plant, generally grows in arid climates with unpredictable rainfall and poor soil. The plant exhibits what is commonly called big bang reproduction. An agave grows for years, accumulating nutrients in its tissues until there is an unusually wet year. Then it sends up a large flowering stalk, produces seeds, and dies.

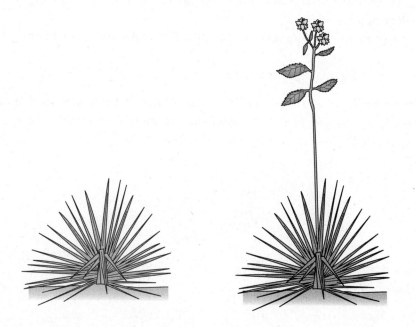

Which of the following statements gives the most likely explanation for how the agave plant has evolved this unusual life strategy?

(A) This strategy is determined by the plant's genes, which are inherited.
(B) This particular life strategy was selected for by the harsh, desert environment.
(C) This strategy is the result of a series of mutations that occurred because of the great heat in the desert.
(D) This strategy is the result of a series of mutations that occurred because of the great heat in addition to the lack of rain in the desert.

61. Which of the following statements is NOT correct about apoptosis?

    (A) Apoptosis, a special type of cell division, requires multiple cell signaling.
    (B) The fact that apoptosis is widespread across several kingdoms is evidence that it evolved early in the evolution of eukaryotes.
    (C) Apoptosis plays a crucial role in the development of fingers in embryonic development.
    (D) Apoptosis prevents an aging and dying cell from damaging neighboring cells.

62. Here are drawings of four amino acids.

    Molecule A is serine and has polar side chains. Molecule B is lysine, which is basic and positively charged. Molecule C, glutamate, is acidic and negatively charged. Molecule D is glycine. It has nonpolar side chains.

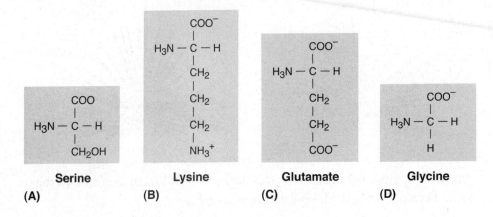

Serine	**Lysine**	**Glutamate**	**Glycine**
(A)	**(B)**	**(C)**	**(D)**

    Which of the following statements is correct about these amino acids?

    (A) Molecule A can readily dissolve through a plasma membrane.
    (B) Only Molecules B and C can readily dissolve through a plasma membrane.
    (C) Molecules A, B, and C can all readily dissolve through a plasma membrane.
    (D) Only Molecule D can readily dissolve through a plasma membrane.

63. Which of the following statements describes the structural level of a protein that is least affected by hydrogen bonding?

    (A) Primary structure depends on the sequence of amino acids.
    (B) Tertiary structure has a shape dependent on the interactions of side chains of amino acids.
    (C) Quaternary structure results from the aggregation of more than one polypeptide unit.
    (D) An α-helix is an example of a secondary structure of a polypeptide.

Directions: Read each of the 6 questions, then provide a numeric answer for each. Units are not required.

Earth's Water Cycle

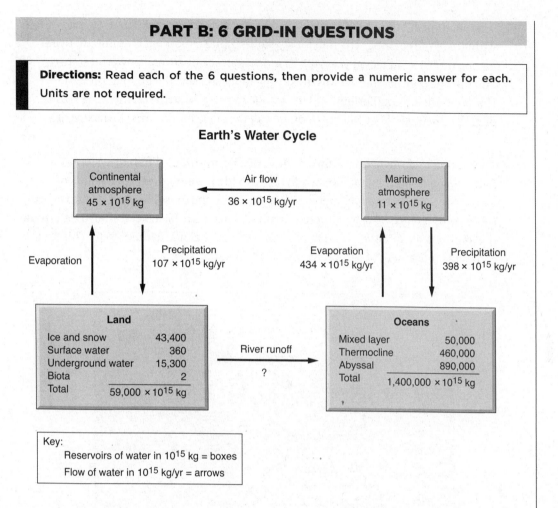

1. Here is a chart of Earth's water cycle. Reservoirs (boxes) are measured in 10^{15} kg. Fluxes or flows (arrows) are shown as 10^{15} kg/yr. Calculate the flux of water from the land to the oceans or rivers. Record your answer as a whole number.

2. A student carried out an experiment with fruit flies, *Drosophila melanogaster*, to determine how the white-eyed trait is inherited. To determine if the trait is sex-linked or autosomal, she initally did two crosses. First she crossed a pure white-eyed female with a pure red-eyed male. She counted 100 F1offspring. All the female offspring had red eyes, and all the males had white eyes. Then she did the reciprocal cross, mating a pure white-eyed male with a pure red-eyed female. All the F1 offspring had red eyes. Because the student found different results in the two trial crosses, she assumed that she had confirmed that the trait is a sex-linked one.

   Next she wanted to determine if the trait is sex-linked dominant or recessive. She hypothesized that since sex-linked recessive is more common, the white-eyed trait is recessive. This time she crossed a male F1 and a female F1 from the first cross. She counted 200 F2 flies. Here are her results:

   There were 46 red-eyed females, 52 white-eyed females, 53 white-eyed males, and 49 red-eyed females.

   Calculate the chi-squared value for the null hypothesis that the white-eyed trait is sex-linked recessive. Give your answer to the nearest tenth.

3. A population of butterflies found in Madagascar is polymorphic. There are two varieties of the coloring. The trait for a yellow stripe on the wing is dominant over having no stripe. In recent years the island was struck by several powerful storms and the butterfly population was drastically reduced. Now the population is in Hardy-Weinberg equilibrium, and the trait for having no stripe on the wing is 21% of the population. What will the frequency of the allele for no stripe be next year?

4. The carbon cycle involves the flux, or flow, of carbon among different systems on Earth. Scientists throughout the world are working to determine the amounts of carbon stored in different components of Earth and the movements of carbon among those components. By using different methods, scientists have been able to determine the approximate quantities and fluxes involved in the global carbon cycle. This sketch shows the movement of carbon measured in gigatons (GT).

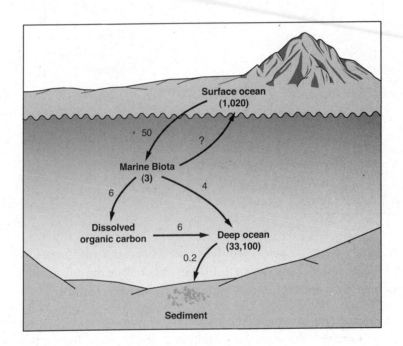

Determine the amount of carbon that reaches the ocean surface from marine biota.

5. Here is a growth curve for a population of rabbits that were accidentally introduced into a region of New Zealand.

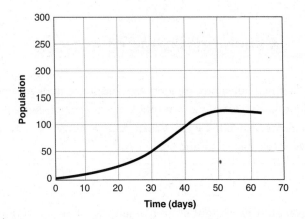

Calculate the mean rate of growth from Day 30 to Day 40. Record your answer to the nearest whole number.

6. Here is a graph showing the amount of product formed during an enyzyme-mediated reaction. Calculate the mean rate of product formed in the first 90 seconds.

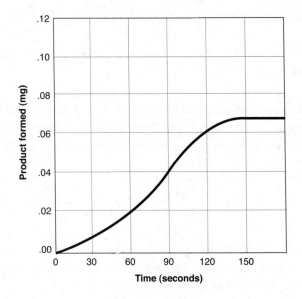

SECTION II

8 Free-Response Questions

Time: 90 minutes

Directions: The first two questions are long, free-response questions. Spend about 20 minutes answering each. Questions 3 through 8 are short, free-response questions. Take about 6 minutes answering each. Write your responses on a separate piece of paper and ask your teacher to "grade" it.

1. All living things require a constant input of energy for their metabolic functions. Special structures within cells have evolved to produce this energy. Discuss the process of cellular respiration in aerobic organisms. Include a discussion of each of the following:

   (a) The structure of a mitochondrion and relate the structure to the particular respiration processes

   (b) The general process of glycolysis and fermentation and why these processes are important

   (c) The general description of the citric acid cycle and its importance in producing energy

   (d) A description of oxidative phosphorylation and the importance of the proton motive force.

2. All living things depend on producers to harness the energy from the sun and make it available for all living things on Earth. The essence of what green plants do is photosynthesis. Here is the simplified equation for photosynthesis:

$$\text{Light}$$
$$6\,CO_2 + 12H_2O \rightarrow C_6H_{12}O_6 + 6\,H_2O + 6O_2$$

   (a) Describe and explain the structure of the chloroplast. You may sketch and label a drawing of it, but you must also explain the function of each part.

   (b) Describe the process of light harvesting in the chloroplast.

   (c) The making of sugar uses a great deal of energy in the form of ATP. Explain how ATP is made in chloroplasts.

   (d) Plants take in water and carbon dioxide and give off water and oxygen. Explain where these chemicals fit into a discussion of photosynthesis. Include a brief explanation of the Calvin cycle.

MODEL TEST 2

3. Your teacher tells you that fruit flies prefer darkness to light and warm to cool temperatures. Your teacher asks you to prove these statements. Devise an experiment to prove your teacher's hypothesis.

4. Polyploidy, a situation of having extra complete sets of chromosomes ($3n$, $4n$, or more) is rare in animals but common in plants. This phenomenon can occur if two individuals mate and the gametes of one or both of them are abnormal because of an error during meiosis (nondisjunction). One gamete might be diploid and one monoploid. Another way in which polyploidy can occur is if individuals from two different species interbreed and produce hybrid offspring. In either case, polyploid individuals may not be able to mate with any individuals that do not exhibit polyploidy. Often these polyploid individuals become a separate species. Briefly explain how polyploidy can bring about the formation of new species.

5. Stomata (stomates) are located on the leaves of most plants to allow the intake of carbon dioxide and the release of water and oxygen. Stomates open and close in response to ambient conditions around the leaf. Most plants have stomates that are open during the day. However, some plants have stomates that are open only at night. The number and placement of stomates on leaves also varies and is related to the habitat and evolutionary history of the plant. Some plants have stomates on their upper leaf surface, while others have stomates on the lower surface. Choose one of these variations of where stomates are located or when they open and close, and give a reasonable explanation for why a particular plant would have stomates with that characteristic.

6. Explain three ways in which the innate immune system differs from the adaptive immune system.

7. Describe the process of ecological succession in a deciduous forest after it is burned down by a fire.

8. Several products are sold in pharmacies for cleaning and preserving contact lenses. The products contain solutions that are all *sterile* and *buffered* and contain *isotonic saline*. Explain what each of those italicized terms mean and why they are required for something that will come in contact with delicate eye tissues.

ANSWER KEY

Section I Part A: Multiple-Choice Answers

1. **(C)**	12. **(D)**	23. **(D)**	34. **(A)**	45. **(D)**	56. **(C)**
2. **(B)**	13. **(D)**	24. **(B)**	35. **(B)**	46. **(C)**	57. **(C)**
3. **(C)**	14. **(C)**	25. **(B)**	36. **(C)**	47. **(C)**	58. **(D)**
4. **(C)**	15. **(D)**	26. **(A)**	37. **(B)**	48. **(C)**	59. **(A)**
5. **(D)**	16. **(A)**	27. **(A)**	38. **(C)**	49. **(D)**	60. **(B)**
6. **(C)**	17. **(C)**	28. **(B)**	39. **(B)**	50. **(D)**	61. **(A)**
7. **(C)**	18. **(B)**	29. **(D)**	40. **(C)**	51. **(D)**	62. **(D)**
8. **(C)**	19. **(B)**	30. **(D)**	41. **(D)**	52. **(B)**	63. **(A)**
9. **(D)**	20. **(C)**	31. **(B)**	42. **(D)**	53. **(B)**	
10. **(C)**	21. **(A)**	32. **(A)**	43. **(B)**	54. **(C)**	
11. **(D)**	22. **(D)**	33. **(C)**	44. **(A)**	55. **(B)**	

Section I Part B: 6 Grid-In Answers

1. 36×10^{15} kg/yr.
2. 0.6
3. 0.46
4. 40 GT
5. 5
6. 0.00044 or 4.4×10^{-4}

Section II Free-Response Answers

See Answer Explanations.

ANSWER EXPLANATIONS

Section I Part A: Multiple-Choice Answers

1. **(C)** This graph is focused on the strength of the bond between Hb and oxygen at pH 7.2 and pH 7.4. At a lower at pH, which is more acidic, the attachment is weaker. This makes sense because as cells carry out cellular respiration, they release CO_2, which makes the blood more acidic. The blood pH changes from normal 7.4 to the more acidic 7.2. At pH 7.2, the Hb can release more of its oxygen where it is needed, at the cells.

2. **(B)** Notice that the question says NOT. In each cell in a woman's body, one of the X chromosomes is inactivated. One X is inactivated in half of her cells and the other X in the other half of her cells. The other statements are correct about Barr bodies.

3. **(C)** Although it is true that during the Krebs cycle pyruvate is completely broken down to CO_2, the most energy from cellular respiration is released during oxidative phosphorylation, when all the energy stored in NADH and $FADH_2$ from previous stages of cellular respiration is released.

4. **(C)** Oxidation is the loss of electrons or of hydrogen. You can see that hydrogen has been lost by the succinate to form fumacate. If this is part of the citric acid cycle, then pyruvate is the starting point of this reaction, not the product. The citric acid cycle produces CO_2 and ATP. Feedback mechanisms are complex. This is certainly not one. The stem of the question states that this is part of the citric acid cycle, which occurs in the inner matrix of the mitochondria, not the cristae membrane.

5. **(D)** The father's genotype is X-Y; the mother's is XX. In this case, the father passes his X-gene to all their daughters. So all the daughters inherit the X-gene. He passes his Y chromosome to his sons.

6. **(C)** Recombination occurs when homologous pairs bind together during synapsis and crossing-over occurs. Picture 1 is in metaphase. Synapsis and crossing-over have already occurred. The homologous pairs are about to separate. Picture 2 shows prophase. Synapsis is occurring, and crossing-over can occur at this point. Picture 3 shows sister chromatids lined up on the metaphase plate. There can be no crossing-over because there is only one chromosome.

7. **(C)** When you see a blending of the parental phenotypes in the offspring, you are dealing with incomplete dominance. The genotype of the black animals is *BB*. The genotype of the white animals is *WW*. The genotype of the gray animals is *BW*.

8. **(C)** A soil with acid pH causes the flowers to be pink; a basic soil gives rise to blue flowers. Polygenic inheritance is characteristic of traits like skin color, hair color, and height that exist in a complete spectrum. ABO blood type is an example of codominance because both A and B genes show.

9. **(D)** By analyzing the F1 results, you see that there is almost a 50-50 ratio of gray normal to black vestigial. The 50-50 ratio of those traits tells you that the traits are linked. If the body is gray, the wings are normal; if the body is black, the wings are vestigial. However, a small variation from the expected values occurred. Those numbers, 190 and 184, are due to crossing-over. Genes that are very close together usually are inherited together. So these genes must be far apart on the same chromosome.

10. **(C)** The two genes are linked and far apart on the same chromosome.

11. **(D)** The coupling of exergonic and endergonic reactions is an important biochemical strategy. The exergonic flow of electrons in the cristae membrane of mitochondria provides the energy to pump protons across the cristae membrane and maintain a gradient.

12. **(D)** A macrophage is an antigen-presenting cells (APC). It will alert the adaptive immune system about a particular invader.

13. **(D)** The action potential is an all-or-nothing response. It either happens or it does not. It doesn't change strength. With a stronger stimulus, only the frequencies of the response increases. The axon does not become hyperpolarized; it depolarizes. If the axon at rest is at –90 mV, the hyperpolarization will measure at –100 mV, for example, not at 50 mV.

14. **(C)** The definition of Hardy-Weinberg equilibrium is that the allelic frequencies will not change. If one trait becomes advantageous and the allelic frequency changes, then the population is no longer in Hardy-Weinberg equilibrium.

15. **(D)** Green plants reflect green light; they do not absorb it and cannot use it as an energy source. Placing a plastic bag over the leaves will decrease transpiration, resulting in Line *C*. Placing a fan near the plant will increase transpiration, causing Line *A*.

16. **(A)** Notice the question states NOT. All the other answer choices are correct. Water moves up the plant through the xylem by passive transport because of the physical properties of water. Choice D refers to the use of a heat sink, which is necessary in this experiment to make sure that the temperature does not rise during the experiment. If the temperature did rise, it would constitute another variable in the experiment.

17. **(C)** The importance of internal organelles is that they compartmentalize the cell so that widely different reactions can all occur at the same time, each in their own space.

18. **(B)** Glycolysis would cease if there were not a constant supply of NAD^+ to bind to hydrogen. Glycolysis produces pyruvate, which is the raw material for the citric acid cycle, not oxidative phosphorylation. Glyocolysis does occur in the cytoplasm, but that is not important.

19. **(B)** Earth is about 4.5 billion years old. The planet took about 1 billion years to cool down and be stable enough for life to evolve, and for the first cell—a prokaryotic cell—to develop. About 2 billion years ago, the first eukaryotic cell evolved. We have fossil evidence of that. Humans, bananas, and all other eukaryotic cells evolved from some ancestral eukaryotic cell. That is why there are basic similarities among all living things.

20. **(C)** Mutations were not discovered until around 1900 when DeVries, while working with polyploid plants, coined the term. Darwin could not explain the origin of all the variation he saw. This lack of mechanism was the weakest part of his theory.

21. **(A)** After deforestation, the nitrate runoff onto the rivers increased. There is no evidence presented that trees were replanted. Choice D is clearly false.

22. **(D)** Even though numbers are mentioned in the question, they are not relevant. There is no way to determine if hawks will begin to eat frogs, which is niche displacement. You can deduce only the obvious.

23. **(D)** Acid rain is caused by the accumulation of sulfuric, sulfurous, nitric, and nitrous acids in the air from the burning of fossil fuels. A pH of 6.0 is 100 times more acidic than pH 8. The hydrogen ion concentration of pH 6 = 1×10^{-6}M.

24. **(B)** The drug Oblimersen is an antisense RNA. Therefore, it binds to the overexpressed RNA and prevents it from translating. Theoretically, Oblimersen should work in preventing cancer.

25. **(B)** The extra nitrate does not contribute any more to the algae growth than the initial increase. Phosphate is the main contributing factor to algae growth.

26. **(A)** This is an accurate description of an autoimmune disease.

27. **(A)** Apoptosis is anything but random. It is controlled by both internal and external signals that lead a class of enzymes called caspases to assist in the digesting of a cell marked for apoptosis.

28. **(B)** Uric acid is the nonsoluble nitrogen waste, not urea. Ammonia is the most toxic nitrogen waste, not urea.

29. **(D)** Protons flowing down a gradient through ATP synthase provide energy to phosphorylate ADP into ATP. Energy from the exergonic flow of electrons to oxygen

provides the energy to pump protons across the cristae membrane and form a gradient.

30. **(D)** The exergonic flow of electrons in the electron transport chain of mitochondria provides the energy to establish a proton gradient. Most CO_2 released from cellular respiration comes from the citric acid cycle. Protons flow through the ATP synthase down a gradient, not by active transport.

31. **(B)** Oxygen is needed to attract electrons down an electron transport chain in the light-dependent reactions. Water entering the plant through the roots ultimately provides hydrogen for the Calvin cycle, but first it is needed in the light-dependent reactions. CO_2 enters the plant through the stomates; it is not released by the light-dependent reactions.

32. **(A)** The only thing this process produces is ATP. No NADPH or oxygen is released. The function of this process is to provide ATP for the Calvin cycle, which uses enormous quantities of ATP.

33. **(C)** Saturated fatty acids are linked to heart disease. Proteins make up most cell surface receptors. Steroids are lipids but are not made of glycerol and fatty acids like typical lipids.

34. **(A)** All the other choices promote outbreeding, the introduction of new genes into a population.

35. **(B)** Innate immunity is the inborn, nonspecific first and second lines of defense. Adaptive immunity is specific and involves antigens and antibodies. Adapative immunity follows innate immunity and is slower. Adaptive immunity involves the humoral (antibody) response. Dendritic cells are antigen-presenting cells.

36. **(C)** In order for transcription of a gene or genes to occur, RNA polymerase must bind to a promoter. In Choices A and B, the repressor is bound to the operator, which excludes RNA polymerase from binding to the promoter and prevents transcription. Choice D does not make any sense because RNA polymerase does not bind to the operator. It's a matter of specificity, like a lock and key.

37. **(B)** A mutation in AUG would have disastrous consequences because translation could not begin. Choice C refers to what happens after transcription. RNA processing occurs after transcription.

38. **(C)** The new strand is built 5′ to 3′, and the Okazaki fragments form, moving away from the replication fork.

39. **(B)** RNA processing occurs before the preRNA transcript leaves the nucleus. Introns, which are intervening sequences, are removed. Exons are the expressed sequences, which are the genes or coding regions.

40. **(C)** Other restriction enzymes are *Bam*HI and *Hind*III. They are all used extensively in biotechnology. The name of the bacteria that lives in the human intestine is *E. coli*. The name of the best-studied bacteriophage is lambda (λ) phage.

41. **(D)** snRNPs stands for small ribonucleoproteins. They form a complex with spliceosomes on a preRNA molecule complex and excise the introns during RNA processing.

42. **(D)** Vaccinations are given to immunize against a particular antigen. They increase the number of a particular white blood cell that recognizes and binds to a particular antigen.

43. **(B)** Without the signal from the gonads to develop into a male, the rabbit remains a female.

44. **(A)** The testosterone would send the signal that the male gonad would have sent.

45. **(D)** The point of meiosis is variation. Daughter cells are different from the parent cells as well as each other. Homologues pair during prophase I, not prophase II. DNA replication occurs before meiosis I only.

46. **(C)** Pioneer organisms colonize an area that is not habitable by other organisms and alter it over time in a way that makes the area habitable for other organisms. Pioneer organisms are producers, such as lichens that are symbionts—an intertwined relationship of a fungus and an algae.

47. **(C)** Almost by definition, the two species do not share a niche because both populations are surviving. However, the environment has limited resources. In fact, the S-shaped growth curve shown in all three of these graphs is called a logistic growth curve. It is characteristic of some microorganisms under conditions of limited resources.

48. **(C)** Oxygen and sunlight do not vary when driving from forest to grasslands. The limiting factor for plant growth is water. Less precipitation occurs in the grasslands than in the forests.

49. **(D)** Choice A refers to hybrid vigor. Choice B refers to polymorphism. Choice C refers to survival of the fittest and sexual selection.

50. **(D)** There was no free oxygen in ancient Earth's atmosphere. In fact, given the strong oxidative nature of oxygen, had it been available, organic molecules that formed in early Earth might not have remained as they did.

51. **(D)** The amount of energy in light is inversely proportional to its wavelength. The shorter the wavelength of light, the more energy it contains. The lines on the graph that are closest to the x-axis represent the greatest amount of reduction and, therefore, the greatest amount of photosynthesis. According to the graph, the most reduction or photosynthesis occurs with two wavelengths of light combined. The reason is that in chloroplasts, two photosystems, P700 and P680, absorb light of different wavelengths.

52. **(B)** There used to be one population of black-bellied seedcracker; now there are two. The single population diverged because of a change in food.

53. **(B)** The y-axis shows time. The change you see in the birds is sudden. It is not gradual and slow. It occured suddenly. The environment directs evolution. So if a species remains the same for a long time, one can assume that the environment has not changed for all that time. The ancestral species is still alive; it did not go extinct. None of the information leads you to think that evolution occurred because of genetic drift.

54. **(C)** The cell cycle consists in order of G1 (the cell is growing), S (DNA replicates), G2 (second gap of growth), and then mitosis (division of the nucleus). Therefore, if G1 is X (before the S phase), then in mitosis (after the S phase), the DNA content must be double X or 2X.

55. **(B)** There is a double hydrogen bond between adenine and thymine in DNA and a triple bond between cytosine and guanine. Breaks in DNA occur more often between adenine and thymine than between cytosine and guanine.

56. **(C)** This is a chloroplast. The light-dependent reactions, where ATP is synthesized and oxygen is released, occurs in the grana (*C*). The light-independent reactions, where PGAL is made, occur in the stroma (*B*). Nothing related to photosynthesis is occurring in *A*, which is the outer membrane of the chloroplast.

57. **(C)** More than anything else, sexual reproduction with recombination of alleles is the greatest source of variation in humans. Genetic drift and mutation are also an important source of variation in humans, but they do not provide nearly as much variation as does recombination of alleles. Changes in the environment provide pressure for evolution, but they are not a source for variation.

58. **(D)** This was a very successful treatment. The patient's hemoglobin (Hb) levels are approaching normal levels. There are no sudden and severe drops in Hb levels, and the patient no longer requires transfusions.

59. **(A)** Line *A* describes the survivorship curve for mammals like humans: few offspring, much parental care, and death at old age. Line *C* is characteristic of external fertilizers, like sea stars, where there is enormous predation of young. Line *B* is characteristic of a population that is vulnerable throughout its life but is not particularly vulnerable at any one stage more than at any others.

60. **(B)** Choices C and D have the theory backward. Mutations may provide variety in a population, but they do not occur to provide a variety or a solution to a problem. Mutations occur. If they are advantageous, that new characteristic may increase in frequency in a population. If they are deleterious, that trait may disappear.

61. **(A)** The question states NOT. Every choice is correct about apoptosis except Choice A. Apoptosis is programmed cell death, not a form of cell division.

62. **(D)** In general, substances that are polar and/or charged cannot diffuse through a membrane; only nonpolar molecules can.

63. **(A)** Hydrogen bonding determines secondary and tertiary structure but not primary.

Section I Part B: 6 Grid-In Answers

1. The answer is 36×10^{15} kg/yr.

   This is the water cycle, and what goes in must come out. Even if humans are consuming and wasting lots of water, the water does not disappear. It flows (fluxes) from one place to another. So the volume of water that flows from the maritime atmosphere to the continental atmosphere equals the volume that fluxes from the land to the oceans or rivers.

2. The answer is 0.6.

Phenotype	Observed (o)	Expected (e)	(o − e)	(o − e)²	(o − e)² ÷ e
Red-eyed female	46	50	−4	16	0.32
White-eyed female	52	50	2	4	0.08
Red-eyed male	49	50	−1	1	0.02
White-eyed male	53	50	3	9	0.18

Use the chi-squared equation found on your reference table.

$$\chi^2 = \Sigma\,(o - e)^2 = 0.32 + 0.08 + 0.02 + 0.18 = 0.6$$

3. The answer is 0.46.

The population is in Hardy-Weinberg equilibrium. So the frequency of the allele next year will be the same as the frequency of the allele this year. The trait for having no stripe is recessive. The frequency for that trait is $21\% = 0.21 = q^2$. Therefore, $q = 0.46$.

4. The answer is 40 GT.

Follow the arrows. You can see that 50 GT enters the biota in the ocean. Of that, 6 GT ends up in dissolved organic carbon and 4 GT ends up in the deep ocean. That means 10 GT does not cycle back to the surface of the ocean but, instead, is captured by the ocean.

$$6\ GT + 4\ GT = 10\ GT$$
$$50\ GT - 10\ GT = 40\ GT$$

5. The answer is 5.

The number of animals increased from 50 on Day 30 to 100 on Day 40. Divide that increase in numbers by the period of time, 10 days.

$$100 - 50 = 50$$
$$50 \div 10 = 5$$

6. The answer is 0.00044 or 4.4×10^{-4}.

Remember that this type of problem can apply to many topics in science. Divide the amount of product formed, 0.04, by the time, 90 seconds.

$$0.04 \div 90 = 0.00044 = 4.4 \times 10^{-4}$$

Section II: Free-Response Answers

1. For ease of studying only, this answer is not written in essay format. Instead, key facts and concepts are simply listed. In your test, you must write your answer in prose.

Cellular respiration is a catabolic process—a reaction in which complex molecules are broken down into simple ones along with the release of energy. Cellular respiration is enzyme mediated and takes place in many small steps.

(a) Structure of a mitochondrion

- Cytoplasm—site of glycolysis
- Inner compartment or inner matrix—site of the citric acid cycle
- Outer double membrane—does not take part in the production of ATP
- Outer compartment—site of the build-up of protons
- Cristae membrane —site of the electron transport chain (respiratory chain)

(b) Glycolysis—here are the main points

- Partial oxidation (breakdown) of glucose into 2 three-carbon pyruvate molecules (pyruvic acid)
- Small amount of energy released; 2 ATP net
- Takes place in the cytoplasm
- The product of glycolysis, pyruvate, is the raw material for the citric acid cycle and the rest of cellular respiration

Fermentation—here are the main points

- Anaerobic process that includes glycolysis plus the production of alcohol or lactic acid. Yeast ferments pyruvic acid, the product of glycolysis, into alcohol and CO_2. Bacteria and skeletal muscle ferment pyruvic into lactic acid.
- Oxidizes NADH back to NAD$^+$, which is required in order for glycolysis to continue.

(c) Citric acid cycle:

- Occurs in the matrix of mitochondrion
- Completes the oxidation of pyruvate to CO_2
- Reduces NAD$^+$ into NADH, which will carry protons and electrons to the electron transport chain (respiratory chain) and to oxidation phosphorylation
- Reduces FAD into FADH$_2$, which will carry protons and electrons to the electron transport chain (respiratory chain) and oxidation phosphorylation
- Releases CO_2 into the atmosphere
- Produces a small amount of ATP

(d) Oxidative phosphorylation

- ATP is formed from ADP + P$_i$ at ATP synthase in the cristae membrane
- Occurs in the cristae membrane
- Couples the exergonic flow of electrons in the electron transport chain with the endergonic pumping of protons across the cristae membrane into the outer compartment

- Formation of a proton gradient in the outer compartment, representing enormous potential energy (a proton motive force) that will power the production of enormous amounts of ATP
- As protons flow down the gradient through the ATP synthase, ADP gets phosphorylated into ATP
- Where the greatest amount of ATP is produced during cell respiration
- Water is released as a byproduct of oxidative phosphorylation

2. (a) *A* Double outer membrane—does not take part in production of sugar
   B Stroma—site of light-independent reactions
   C Grana—site of light-dependent reactions; consists of membranes called thylakoids
   D Inner plasma membrane—does not take part in production of glucose

   (b) Light is harvested in the thylakoid membranes in the grana of chloroplasts by photosystems. Each photosystem consists of a reaction center containing the photosynthetic pigment chlorophyll *a* and several hundred other photosynthetic antenna pigments that funnel light energy to chlorophyll *a*. There are two photosystems (PS). PS I absorbs light in the 700 nm range. PS II absorbs light in the 600 nm range. The role of the antenna pigments is to broaden the spectrum of light absorbed by chlorophyll *a*. Light is energy, and the energy absorbed by chlorophyll excites electrons within the molecule. These excited electrons are captured by an electron transport chain that will ultimately transfer the energy to ATP.

   (c) When excited electrons from chlorophyll *a* are captured by an electron transport chain in the thylakoid membrane, they enter at a high energy level and then drop down to successively lower energy levels, releasing energy as they fall. The energy released from this exergonic flow of electrons is coupled with an endergonic reaction and used to pump protons across the thylakoid membrane. This process creates a proton gradient in the thylakoid space (lumen). The thylakoid membrane is impermeable to protons except at ATP synthase structures. As protons flow down the gradient and through the ATP synthase molecules, ADP is phosphorylated into ATP. This process is similar to what happens in oxidative phosphorylation during cellular respiration. However, because the movement of electrons is powered by solar energy, this process is called photophosphorylation. Finally, NADPH and $FADH_2$ carry spent protons and electrons from the light-dependent reactions to the Calvin cycle.

   (d) The business of the light-independent reactions is to fix carbon into a molecule of sugar—PGAL, also known as 3-PGA. The Calvin cycle is the process by which carbon fixation occurs. The raw material for carbon fixation is CO_2, which enters the Calvin cycle as it combines with the enzyme RuBP. These reactions do not depend on the light directly, but rather on the products of the light reactions, protons and electrons carried by NADPH and $FADH_2$. The Calvin cycle occurs in the stroma of chloroplasts.

3. Choose a large enough sample of fruit flies. Thirty flies is a good-sized sample. Ten flies is too small, and 100 may be too difficult to manage and might result in errors.

   Build a choice chamber. Use two 4" long vials that the flies were transported in, or use two graduated cylinders.

   To test the hypothesis that flies prefer darkness, put the flies to sleep. Then place them into the cylinders, and seal the cylinders with clear tape. Cover one side of the choice chamber to prevent light from entering it. Leave the choice chamber undisturbed. After 10 minutes, remove the cover and place a piece of cardboard over each side of the chamber to keep the flies in each chamber. Count the flies in each cylinder.

   To test the hypothesis that flies prefer one temperature over another, warm one half of the choice chamber. Before you put the flies into the choice chamber, make sure that you can control and measure the temperature of each chamber. Be careful that the temperature is not too warm or too cold. Carry out the experiment in the same way as before. Carefully count the flies in each chamber.

   You must always include a control in any experiment. In this case, your control is a choice chamber just like the experimental one except without altered light or temperature. After all, you have to make sure that the flies are not moving from one part of the chamber to the other solely by chance.

4. If polyploid individuals can mate only with other polyploid individuals, they are isolated within their population. They have become separated by reproductive isolation. The two groups ($2n$ and polyploid) cannot mate and cannot exchange genes. It is as if they were separated by thousands of miles or a giant sea. The offspring will experience separate evolutionary histories and will evolve separately. Ultimately, the offspring will differ so much from the original population that they will become a separate species.

5. Include only one of the following.

   Stomates must exchange photosynthetic gases but must also minimize excessive water loss from transpiration. A plant that floats on water, such as a lily pad, would have stomates only on the top surface because that is the only place where gas exchange can occur. Plants that live in very hot and dry places, such as deserts, might have stomates that open only at night or are only on the undersurface, protected from intense sun and heat.

6. For ease of studying only, this answer is not written in essay format. Instead, key facts and concepts are simply listed. In your test, you must write your answer in prose.

Innate Immunity Includes	Adaptive Immunity Includes
First and second lines of defense	Third line of defense
Prevents germs from getting into the body	Is specific
Is nonspecific	B lymphocytes begin to produce antibodies to antigens; the humoral response
Includes skin, mucous membranes, stomach acid, cilia in respiratory system	T lymphocytes kill infected body cells; cell-mediated response
Activates inflammatory response: histamine brings greater blood supply to the area	Helper T cells announce that invaders have infected the body; they attract other specific fighting cells to the battle
Phagocytes gobble up invaders indiscriminately	Macrophages and dendritic cells are antigen-presenting cells that help the system identify what the antigen looks like
Interferons block viral infections	Clonal selection is fundamental to the immune response; the correct B and T cells are chosen to make copies of themselves and provide a defense

7. Although a fire can destroy all the trees and plants in a forest, it leaves the soil intact. What occurs after this type of destruction is called secondary ecological succession. Some trees may benefit from or even require a fire to germinate their seeds. Scrub pine is one example of a tree whose seeds germinate only after a high-intensity fire. A variety of plants will grow back rapidly because adequate sunlight is now available, where once there were mature trees to compete with and the forest floor was dark. Prior to the fire, there probably were only one or two dominant trees in the forest. Those trees were the fittest in that environment. After the fire, many new trees will have an opportunity to grow. As a result, after the fire the forest will have greater diversity than before the fire. New bushes and trees will replace others in a succession that is characteristic of that biome. For example, no cactus will germinate and grow in a Canadian evergreen forest.

8. The solutions are sterile, meaning that all germs have been killed. Obviously, you would not wish to introduce any pathogens into your eyes.

A buffer is a solution that maintains a certain pH. Buffers work by absorbing or releasing protons that would make a solution less or more acidic. Your eye is delicate and would be harmed if exposed to a strong acid or to a strong base. That is why one must wear proper eye protection in the lab when working with chemicals.

Isotonic saline is a solution with the same concentration of salt as body cells. If body cells were exposed to a hypertonic solution, they would shrink, a process known as plasmolysis. Water would diffuse from an area with high water potential (cells of the eye) to an area with low water potential (hypertonic solution). In contrast, if the ophthalmic solution were hypotonic, the cells would swell because water from the saline solution would diffuse into the eye for the same reason as just described. An isotonic solution contains the same solute concentration as body cells.

Appendix A

BIBLIOGRAPHY

These textbooks are valuable resources for any basic college biology course.

Campbell, Neil A., and Jane Reece. *Biology*, 9th ed., San Francisco, California: Pearson, Benjamin Cummings, 2009

Freeman, Scott. *Biological Science*, 2nd ed., Upper Saddle River, New Jersey: Pearson Prentice Hall, 2005

Hillis, David M. and Heller, Craig H. *Principles of Life, High School Edition* Sinauer Assoc., W. H. Freeman and Co., 2012

Mader, Sylvia S. *Biology*, 9th ed., Dubuque, Iowa: McGraw-Hill, 2007

Micklos, David A. and Freyer, George A. *DNA Science, A First Course*, 2nd ed., Cold Spring Harbor, New York: Cold Spring Harbor Press, 2003

Nelson, David L., and Cox, Michael M. *Lehninger Principles of Biochemistry*, 4th ed., New York: W. H. Freeman, 2005

Purves, William K., Gordon H. Orians, and H. Craig Heller. *Life: The Science of Biology*, 6th ed., New York: W. H. Freeman, 2000

Raven, Peter H. and George B. Johnson, *Biology*, 9th ed., Dubuque, Iowa: William C. Brown/McGraw Hill Publishers, 2011

Starr, Cecie, and Ralph Taggart. *Biology: The Diversity of Life*, 12th ed., Belmont, California: Wadsworth Publishing, 2009

Tobin, Allan J. and Jennie Dushek. *Asking About Life*, 2nd ed. Orlando, Florida: Harcourt College Publishers, 2001

Wallace, Robert A., Gerald P. Sanders, and Robert J. Ferl. *Biology: The Science of Life*, 4th ed., New York: Addison-Wesley Publishing Co., 1996

Appendix B

MEASUREMENTS USED IN BIOLOGY

Measurements			
Quantity	Name of Unit	Symbol	Conversion
length	meter	m	1 m = 1,000 mm = .001 km = 100 cm
	kilometer	km	1 km = 1,000 m
	centimeter	cm	1 cm = 1/100 m
	millimeter	mm	1 mm = 1,000 µm
	micrometer	µm	1 µm = 1,000 nm
	nanometer	nm	1 nm = 1/1,000 µm
area	square meter	m^2	area encompassed by a square, each side of which is 1 meter in length
	hectare	ha	1 ha = 10,000 m^2 = 2.47 acres
	square centimeter	cm^2	1 cm^2 = 1/10,000 m^2
volume	liter	L	1 liter = 1/1,000 m^3 = 1.057 quarts
	milliliter	mL	1 mL = 0.0001 L
	microliter	µL	1 µL = 0.0001 mL
mass	kilogram	kg	1 kg = 1,000 g
	gram	g	1 g = 1,000 mg
	milligram	mg	1 mg = 1,000 mg
	microgram	µg	1 µg = 1/1,000 mg
temperature	Kelvin	K	0 K = absolute zero
	degrees Celsius	°C	°C + 273 = Kelvin
heat, work	calorie	cal	1 calorie = the amount needed to raise 1 gram of pure water 1° Celsius (from 14.5°C to 15.5°C = 4.184 Joules)
	kilocalorie	kcal	1 kcal = 1,000 cal
	joule	J	1 cal = 4.184 J
electric potential	volt	V	a unit of potential difference or electromotive force
	millivolt	mV	1 mV = 1/1,000 V = 10^{-3} V

Measurements			
Quantity	Name of Unit	Symbol	Conversion
time	second	s.	1 s. = 1/60 min.
	minute	min.	1 min. = 1/60 hr.
	hour	hr.	1 hr. = 1/24 d.
	day	d.	1 d. = 24 hr.

Glossary

ABA (abscisic acid) Plant hormone that inhibits growth, closes stomates during times of water stress and counteracts breaking of dormancy.

Abiotic Nonliving and includes temperature, water, sunlight, wind, rocks, and soil.

Abscission The process of leaves falling off a tree or bush.

Acetylcholine One of many neurotransmitters.

Acid rain Caused by pollutants in the air from combustion of fossil fuels. The pH is less than 5.6.

Actin Thin protein filaments that interact with myosin filaments in the contraction of skeletal muscle.

Action potential A rapid change in the membrane of a nerve or muscle cell when a stimulus causes an impulse to pass.

Active immunity The type of immunity when an individual makes his or her own antibodies after being ill and recovering or after being given an immunization or vaccine.

Adaptive radiation The emergence of numerous species from one common ancestor introduced into an environment.

Adenine A nucleotide that binds to thymine and uracil. It is a purine.

Adipose Fat tissue.

Allopatric speciation The formation of new species caused by separation by geography, such as mountain ranges, canyons, rivers, lakes, glaciers, altitude, or longitude.

Allosteric A type of enzyme that changes its conformation and its function in response to a modifier.

Amoebocytes Found in sponges, these cells are mobile and perform numerous functions, including reproduction, transport of food particles to nonfeeding cells, and secretion of material that forms the spicules.

Amphipathic A molecule with both a positive and negative pole.

Anaerobic respiration The anaerobic breakdown of glucose into pyruvic acid with the release of a small amount of ATP.

Analogous structures Structures, such as a bat's wing and a fly's wing, that have the same function, but the similarity is superficial and reflects an adaptation to similar environments, not a common ancestry.

Aneuploidy Any abnormal number of a particular chromosome.

Angiosperms Flowering plants.

Anode The positive pole in an electrolytic cell.

Antenna pigment Accessory photosynthetic pigment that expands the wavelengths of light that can be used to power photosynthesis.

Anterior pituitary Gland in the brain that releases many hormones, including growth hormone, luteinizing hormone, thyroid-stimulating hormone, adrenocorticotropic hormone, and follicle-stimulating hormone.

Anther Part of a flowering plant that produces male gametophytes.

Antheridia Structures in plants that produce male gametes.

Antibodies Produced by B lymphocytes and destroy antigens.

Anticodon The three-base sequence of nucleotides at one end of a tRNA molecule.

Antidiuretic hormone Released by the posterior pituitary, its target is the collecting tube of the nephron.

Apical dominance The preferential growth of a plant upward (toward the sun), rather than laterally.

Apoplast The network of cell walls and intercellular spaces within a plant body that permits extensive extracellular movement of water within a plant.

Apoptosis Programmed cell death.

Aposematic coloration The bright, often red or orange coloration of poisonous animals as a warning that predators should avoid them.

Archegonia Structures in plants that produce female gametes.

Artificial selection The intentional selection of specific individuals with desired traits for breeding.

Associative learning One type of learning in which one stimulus becomes linked, through experience, to another.

ATP-synthase channels Located in the cristae of mitochondria and thylakoids of chloroplasts, these are membrane channels that allow protons to diffuse down a gradient in the production of ATP.

Autonomic nervous system The branch of the vertebrate peripheral nervous system that controls involuntary muscles.

Autotrophs Organisms that synthesize their own nutrients.

Auxin A plant hormone that stimulates stem elongation and growth, enhances apical dominance, and is responsible for tropisms.

Back cross See testcross.

Bacteriophage A virus that attacks bacteria.

Balanced polymorphism The presence of two or more phenotypically distinct forms of a trait in a single population, such as two varieties of peppered moths, black ones and white ones.

Barr body An inactivated X chromosome seen as a condensed body lying just inside the nuclear envelope.

Batesian mimicry The copycat coloration where one harmless animal mimics the coloration of one that is poisonous. An example is the viceroy butterfly, which is harmless but looks similar to the monarch butterfly.

Binomial nomenclature A scientific naming system where every organism has a unique name consisting of two parts: a genus name and a species name.

Biological magnification A trophic process in which substances in the food chain become more concentrated with each link of the food chain.

Biomes Very large regions of the earth, named for the climatic conditions and for the predominant vegetation. Examples are marine, tropical rain forest, and desert.

Biosphere The global ecosystem.

Biotic potential The maximum rate at which a population could increase under ideal conditions.

B lymphocyte A lymphocyte that produces antibodies.

Bottleneck effect An example of genetic drift that results from the reduction of a population, typically by natural disaster. The surviving population is no longer genetically representative of the original one.

Botulinum The genus name for the bacterium that produces botulism, a very serious form of food poisoning.

Bryophytes Nonvascular plants like mosses.

Bulk flow The general term for the overall movement of a fluid in one direction in an organism, such as sap flowing in a tree or blood flowing in a human.

Bundle sheath cell A type of photosynthetic plant cell that is tightly packed around the veins in a leaf.

C-3 plant The common type of plant, different from C-4 and CAM plants.

C-4 plant A plant with anatomical and biochemical modifications for a dry environment that differ from C-3 and CAM plants. Examples are sugarcane and corn.

Calvin cycle A cyclical metabolic pathway in the dark reactions of photosynthesis that fixes or incorporates carbon into carbon dioxide and produces phosphoglyceraldehyde (PGAL), a three-carbon sugar.

CAM (crassulacean acid metabolism) A form of photosynthesis that is an adaptation to dry conditions; stomates remain closed during the day and open only at night.

Capsid The protein shell that encloses viral DNA or RNA.

Carbon fixation Carbon becomes fixed or incorporated into a molecule of PGAL. This happens during the Calvin cycle.

Carbonic acid anhydrase An enzyme found in red blood cells that catalyzes the conversion of carbon dioxide and water into carbonic acid as part of the system that maintains blood pH at 7.4.

Carotenoid Accessory photosynthetic pigment that is yellow or orange.

Carrying capacity The limit to the number of individuals that can occupy one area at a particular time.

Catalase An enzyme produced in all cells to decompose hydrogen peroxide, a by-product of cell respiration.

Cathode The negative pole in an electrolytic cell.

CDKs (cyclin-dependent kinases) A kinase whose activity depends on the level of cyclins and that controls the timing of cell division.

Cell plate A double membrane down the midline of a dividing plant cell between which the new cell wall will form.

Centriole One of two structures in animal cells involved with cell division.

Centromere   A specialized region in a chromosome that holds the two chromatids together.

Chemiosmosis   The process by which ATP is produced from the flow of protons through an ATP-synthetase channel in the thylakoid membrane of the chloroplast during the light reactions of photosynthesis and in the cristae membrane of the mitochondria during cell respiration.

Chemokines   A chemical secreted by blood vessel endothelium and monocytes during an immune response to attract phagocytes to an area.

Chiasma/chiasmata   The site at which a crossover and recombination occurs.

Chitin   A structural polysaccharide found in the cell walls.

Chlorophyll _a_   One type of chlorophyll that participates directly in the light-dependent reactions of photosynthesis.

Chlorophyll _b_   One type of chlorophyll that acts as an antenna pigment, expanding the wavelengths of light that can used to power photosynthesis.

Chloroplast   The site of photosynthesis in plant cells.

Choanocytes   Collar cells that line the body cavity and have flagella that circulate water in sponges.

Chromatid   Either of the two strands of a replicated chromosome joined at the centromere.

Chromatin network   The complex of DNA and protein that makes up a eukaryotic chromosome. When the cell is not dividing, chromatin exists as long, thin strands and is not visible with the light microscope.

Cilia   Hairlike extensions from the cytoplasm used for cell locomotion.

Citric acid cycle   Another name for the Krebs cycle.

Cladogenesis   Branching evolution occurs when a new species branches out from a parent species.

Classical conditioning   One type of associative learning that is widely accepted because of the ingenious work of Ivan Pavlov associating a novel stimulus with an innately recognized one.

Cleavage furrow   A shallow groove in the cell surface in an animal cell where cytokinesis is taking place.

Climax community   The final, stable community in an ecosystem.

Cline   A variation in some trait of individuals coordinated with some gradual change in temperature or other factor over a geographic range.

Clonal selection   A fundamental mechanism in the development of immunity. Antigenic molecules select or bind to specific B or T lymphocytes, activating them. The B cells then differentiate into plasma cells and memory cells.

Cnidocytes   Stinging cells in all cnidarians.

Codominance   The type of inheritance when there is no trait that dominates over another; both traits show.

Codons   The three-base sequence of nucleotides in mRNA.

Coelom   The body cavity that arises from within the mesoderm and is completely surrounded by mesoderm tissue.

Coevolution   Evolution that is caused by two species that interact and influence each other. All predator-prey relationships are examples.

Cohesion tension   Force of attraction between molecules of water due to hydrogen bonding.

Collaboration   Two genes interact to produce a novel phenotype.

Collenchyma cells   Plant cells with unevenly thickened primary cell walls that are alive at maturity and that function to support the growing stem.

Commensalism   A symbiotic relationship where one organism benefits and one is unaware of the other organism (+/o).

Community   All the organisms living in one area.

Companion cell   Connected to each sieve tube member in the phloem and nurtures the sieve tube elements.

Complement   An important part of the immune system, a group of about twenty proteins that assists in lysing cells.

Complementary genes   The expression of two or more genes where each depends upon the alleles of the other in order for a trait to show.

Conformation   The particular three-dimensional shape of a protein molecule.

Conjugation   A primitive form of sexual reproduction that is characteristic of bacteria and some algae.

Convergent evolution   Evolution that occurs when unrelated species occupy the same environment and are subjected to similar selective pressures and show similar adaptations.

Countercurrent mechanism   A mechanism or strategy to maximize the rate of diffusion. This is a major strategy to transport substances across membranes passively, such as in the nephron.

Cristae   The internal membranes of mitochondria that are the site of the electron transport chain.

Crop   Part of the digestive tract of many animals where food is temporarily stored until it can continue to the gizzard.

Crossing-over   The reciprocal exchange of genetic material between nonsister chromatids during synapsis of meiosis I.

Cutin   The main component of the waxy cuticle covering leaves to minimize water loss.

Cyclic photophosphorylation   Part of the light-dependent reactions in photosynthesis where electrons travel on a short-circuit pathway to replenish ATP levels only.

Cyclin   A regulatory protein whose levels fluctuate cyclically in a cell, in part, related to the timing of cell division.

Cystic fibrosis   The most common lethal genetic disease in the United States; characterized by a buildup of extracellular fluid in the lungs and digestive tract.

Cytochrome   An iron-containing pigment present in the electron transport chain of all aerobes.

Cytokines   Chemicals that stimulate helper T cells, B cells, and killer T cells.

Cytokinesis   Division of the cytoplasm.

Cytokinins   Plant hormone that stimulates cell division and delays senescence (aging).

Cytosine   A nucleotide that binds with guanine. A pyrimidine.

Cytoskeleton   A complex network of protein filaments that gives a cell its shape and helps it move.

Cytotoxic T cells   A type of lymphocyte that kills infected body cells and cancer cells.

Decomposers   Organisms, like bacteria and fungi, that recycle nutrients back to the soil.

Deletion   A chromosomal mutation where a fragment is lost during cell division.

Dendrites   The sensory processes of a neuron.

Denitrifying bacteria   Convert nitrates (NO_3) into free atmospheric nitrogen.

Density-dependent factors   Factors, such as starvation, that increase directly as the population density increases.

Density-dependent inhibition   A characteristic of normal cells grown in culture that causes cell division to cease when the culture becomes too crowded.

Density-independent factors   Factors, such as earthquakes, whose occurrence is unrelated to the population density.

Depolarization   An electrical state where the inside of an excitable cell is made less negative compared with the outside. If an axon is depolarized, an impulse is passing.

Detrivores   Consumers that derive their nutrition from nonliving, organic matter.

Deuterostomes   Animals in which the blastopore becomes the anus during early embryonic development.

Dicotyledon   A subdivision of flowering plants whose members possess an embryonic seed leaf made of two halves or cotyledons.

Dihybrid cross   A cross between individuals that are hybrid for two different traits, such as height and seed color.

Diploblastic   An organism whose body is made of only two cell layers, the ectoderm and the endoderm. The two are connected by a noncellular layer called the mesoglea. Animal phyla that are diploblastic are the Porifera (sponges) and the Cnidaria (jellyfish and hydra).

Directional selection   Selection where one phenotype replaces another in the gene pool.

Disruptive selection   Selection that increases the extreme types in a population at the expense of intermediate forms.

Divergent evolution   Evolution that occurs when a population becomes isolated (for any reason) from the rest of the species, becomes exposed to new selective pressures, and evolves into a new species.

DNA ligase   An enzyme that permanently attaches pieces of DNA together.

Dopamine   A neurotransmitter.

Down syndrome   A genetic condition caused by trisomy 21.

Duodenum   The first 12 inches (30 cm) of the human small intestine.

Ecdysone A hormone that helps control metamorphosis in insects.

Ecological succession   The sequential rebuilding of an entire ecosystem after a disaster.

Ecosystem   All the organisms in a given area as well as the abiotic (nonliving) factors with which they interact.

Ectoderm   The germ layer that gives rise to the skin and nervous system.

Effectors   Muscles or glands.

Electron transport chain   A sequence of membrane proteins that carry electrons through a series of redox reactions to produce ATP.

Endergonic   Any process that absorbs energy.

Endoderm   The embryonic germ layer that gives rise to the viscera, the digestive tract, and other internal organs.

Endodermis   The tightly packed layer of cells that surrounds the vascular cylinder in the root of a plant.

Endoplasmic reticulum   A system of transport channels inside a eukaryotic cell.

Endosperm   The food source for the growing embryo in monocots.

Endosymbiosis   This theory states that mitochondria and chloroplasts were once free-living prokaryotes that took up residence inside larger prokaryotic cells in a permanent, symbiotic relationship.

Endotherms   Animals that can raise their body temperature, although they cannot maintain a stable body temperature.

Envelope   Cloaks the capsid of a virus and aids the virus in infecting the host. The envelope is derived from membranes of host cells.

Enzyme   A protein that serves as a catalyst.

Epicotyl   Part of the developing embryo that will become the upper part of the stem and the leaves of a plant.

Epinephrine   A neurotransmitter.

Epiphytes   Photosynthetic plants that grow on other trees rather than supporting themselves.

Epistasis   Two separate genes control one trait, but one gene masks the expression of the other gene.

Esterase   An enzyme that breaks down excess neurotransmitter.

Ethylene   A gaseous plant hormone that promotes fruit ripening and opposes auxins in its actions.

Eukaryotes   Cells with internal membranes.

Eutrophication   Translates as "true feeding." A process begun by the entrance of large amounts of nutrients into a lake, ultimately ending with the death of the lake.

Exocytosis   The process by which cells expel substances.

Exons   Stands for expressed sequences of DNA. These are genes.

Exothermic   Any process that gives off energy.

Expressivity   The range of expression of mutant genes.

Extranuclear genes   Genes outside the nucleus, in the mitochondria and chloroplasts.

Facultative anaerobes   Organisms that can live without oxygen in the environment.

FAD   (flavin adenine dinucleotide)   A coenzyme that carries protons or electrons from glycolysis and the Krebs cycle to the electron transport chain.

Fermentation   A synonym for anaerobic respiration. The anaerobic breakdown of glucose into pyruvic acid.

Fixed action pattern   An innate, highly stereotypic behavior, which when begun, is continued to completion, no matter how useless.

Flagella   The tail-like structure that propels some single-celled organisms. Flagella consist of microtubules.

Follicle-stimulating hormone (FSH)   A hormone released from the anterior pituitary that stimulates the ovarian follicle.

Food chain   The pathway along which food is transferred from one trophic level to the next.

Food pyramid A model of the food chain that demonstrates the interaction of the organisms and the loss of energy.

Food web The interconnected feeding relationships of organisms in an ecosystem.

Founder effect An example of genetic drift, when a small population breaks away from a larger one to colonize a new area; it is most likely not genetically representative of the original larger population.

Frameshift One type of mutation caused by a deletion or addition where the entire reading sequence of DNA is shifted. AAA TTT CCC GGG could become AAT TTC CCG GG.

Frequency-dependent selection A form of selection that acts to decrease the frequency of the more-common phenotypes and increase the frequency of the less-common ones.

Fruit A ripened ovary of a flowering plant.

Fungi The kingdom that consists of heterotrophs that carry out extracellular digestion and have cell walls made of chitin; includes mushrooms and yeast.

GABA (gamma-aminobutyric acid) A neurotransmitter.

Gametangia A protective jacket of cells that prevents some plants' gametes and zygotes from drying out.

Gametophyte The monoploid generation of a plant.

Gastrodermis Cells that line the gastrovascular cavity in cnidarians.

Gastrovascular cavity A digestive cavity with only one opening, characteristic of cnidarians.

Gated-ion channel A channel in a plasma membrane for one specific ion, such as sodium or calcium. In the terminal branch of a neuron, it is responsible for the release of neurotransmitter into the synapse.

Gene flow The movement of alleles into or out of a population.

Genetic drift Change in the gene pool due to chance.

Genetic engineering The technology of manipulating genes for practical purposes.

Genomic imprinting Certain traits whose expression varies, depending on the parent from which they are inherited. Diseases that result from imprinting are Prader-Willi and Angelman syndromes.

Genotype The types of genes an organism has.

Gibberellins Plant hormone that promotes stem elongation.

Gizzard Part of the digestive tract of many animals. It is the site of mechanical digestion.

Glial cells Cells that nourish neurons.

Gluteraldehyde A chemical fixative often used in the preparation of tissue for electron microscopy.

Glycocalyx The external surface of a plasma membrane that is important for cell-to-cell communication.

Glycolysis A nine-step, anaerobic process that breaks down one glucose molecule into two pyruvic acid molecules and four ATP.

Golgi apparatus An organelle in eukaryotes that lies near the nucleus and that packages and secretes substances for the cell.

Gonadotropic-releasing hormone (GgRH) A hormone released by the hypothalamus that stimulates other glands to release their hormones.

Gradualism The theory that organisms descend from a common ancestor gradually, over a long period of time, in a linear or branching fashion.

Grana   Membranes in the chloroplast where the light reactions occur.

Greenhouse effect   The warming of the planet because of the accumulation of atmospheric carbon dioxide.

Ground tissue   The most common tissue type in a plant, functions mainly in support and consists of parenchyma, collenchyma, and sclerenchyma cells.

GTP (guanosine triphosphate)   A molecule closely related to ATP that provides the energy for translation.

Guanine   A nucleotide that binds with cytosine. A purine.

Guttation   Due to root pressure, droplets of water appear in the morning on the leaf tips of some herbaceous leaves.

Gymnosperms   Conifers or cone-bearing plants.

Habitat isolation   Separation of two or more organisms of the same species living in the same area but in separate habitats, such as in the water and on land.

Habituation   One of the simplest forms of learning in which an animal comes to ignore a persistent stimulus.

Halophiles (halobacteria)   Aerobic bacteria that thrive in environments with very high salt concentrations.

Hatch-Slack pathway   An alternate biochemical pathway found in C-4 plants; its purpose is to remove CO_2 from the airspace near the stomate.

Head-foot   The part of the body of mollusks that contains both sensory and motor organs.

Helicase   An enzyme that untwists the double helix at the replication fork.

Helper T cells   One type of T lymphocyte that activates B cells and other T lymphocytes.

Hemocoels   Blood-filled cavities within the body of arthropods and mollusks with open circulatory systems.

Hemophilia   An inherited genetic disease caused by the absence of one or more proteins necessary for normal blood clotting.

Hermaphrodites   Organisms possessing both male and female sex organs.

Heterosis   See hybrid vigor.

Heterosporous   A plant that produces two kinds of spores, male and female.

Heterotroph hypothesis   This theory states that the first cells on earth were heterotrophic prokaryotes.

Heterotrophs   Organisms that must ingest nutrients rather than synthesize them.

Histamine   A chemical released by the body during an inflammatory response that causes blood vessels to dilate.

Homeotherms   Organisms that maintain a consistent body temperature.

Homologous structures   Structures in different species that are similar because they have a common origin.

Homosporous   A plant that produces a single bisexual spore.

Huntington's disease   A degenerative, inherited, dominant disease of the nervous system that results in certain and early death.

Hybrid vigor   A phenomenon in which the hybrid state is selected because it has greater survival and reproductive success. Also known as heterosis.

Hydrophilic   Having an affinity for water.

Hyperpolarized An electrical state where the inside of the excitable cell is made more negative compared with the outside of the cell and the electric potential of the membrane increases (gets more negative).

Hypertonic Having a greater concentration of solute than another solution.

Hypocotyl Part of the developing embryo that will become the lower part of the stem and roots.

Hypothalamus Gland located in the brain above the pituitary that is the bridge between the endocrine and nervous systems.

Hypotonic Having a lesser concentration of solute than another solution.

Immunoglobins See antibodies.

Immunological memory The capacity of the immune system to generate a secondary immune response against a specific antigen for a lifetime.

Imprinting A type of learning that is responsible for the bonding between mother and offspring. Common in birds, it occurs during a sensitive or critical period in early life.

Incomplete dominance The type of inheritance that is characterized by blending traits. For instance, one gene for red plus one gene for white results in a pink four o'clock flower.

Indoleacetic acid IAA. A naturally occurring auxin.

Inflammatory response A nonspecific defensive reaction of the body to invasion by a foreign substance that is accompanied by the release of histamine, fever, and red, itchy areas.

Interferons A class of chemicals that block viral infections.

Interneuron Also known as association neuron, resides within the spinal cord and receives sensory stimuli and transfers the information directly to a motor neuron or to the brain for processing.

Interphase The longest stage of the life cycle of a cell; it consists of G_1, S, and G_2.

Introns Intervening sequences, the noncoding regions of DNA, that are sometimes referred to as junk.

Inversion A chromosome mutation where a chromosomal fragment reattaches to its original chromosome but in the reverse orientation.

In vitro In the laboratory.

In vivo In the living thing.

Isotonic Two solutions containing equal concentrations of solutes.

Karyotype A procedure that analyzes the size, number, and shape of chromosomes.

Kinase An enzyme that transfers phosphate ions from one molecule to another.

Kinetochore A disc-shaped protein on the centromere that attaches the chromatid to the mitotic spindle during cell division.

Klinefelter's syndrome A genetic condition in males in which there is an extra X chromosome; the genotype is XXY.

Kranz anatomy Refers to the structure of C-4 leaves and differs from C-3 leaves. In C-4 leaves, the bundle sheath cells lie under the mesophyll cells, tightly wrapping the vein deep within the leaf, where CO_2 is sequestered.

Krebs cycle Also known as the citric acid cycle, it completes the breakdown of pyruvic acid into carbon dioxide, with the release of a small amount of ATP.

Lactic acid fermentation   The process by which pyruvate from glycolysis is reduced to form lactic acid or lactate. This is the process that the dairy industry uses to produce yogurt and cheese.

Lateral meristem   Growth region of a plant that provides secondary growth, increase in girth.

Law of dominance   One of Mendel's laws. It states that when two organisms, each pure for two opposing traits, are crossed, the offspring will be hybrid but will exhibit only the dominant trait.

Law of independent assortment   States that each allelic pair separates during gamete formation. Applies when genes for two traits are not on the same chromosome.

Law of segregation   During the formation of gametes, allelic pairs for two traits separate.

Learning   A sophisticated process in which the responses of the organism are modified as a result of experience.

Linked genes   Genes that are on the same chromosome.

Luteinizing hormone   Triggers the ovulation of the secondary oocyte from the ovary.

Lysosomes   Sacs of hydrolytic enzymes and the principal site of intracellular digestion.

Lytic cycle   A type of viral infection that results in the lysing of the host cell and the release of new phages that will infect other cells.

Macroevolution   The development of an entirely new species.

Macrophage   While acting as an antigen-presenting cell, it engulfs bacteria by phagocytosis and presents a fragment of the bacteria on the cell surface by an MHC II molecule.

Malpighian tubules   The organ of excretion in insects.

Mantle   The part of the body of mollusks that contains specialized tissue that surrounds the visceral mass and secretes the shell.

Map unit   The distance on a chromosome within which recombination occurs 1 percent of the time.

Marsupials   Animals whose young are born very early in embryonic development and where the joey completes its development nursing in the mother's pouch. Includes kangaroos.

Matrix   The inner region of a mitochondrion, where the Krebs cycle occurs.

Medusa   The free-swimming, upside-down, bowl-shaped stage in the life cycle of the cnidarians. An example is jellyfish.

Megaspores   In flowering plants, these produce the ova.

Meiosis   Occurs in sexually reproducing organisms and results in cells with half the chromosome number of the parent cell.

Membrane potential   A measurable difference in electrical charge between the cytoplasm (negative ions) and extracellular fluid (positive ions).

Memory cells   A long-lived form of a lymphocyte that bear receptors to a specific antigen and that remains circulating in the blood in small numbers for a lifetime.

Meristem   Actively dividing cells that give rise to other cells such as xylem and phloem.

Mesoderm   The germ layer that gives rise to the blood, bones, and muscles.

Methanogens   Prokaryotes that synthesize methane from carbon dioxide and hydrogen gas.

MHC (major histocompatibility complex)   A collection of cell surface markers that identify the cells as self. No two people, except identical twins, have the same MHC markers. Also known as HLA (human leukocyte antigens).

Microevolution   Refers to the changes in one gene pool of a population.

Microfilaments   Solid rods of the protein actin that make up part of the cytoskeleton.

Micropyle   The opening to the ovule in a flowering plant.

Microspores   In flowering plants, these produce sperm.

Microtubules   A hollow rod of the protein tubulin in the cytoplasm of all eukaryote cells that make up cilia, flagella, spindle fibers, and other cytoskeletal structures of cells.

Middle lamella   A distinct layer of adhesive polysaccharides that cements adjacent plant cells together.

Minority advantage   See frequency-dependent selection.

Mitchell hypothesis   An attempt to explain how energy is produced during the electron transport chain by oxidative phosphorylation.

Mitochondria   The site of cell respiration and ATP synthesis in all eukaryotic cells.

Mitosis   Produces two genetically identical daughter cells and conserves the chromosome number ($2n$).

Monera   No longer used as the name of the kingdom that contains all the prokaryotes, including bacteria.

Monoclonal antibodies   Antibodies produced by a single B cell that produces a single antigen in huge quantities. They are important in research and in treating and diagnosing certain diseases.

Monocotyledon   A subdivision of flowering plants whose members possess one embryonic seed leaf or cotyledon.

Monocytes   A type of white blood cell that transforms into macrophages, extends pseudopods, and engulfs huge numbers of microbes over a long period of time.

Monohybrid cross   This is the cross between two organisms that are each hybrid for one trait.

Monotremes   Egg-laying mammals where the embryo derives nutrition from the yolk, like the duck-billed platypus.

Motor neuron   A neuron that stimulates effectors (muscles or glands).

Mucosa   The innermost layer of the human digestive tract. In some parts of the digestive system, it contains mucus-secreting cells and glands that secrete digestive enzymes.

Müllerian mimicry   Two or more poisonous species resemble each other and gain an advantage from their combined numbers. Predators learn more quickly to avoid any prey with that appearance.

Multiple alleles   More than two allelic forms of a gene.

Mutagenic agents   Substances that cause mutations.

Mutation   Changes in DNA.

Mutualism   Asymbiotic relationship where both organisms benefit (+/+).

Mycorrhizae   The symbiotic structures consisting of the plant's roots intermingled with the hyphae (filaments) of a fungus that greatly increase the quantity of nutrients that a plant can absorb.

Myosin   Thick protein filaments that interact with actin filaments in the contraction of skeletal muscle.

NAD (nicotinamide adenine dinucleotide)   A coenzyme that carries protons or electrons from glycolysis and the Krebs cycle to the electron transport chain.

NADP **(nicotinamide nucleotide phosphate)** Carries hydrogen from the light reactions to the Calvin cycle in the dark reactions of photosynthesis.

Natural killer (NK) cells Part of the nonspecific immune response that destroys virus-infected body cells (as well as cancerous cells).

Natural selection A theory that explains how populations evolve and how new species develop.

Neuromuscular junction The place where a neuron synapses on a muscle.

Neurotransmitter The chemical held in presynaptic vesicles of the terminal branch of the axon that are released into a synapse and that excite the postsynaptic membrane.

Neutrophils A type of white blood cell that engulfs microbes by phagocytosis.

Niche Organisms that live in the same area and use the same resources.

Nitric oxide Acts as a local signaling molecule.

Nitrifying bacteria Convert the ammonium ion into nitrites and then into nitrates.

Nitrogen-fixing bacteria Convert free nitrogen into the ammonium ion.

Nondisjunction Homologous chromosomes fail to separate as they should during meiosis.

Norepinephrine A neurotransmitter.

Notochord A rod that extends the length of the body and serves as a flexible axis in all chordates.

Nucleoid Nuclear region in prokaryotes.

Nucleolus Located in the nucleus and is the site of protein synthesis.

Nucleotides The building blocks of nucleic acids. They consist of a five-carbon sugar, a phosphate, and a nitrogenous base: adenine, thymine (in DNA), cytosine, guanine, or uracil (in RNA).

Obligate anaerobes Prokaryotes that cannot live in the presence of oxygen.

Okazaki fragments In DNA replication, the segments in which the 3′ to 5′ lagging strand of DNA is synthesized.

Omnivores Organisms, like humans, that eat both plants and animals.

Operant conditioning A type of associative learning where an animal learns to associate one of its own behaviors with a reward or punishment and then repeats or avoids that behavior. Also called trial-and-error learning.

Operator In an operon, the binding site for the repressor.

Operon Functional genes and their switches that are found in bacteria.

Osmotic potential The tendency of water to move across a permeable membrane into a solution.

Outbreeding Mating of organisms that are not closely related; it is a major mechanism of maintaining variation within a species.

Oxidative phosphorylation The production of ATP using energy derived from the electron transport chain.

Oxytocin Hormone released by the posterior pituitary that stimulates labor and the production of milk from mammary glands.

Parallel evolution Evolution that occurs when two related species have made similar evolutionary adaptations after their divergence from a common ancestor.

Parasitism A symbiotic relationship (+/−) where one organism, the parasite, benefits while the other organism, the host, is harmed.

Parasympathetic One of two branches of the autonomic nervous system that has a relaxing effect.

Parenchyma cells Traditional plant cells with primary cell walls that are thin and flexible and that lack secondary cell walls.

Passive immunity Immunity is transferred to an individual from someone else.

Pathogens Organisms that cause disease.

Pedigree A family tree that indicates the phenotype of one trait being studied for every member of a family and will help determine how a particular trait is inherited.

Peroxisomes Organelles in both plants and animals that break down peroxide, a toxic by-product of cell respiration.

Phage Short form of bacteriophage, the virus that attacks bacteria.

Phagocytes A type of white blood cell that ingests invading microbes.

Phagocytosis The process by which a cell engulfs large particles using pseudopods.

Phenotype The appearance of an organism.

Phenylketonuria An inborn inability to break down the amino acid phenylalanine. It requires elimination of phenylalanine from the diet, otherwise serious mental retardation will result.

Phloem Transport vessels in plants that carry sugars from the photosynthetic leaves to the rest of the plant by active transport.

Phosphodiester linkages The bonds that join the nucleotides in DNA.

Phosphofructokinase (PFK) An allosteric enzyme important in glycolysis.

Phosphoglyceraldehyde (PGAL) A three-carbon sugar, the first stable carbohydrate to be produced by photosynthesis.

Photolysis The process of splitting water, providing electrons to replace those lost from chlorophyll *a* in P680. This is powered by the light energy absorbed during the light-dependent reactions.

Photoperiod The environmental stimulus a plant uses to detect the time of year and the relative lengths of day and night.

Photophosphorylation The process of generating ATP by means of a proton motive force during the light reactions of photosynthesis.

Photorespiration A process that occurs when rubisco binds with O_2 instead of CO_2. It is a dead-end process because no ATP is produced and no sugar is formed.

Photosynthesis The process by which light energy is converted to chemical bond energy.

Photosystem I (P700) Energy, with an average wavelength of 700nm, is absorbed in this photosystem and transferred to electrons that move to a higher energy level.

Photosystem II (P680) Energy, with an average wavelength of 680nm, is absorbed in this photosystem and transferred to electrons that move to a higher energy level.

Photosystems Light-harvesting complexes in the thylakoid membranes of chloroplasts. They consist of a reaction center containing chlorophyll *a* and a region containing several hundred antenna pigment molecules that funnel energy into chlorophyll *a*.

Phycobilin A photosynthetic antenna pigment common in red and blue-green algae.

Phytochrome   The photoreceptor responsible for keeping track of the length of day and night. There are two forms of phytochrome, Pr (red light absorbing) and Pfr (infrared light absorbing).

Pili   Cytoplasmic bridges that connect one cell to another and that allow DNA to move from one cell to another in a form of primitive sexual reproduction called conjugation.

Pinocytosis   A type of endocytosis in which a cell ingests large, dissolved molecules.

Pioneer organisms   The first organisms, such as lichens and mosses, to inhabit a barren area.

Pistil   Part of a flowering plant that produces female gametes.

Placental mammals   Animals whose young are born and where the embryo develops internally in a uterus connected to the mother by a placenta where nutrients diffuse from mother to embryo. Also called eutherians.

Plasma cells   A short-lived form of a lymphocyte that secretes antibodies.

Plasmid   Foreign, small, circular, self-replicating DNA molecule that inhabits a bacterium and imparts characteristics to the bacterium such as resistance to antibiotics.

Plasmodesmata   An open channel in the cell walls of plant cells allowing for connections between the cytoplasm of adjacent cells.

Plasmolysis   Cell shrinking.

Plastids   Organelles in plant cells, including chloroplasts, chromoplasts, and leucoplasts.

Plastoquinone   A proton and electron carrier in the electron transport chain during the light reactions of photosynthesis.

Pleiotropy   One single gene affects an organism in several or many ways.

Poikilotherms   Cold-blooded animals.

Point mutation   A change in one nucleotide in DNA.

Polarized membrane   An axon membrane at rest where the inside of the cell is negative compared with the outside of the cell.

Pollen   One pollen grain contains three monoploid nuclei, one tube nucleus, and two sperm nuclei.

Pollination   The transfer of pollen from the stamen to the pistil.

Polygenic   Genes that vary along a continuum, like skin color or height.

Polymerase chain reaction (PCR)   A cell-free, automated technique by which a piece of DNA can be rapidly copied or amplified.

Polyp   A vase-shaped body or the sessile phase in the life cycle of cnidarians.

Polyploid   A chromosome mutation in which the organism possesses extra sets of chromosomes; the cell becomes $3n$, $4n$, $5n$, and so on.

Population   A group of individuals of one species living in one area.

Predation   One animal eating another animal, or it can also refer to animals eating plants.

Primary consumer   The animal that eats the producer.

Primary immune response   The initial immune response to an antigen.

Primase   An enzyme that joins RNA nucleotides to make the primer.

Prions   Infectious proteins that cause several brain diseases: scrapie in sheep, mad cow disease in cattle, and Creutzfeldt-Jakob disease in humans.

Producer   Those photosynthetic organisms at the bottom of any food chain.

Prokaryotes Cells with no internal membranes. Bacteria are one example.

Promoter The binding site of RNA polymerase in an operon.

Prophage A phage genome that has been inserted into a specific site in a bacterial chromosome.

Prostaglandin A hormone that promotes blood supply to an area.

Protista The kingdom that consists of single-celled and primitive multicelled organisms, such as paramecium and amoeba.

Proton pump A mechanism in cells that uses ATP to pump protons across a membrane to generate a membrane or electric potential.

Protostome An animal in which the blastopore becomes the mouth during early embryonic development. Literally, first opening.

Pseudocoelomate A body cavity with mesoderm on only one side, characteristic of nematodes.

Pseudopods Cellular extensions of amoeboid cells used in moving and feeding.

Punctuated equilibrium A theory that proposes that new species appear suddenly after long periods of stasis.

Purines A class of nucleotides that includes adenine and guanine.

Pyrimidines A class of nucleotides that includes cytosine, thymine, and uracil.

Pyrogens A chemical released by certain leukocytes that increases body temperature to speed up the immune system and make it more difficult for microbes to function.

Pyruvate A variant of pyruvic acid.

Pyruvic acid A three-carbon molecule that is the product of glycolysis and is the raw material for the Krebs cycle.

Radicle In the embryonic root, the first organ to emerge from the germinating seed.

Radula A movable, tooth-bearing structure that acts like a tongue in mollusks.

Receptor-mediated endocytosis The uptake of specific molecules by a cell's receptors.

Recessive trait The trait that remains hidden in the hybrid state.

Recognition sequence A specific sequence of nucleotides at which a restriction enzyme cleaves a DNA molecule.

Recombinant chromosomes Chromosomes that combine genes from both parents due to crossing-over.

Recombination The result of a cross-over.

Reflex arc The simplest nerve response; it is inborn, automatic, and protective.

Refractory period The period of time during which a neuron cannot respond to another stimulus because the membrane is returning to its polarized state.

Replication bubbles There are thousands of replication bubbles along the DNA molecule that speed up the process of replication.

Replication fork A Y-shaped region where the new strands of DNA are elongating.

Repressor Binds to the operator of an operon and prevents RNA polymerase from binding to the promoter, thus blocking transcription.

Resolution A measure of the clarity of an image; the ability to see two objects as separate.

Resource partitioning The exploitation of environmental resources by organisms living in the same area so that each group of organisms can occupy a different niche.

Restriction enzymes Enzymes, naturally occurring in bacteria, that cut DNA at certain specific recognition sites.

Restriction fragment length polymorphisms (RFLPs) Noncoding regions of human DNA that vary from person to person. They can be used to identify a single individual. Pronounced "riflips."

Restriction fragments Fragments of DNA that result from the cuts made by restriction enzymes.

Reverse transcriptase An enzyme found in retroviruses that facilitates the production of DNA from RNA.

Rhizobium A symbiotic bacterium that lives in the nodules on roots of specific legumes and that incorporates nitrogen gas from the air into a form of nitrogen the plant requires.

Ribosomes The site of protein synthesis in the cytoplasm.

RNA polymerase The enzyme that binds to the promoter in DNA and that begins transcription.

RNA primer An already existing chain of RNA attached to DNA to which DNA polymerase adds nucleotides during DNA synthesis.

Rubisco (ribulose biphosphate carboxylase) The enzyme that catalyzes the first step in the Calvin cycle: the addition of RuBP (ribulose biphosphate) to CO_2.

Sarcolemma The modified plasma membrane surrounding a skeletal muscle cell and that can propagate an action potential.

Sarcomere The basic functional unit of skeletal muscle.

Sarcoplasmic reticulum Modified endoplasmic reticulum in skeletal muscle cells.

Satellite DNA Short sequences of DNA that are tandemly repeated as many as 10 million times in the DNA. Much of it is located at the telomeres.

Schwann cells Glial cells that are located in the peripheral nervous system and that form the myelin sheath around the axon of a neuron.

Sclerenchyma cells Plants cells with very thick primary and secondary cell walls fortified with lignin.

Secondary consumer The animal that eats the primary consumer.

Seed After fertilization, the ovule becomes the seed.

Semiconservative replication The way DNA replicates, each double helix separates and forms two new strands of DNA. Each new molecule of DNA consists of one old strand and one new strand.

Senescence Aging.

Sertoli cells The cells found in the mammalian testes that nourish the developing sperm cells, which contain no cytoplasm.

Sessile Nonmotile.

Sex-influenced trait The inheritance of a trait influenced by the sex of the individual carrying the trait.

Sexual selection Selection based on variation in secondary sexual characteristics related to competing for and attracting mates.

Sieve tube members Along with companion cells, these make up the phloem.

Single-stranded binding proteins Proteins that act as scaffolding, holding two DNA strands apart during replication.

snRNPs (small nuclear ribonucleoproteins) Help to process RNA after it is formed and before it moves to the ribosome.

Sodium-potassium pump A protein pump within a plasma membrane of an axon that restores the membrane to its original polarized condition by pumping sodium and potassium ions across the membrane.

Solute The substance dissolved.

Solvent The substance doing the dissolving.

Somatic cell A body cell.

Somatic nervous system The branch of the vertebrate peripheral nervous system that controls skeletal (voluntary) muscles.

Sori Structures on the underside of the fern leaves that are clusters of sporangia containing monoploid spores.

Specific heat The amount of heat a substance must absorb to increase 1 gram of the substance by 1°C.

Spicules Found in sponges, these consist of inorganic materials and support the animal.

Spindle fibers Made of microtubules that connect centrioles to kinetochores of chromosomes and that separate sister or homologous chromosomes during cell division.

Spiracles Openings in the exoskeleton of arthropods, such as the grasshopper, that connect to internal cavities called hemocoels where respiratory gases are exchanged.

Splicesomes Enzymes that (along with SnRPs) help process RNA after it is formed and before it moves to the ribosome.

Spongocoel Found in sponges, it is the central cavity into which water is drawn to filter nutrients.

Sporopollenin A tough polymer that protects plants in a harsh terrestrial environment.

Stabilizing selection Selection that eliminates the extremes and favors the more common intermediate forms.

Stele The vascular cylinder of the root, consisting of vascular tissue.

Stroma The site of the light-independent (dark) reactions in chloroplasts.

Submucosa A layer of the human digestive system that contains nerves, blood vessels, and lymph vessels.

Survivorship curves Show the size and composition of a population.

Sympathetic nervous system One of two branches of the autonomic nervous system that is generally excitatory.

Sympatric speciation The formation of new species without geographic isolation; such as polyploidy or behavioral isolation.

Symplast A continuous system of cytoplasm of cells interconnected by plasmodesmata.

Synapsis The process of pairing replicated homologous chromosomes during prophase I of meiosis.

Systematics Scientific study of the classification of organisms and their relationships to one another.

Taq polymerase A heat-stable form of DNA polymerase extracted from bacteria that live in hot environments, such as hot springs, that is used during PCR technique.

Taxa A particular group at a category level; such as kingdom or genus.

Taxonomy The study of classification of organisms.

Tay-Sachs disease An inherited genetic disease that is caused by lack of an enzyme necessary to break down lipids necessary for normal brain function and results in seizures, blindness, and early death. Common in Ashkenazi Jews.

Telomerase An enzyme that catalyzes the lengthening of the telomeres at the ends of eukaryotic chromosomes.

Telomeres The protective ends of eukaryotic chromosomes.

Tertiary consumer The third trophic level of consumer in a food chain.

Testcross A cross done to determine whether an individual plant or animal showing the dominant trait is homozygous dominant (*B/B*) or heterozygous (*B/b*). The individual in question (*B/_*) is crossed with a homozygous recessive individual.

Tetanus The smooth, sustained contraction of a skeletal muscle.

Thermophiles Prokaryotes that thrive in very high temperatures.

Theta replication The way in which prokaryotes replicate their DNA.

Thylakoids Membranes in chloroplasts that make up the grana, the site of the light reactions.

Thymine A nucleotide that binds with adenine. It is a pyrimidine and is not present in RNA.

T lymphocytes One type of lymphocyte that fights pathogens by cell-mediated response.

Tracheids Long, thin cells that overlap and are tapered at the ends and that, along with vessel elements, make up xylem in a plant.

Tracheophytes Plants that have transport vessels, xylem and phloem.

Transcription The process by which DNA makes RNA.

Transduction Transfer of bacterial DNA by phages from one bacterium to another.

Transformation The transfer of genes from one bacterium into another.

Translation The process by which the codons of an mRNA sequence are changed into an amino acid sequence.

Translocation A chromosome mutation where a fragment of a chromosome becomes attached to a nonhomologous chromosome; the transport of sugar in a plant from source to sink.

Transpiration Loss of water from stomates in leaves.

Transpirational pull–cohesion tension theory This theory describes the passive transport of water up a tree. For each molecule of water that evaporates from a leaf by transpiration, another molecule of water is drawn in at the root to replace it.

Transposons Transposable genetic elements, sometimes called jumping genes.

Triploblastic Having three cell layers: ectoderm, mesoderm, and endoderm.

Triploid A chromosome mutation where an organism has three sets of chromosomes ($3n$) instead of two ($2n$).

Trisomy A chromosome condition in which a cell has an extra copy of one chromosome. The cell has three of that chromosome, instead of two.

Trophic level Any level of a food chain based on nutritional source.

Tropic hormones Hormones released by one endocrine gland that stimulate other endocrine glands to release their hormones.

Tropism The growth of a plant toward or away from a stimulus, for example, phototropism.

T system   A set of tubules that traverse the skeletal muscle, conduct the action potential deep into the cell, and stimulate the sarcoplasmic reticulum to release calcium ions.

Turgid   Firm. Plant cells swollen because they have absorbed water.

Turner's syndrome   A genetic condition in females caused by a deletion of one of the two X chromosomes.

Typhlosole   A large fold in the upper surface of the intestine of the earthworm that increases surface area to increase absorption.

Ultramicrotome   An instrument used to cut very thin sections of tissue for use in the transmission electron microscope.

Uracil   A nucleotide in only RNA that binds with adenine. It is a pyrimidine.

Vacuoles   A membrane-enclosed sac for storage in all cells, particularly in plant cells.

Vasodilation   The enlargement of blood vessels to increase blood supply.

Vegetative propagation   Plants can clone themselves or reproduce asexually from any vegetative part of the plant; the root, stem, or leaf.

Vesicles   Small vacuoles.

Vessel elements   Wide, short tubes that, along with tracheids, make up the xylem.

Vestigial structures   Structures of no importance, such as the appendix, that were once important to ancestors.

Visceral mass   The part of the body of mollusks that contains the organs of digestion, excretion, and reproduction.

Water potential   The tendency of water to move across a semipermeable membrane, ψ. The water potential of pure water is zero. Any solution has a water potential less than zero.

Wave of depolarization   The wavelike reversal of the polarity of the membrane when an impulse passes.

Wobble   Refers to the translation of mRNA to protein. The relaxation of base-pairing rules where the pairing rules for the third base of a codon are not as strict as they are for the first two bases. UUU and UUA both code for the amino acid phenylalanine.

Xanthophyll   A photosynthetic antenna pigment common in algae that is a structural variant of a carotenoid.

Xylem   Transport vessels in plants that carry water and minerals from the soil to the leaves.

Z lines   These define the edges of the sarcomere in a muscle cell.

Zone of cell division   The region of a plant's root with actively dividing cells that grow down into the soil.

Zone of differentiation   The region of root tip where cells undergo specialization.

Zone of elongation   The region of root tip where cells elongate and that are responsible for pushing the root cap downward, deeper into the soil.

Index

Organic compounds, 14–20
Organogenesis, 287–288
Osmoregulation, 240–241
Osmosis, 41–42, 341–344
Osmotic potential, 42
Outbreeding, 184–185
Oxidative phosphorylation, 64
Oxyhemoglobin, 229
Oxytocin, 236

P

Parallel evolution, 192
Parenchymal cells, 209
Parthenogenesis, 286
Particulate inheritance, 105
Passive immunity, 273
Passive transport, 40–44
Pedigree, 116–117
Penetrance, 113–114
Peptidoglycan, 161
Peripheral nervous system, 244
Peristalsis, 226
Peroxisomes, 36, 80
Pesticides, 316
pH, 12–13
Phagocytes, 268
Phagocytosis, 45
Phenotype, 107
Phenotype ratio, 110
Phenylketonuria, 118
Pheromones, 234
Phloem, 207–208
Phosphoenolpyruvate, 80
Phosphoglyceraldehyde, 79
Phospholipids, 16–17
Phosphorylation, 64
Photolysis, 77
Photoperiodism, 217–218
Photosynthesis, 73–88, 344–346
Photosystems, 76
Phototropisms, 216–217
Phycobilins, 73
Phylogenetic trees, 170–171
Phytochrome, 218
Pigments, photosynthetic, 73–75
Pinocytosis, 44
Pituitary hormones, 236
Plants, 162, 203–204, 207–218, 344–346
Plasma, 39–40
Plasma cells, 270
Plasmid, 142, 183
Plasmodesmata, 47
Plasmolysis, 43
Plate tectonics, 179
Platyhelminthes, 167
Pleiotropy, 112
Poikilotherms, 240
Point mutation, 139
Polar bond, 10
Polar molecules, 10

Polydactyly, 186
Polygenic inheritance, 113
Polymerase chain reaction, 147
Polyploidy, 119–120, 190
Polysaccharides, 14–15
Population, 184–187, 299–305, 337–339
Porifera, 166
Positive feedback, 236, 274
Positive gene regulation, 144
Postzygotic barriers, 191
Predation, 305
Predator-prey relationships, 192
Prezygotic barriers, 191
Primary productivity, 305–306, 353
Primates, 170
Prions, 145
Probability, 106
Prokaryotes, 33–34
Prop roots, 210
Prostaglandins, 268
Protein, 17–19, 237
Protein-folding problem, 19
Protista, 162
Protons, 9
Pulmonary circulation, 234
Punctuated equilibrium, 193
Pyrimidines, 131
Pyrogens, 268
Pyruvic acid, 58

R

Radioactive iodine, 10
Radioisotopes, 10
Random fertilization, 95
Receptor-mediated endocytosis, 45
Recessive allele, 188
Recessive mutation, 114
Recessive trait, 106
Recombinant chromosomes, 95
Recombinant DNA, 145–148
Recombination, 115
Reflex arc, 245–246
Regulatory T cells, 270
Reproduction, 184, 213–215, 281–283
Reproductive isolation, 189–191
Resource partitioning, 304
Resting potential, 246
Restriction enzymes, 146
Restriction fragment, 352
Restriction fragment length polymorphisms, 147–148
Retroviruses, 141
Reverse transcriptase, 141, 148
Rh factor, 274

Ribose, 132
Ribosome, 36
Ribulose biphosphate, 79
RNA, 19, 132–133, 135–136
RNA processing, 136
Roots, 209–211
Rough endoplasmic reticulum, 37

S

Sarcomere, 252
Sarcoplasmic reticulum, 252
Saturated fatty acid, 15
Schwann cells, 245
Sclerenchymal cells, 209
Seed, 214–215
Self-tolerance, 270
Sensory neurons, 245
Sex chromosomes, 114
Sex-influenced inheritance, 114
Sexual dimorphism, 183
Sexual reproduction, 184, 213–215, 282–286
Sexual selection, 183
Short-day plants, 217
Sickle cell anemia, 139
Signal transduction pathway, 47–49, 216, 250
Single bonds, 16
Sinoatrial node, 232
Skeletal muscle, 251
Small intestine, 227
Smooth endoplasmic reticulum, 37
Smooth muscle, 251
Social behavior, 328–329
Sodium-potassium pump, 44
Speciation, 189–191
Species, 189
Specific heat, 11
Spermatogenesis, 285
Spindle fibers, 38
Splicesomes, 136
Sponges, 163, 166
Sporophyte, 215
Sporopollenin, 206
Spring overturn, 12
Starch, 15
Steroids, 16
Stomach, 226–227
Stomates, 212–213
Stroma, 76
Structure-function relationships, 362–363
Suberin, 209
Summation, 252–253
Survivorship curves, 300–301
Sympatric speciation, 190–191
Synapse, 248
Synapsis, 92
Systematics, 162
Systolic blood pressure, 233

T

T cells, 268–270
Tandem repeats, 145
Taq polymerase, 147
TATA box, 136
Taxonomy, 159, 180
Telomerase, 133
Telomeres, 133, 145
Temperature regulation, 239–240
Temporal isolation, 191
Termination, 136, 138
Territoriality, 329
Testcross, 107–108
Tetanus, 252–253
Thermophiles, 161
Thigmotropisms, 217
Thylakoids, 76
Thymine, 131
Tight junctions, 46
Tissue, 163
Topoisomerases, 133
Toxins, 314
Tracheids, 208
Tracheophytes, 205
Transcription, 135–136
Transduction, 141
Transfer RNA, 135
Translation, 137–138
Translocation, 119
Transpiration, 213, 354–355
Transpirational-pull cohesion tension, 11, 213, 354
Transposons, 145
Triploid, 120
Tropic hormones, 234
Tropism, 217
Tropomyosin, 252
Troponin, 252
Tryptophan operon, 142–143
Turner's syndrome, 119

U

Uracil, 132
Urea, 241
Uric acid, 241
Urine, 241
Uterus, 282

V

Vacuoles, 38
Vagina, 282
Vector, 145–146
Vesicles, 38
Vessel elements, 208
Viruses, 140–141
Vision, 250

W

Water, 11–13, 42, 312, 341
Wobble, 137, 139

X

Xanthophyll, 73
Xylem, 207–208, 213

Minimum System Requirements

Windows®

Pentium 4 2GHZ or faster processor
Windows 7, Windows Vista®, Windows XP
512MB of RAM (1GB recommended)
1024 × 768 resolution display
Requires Adobe Air

Mac OS®

Intel Core Duo 1.83 GHz or faster processor
Mac OS X 10.5 or greater
512MB of RAM (1GB recommended)
1024 × 768 resolution display
Requires Adobe Air

Note: Mac computers with PowerPC® processors are not supported.

Installing the Application

Windows

1. Insert the CD into your CD-ROM drive.
2. After a few moments, the Installer will open automatically.[†]
3. Follow the onscreen instructions to install the application.

[†]*If the Installer does not automatically open, from the Start menu, select "Run..." and enter: X:\Installer.exe (where "X" is the letter of your CD-ROM drive). Then click OK.*

Mac OS

1. Insert the CD into your CD-ROM drive.
2. After a few moments, the contents of the CD will be displayed.
3. Double click on Installer to begin installation.
4. Follow the onscreen instructions to install the application.

Running the Application

Windows

After installing, you can start the application by double-clicking the AP_Biology icon on your desktop.

Mac OS

Open the Mac's application folder and double-click AP_Biology icon to run the application.